Saas-Fee Advanced Course 20
Lecture Notes 1990

R. D. Blandford   H. Netzer
L. Woltjer

Active
Galactic Nuclei

Saas-Fee Advanced Course 20
Lecture Notes 1990
Swiss Society for Astrophysics
and Astronomy

Edited by T. J.-L. Courvoisier and M. Mayor

With 97 Figures

Springer-Verlag Berlin Heidelberg GmbH

Professor R. D. Blandford

Division of Physics and Astronomy, Theoretical Astrophysics, 130-133 CALTECH, Pasadena, CA 91125, USA

Professor H. Netzer

School of Physics and Astronomy, Tel Aviv University, Ramot Aviv, 69978 Tel Aviv, Israel

Professor L. Woltjer

CNRS, Observatoire de Haute-Provence, F-04870 Saint-Michel-L'Observatoire, France

Volume Editors:

Dr. T. J.-L. Courvoisier
Professor M. Mayor

Observatoire de Genève, CH. des Maillettes 51, CH-1290 Sauverny, Switzerland

This series is edited on behalf of the Swiss Society for Astrophysics and Astronomy:

Société Suisse d'Astrophysique et d'Astronomie

Observatoire de Genève, CH. des Maillettes 51, CH-1290 Sauverny, Switzerland

ISBN 978-3-662-38888-4   ISBN 978-3-662-39816-6 (eBook)
DOI 10.1007/978-3-662-39816-6

Library of Congress Cataloging-in-Publication Data. Blandford, Roger D. Active galactic nuclei: Saas-Fee advanced course 20 lecture notes, 1990, Swiss Society for Astrophysics and Astronomy / R. D. Blandford, H. Netzer, L. Woltjer; edited by T. Courvoisier and M. Mayor. p. cm. Lectures presented at the 20th Advanced Course in Astrophysics held in Les Diablerets, Switzerland, Apr. 1–6, 1990. Includes index. ISBN 3-540-53285-4 (Springer-Verlag Berlin Heidelberg New York: alk. paper). – ISBN 0-387-53285-4 (Springer-Verlag New York Berlin Heidelberg: alk. paper). 1. Galactic nuclei – Congresses. 2. Active galaxies – Congresses. I. Netzer, Hagai. II. Woltjer, Lodewijk, 1930–. III. Courvoisier, T., 1953–. IV. Mayor, M. (Michel) V. Schweizerische Gesellschaft für Astrophysik und Astronomie. VI. Advanced Course in Astrophysics (20th: 1990: Les Diablerets, (Switzerland) VII. Title. QB5858.3.B53 1990 523.1'12 – dc20 90-23472

© Springer-Verlag Berlin Heidelberg 1990
Originally published by Springer-Verlag Berlin Heidelberg New York in 1990.
Softcover reprint of the hardcover 1st edition 1990

Foreword

The Swiss Society for Astrophysics and Astronomy organizes each year in the late winter or early spring an advanced course. The format of the school is always identical: three leading lecturers are invited to cover the subject in nine or ten lectures each and to deliver a written version of their lecture notes. Lectures are held in the morning and late afternoon, thus leaving ample time for discussion and skiing. These arrangements prove very convivial and lead to an excellent atmosphere in which to learn exciting new subjects and establish contacts with colleagues. A wide variety of people attend the school, including many young students, mostly from Europe, and some experienced researchers.

The 20th Advanced Course of the Swiss Society for Astrophysics and Astronomy took place in Les Diablerets from 1 to 6 April 1990. It was devoted to observational and theoretical aspects of active galactic nuclei. The previous advanced courses of the Swiss Society for Astrophysics and Astronomy have regularly taken place in Saas-Fee, a small resort in the Swiss Alps, hence the name "Saas-Fee" used to describe the courses and lecture notes. In the last three years, however, the course was organized in Leysin and in Les Diablerets, both also situated in the Swiss Alps.

This volume contains the three brilliant sets of lectures delivered by R. Blandford on "Physical Processes in Active Galactic Nuclei", H. Netzer on "AGN Emission Lines" and L. Woltjer on "Phenomenology of Active Galactic Nuclei". The three lecturers also produced a set of notes that will certainly prove extremely valuable, not only to the participants of the school but also to all students of nuclear activity in galaxies. We are very indebted to them for their excellent work.

This course is the first to be published and distributed by Springer-Verlag, breaking with a long period during which Geneva Observatory took responsibility for the printing and distribution of the lecture notes. The 19th Course is co-published by University Science Books (Mill Valley) and Geneva Observatory. The present change was motivated by the wish heard many times in recent years to improve the world-wide distribution of the Saas-Fee Lecture Notes. Many of the previous years' volumes can still be ordered from Geneva Observatory.

We would like to extend our thanks to those who helped us in the organization of this year's school – among them our secretary, Mrs. E. Teichmann, and our students A. Orr and R. Walter.

Geneva, November 1990   *T. J.-L. Courvoisier and M. Mayor*

Contents

List of Previous Saas-Fee Advanced Courses:

---

* Out of print

The Milky Way as a Galaxy, 1989   *G. Gilmore, I. King, P. van der Kruit*

Active Galactic Nuclei, 1990   *R. Blandford, H. Netzer, L. Woltjer*

Books from 1971 on may be ordered from: SAAS-FEE COURSES
GENEVA OBSERVATORY
chemin des Maillettes 51
CH-1290 Sauverny
Switzerland

Books from 1990 on may be ordered from Springer-Verlag

Phenomenology of Active Galactic Nuclei

L. Woltjer

With 13 Figures

INTRODUCTION

Active galactic nuclei constitute a somewhat vaguely defined class of objects. It is very well possible that most galaxies have nuclei and that these are active in the sense that there is an energy source in addition to the thermonuclear sources inside the constituent stars. Something is called an AGN if this activity is "substantial" in some characteristic, but the quantitative meaning of this is unclear and tied up with experimental or observational possibilities. For example, in the nucleus of the "normal" galaxy M 81 very faint broad wings (± 3000 km/sec) have been detected in Hα and Hβ (Peim' ᵣt and Torres-Peimbert 1981, Shuder and Osterbrock 1981, Filippenko and Sargent 1988). Such a detection is important because it shows that high velocity material is present, which either has been accelerated by some process or which is moving in a very deep potential well. It cannot be excluded that some such high velocity gas is present in all "normal" nuclei, but that in most cases the quantity is still smaller than in M 81. This example shows that it is not enough to define certain phenomena as corresponding to AGN, but that it is necessary to quantify the activity by measurable criteria. In addition, we have to define the meaning of "nuclear". Is it a region 1 pc or 1 kpc across ? For the moment, we shall take a rather generous definition and not inquire too closely about its precise meaning.

The following classes of AGN or related objects appear in the literature:

1. Radio Galaxies
2. Radio Quasars
3. BL Lac Objects
4. Optically Violent Variables (OVV's)
5. Radio Quiet Quasars
6. Seyfert 1 Galaxies (Sy 1)
7. Seyfert 2 Galaxies (Sy 2)
8. Low Ionization Nuclear Emission-Line Regions (LINERS)
9. Nuclear H II Regions
10. Star Burst Galaxies
11. Strong IRAS Galaxies

1. *Radio Galaxies (RG).* While weak radio emission with[1] $P_{1.4\,GHz} < 10^{23.3}$ W Hz^{-1} occurs in many galaxies and is particularly common in spirals where it may be related to the relativistic electron production by supernovae, the term "Radio Galaxy" is mainly reserved

---

[1] In these lectures we shall always use a Hubble constant H_o = 50 km s^{-1} Mpc^{-1} and a deceleration parameter q_o = 0.

for stronger radio emission. We shall distinguish Powerful Radio Galaxies (PRG) and Weak Radio Galaxies (WRG) and place the separation at $P_{1.4\,GHz} = 10^{25}$ W Hz^{-1}. The PRG tend to be associated with luminous elliptical galaxies, frequently with pronounced peculiarities and with strong emission lines; they show a strong cosmological evolution, their number per unit of comoving volume being much larger at redshift z = 2 than locally at z = 0. The WRG are statistically associated with less luminous ellipticals, with emission lines weak or absent, and there is no evidence for much cosmological evolution.

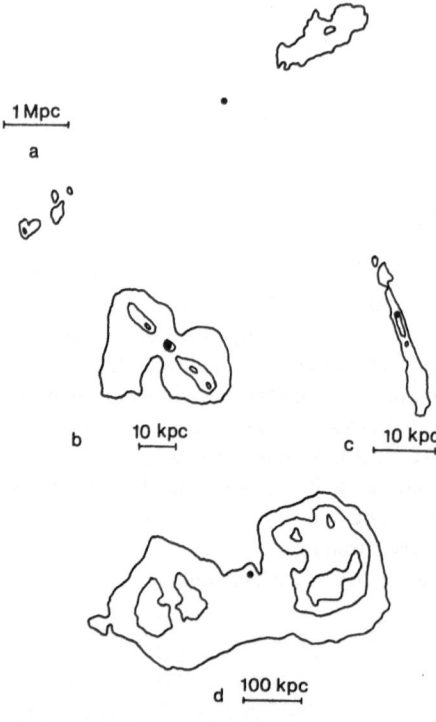

Fig. 1 : Radio isophotes of some radio galaxies

(a) The largest known radio galaxy 3 C 236 at 1.4 GHz. In the powerful nucleus which coincides with an elliptical galaxy there is radio structure with a kpc scale. A compact core in this structure has structure on a scale of 10 pc. Both of these structures are elongated in the same direction as the large scale structure; apparently the energy supply has had a remarkably stable directionality (Barthel et al., 1985).

(b) B 2 2236+35 and **(c)** B 2 2116+26, two weak radio galaxies at 4.86 GHz. The more irregular structure of the former and the stronger concentration of the radio emission near the center are typical for weaker radio galaxies. Bisymmetrical jets are also rather common (Morganti et al., 1987).

(d) The nearby radio galaxy Fornax A at 1.4 GHz. The weak nucleus (0.2 % of the total) of this intermediate power source is located in the galaxy NGC 1316. The eccentric location of the galaxy suggests that the source has lived long enough ($\sim 10^8$ y) for the galaxy to have moved to its present position with respect to the intergalactic medium which restrains the radio structure (Ekers et al., 1983).

On the basis of the spectral properties a subdivision is made into "steep spectrum" and "flat spectrum" sources. If $F_\nu \propto \nu^\alpha$ around 1 GHz the separation generally is taken somewhere near $\alpha = -0.4$. Flat spectrum sources tend to be compact and variable with new components which are partly optically thick in synchrotron radiation appearing from time to time. The steep spectrum sources are usually optically thin and extended. Such radio galaxies may be very large. Saripalli et al. (1986) give a list of 12 sources with a total extent in excess of 1.5 Mpc. At least 2 % of the bright radio galaxies appear to have such large sizes and therefore ages which cannot be less than several million years ($> R/c$).

Maps of the radio emission around 1 GHz show classical doubles relatively frequently among the nearby PRG, and more complex structures among weaker sources (Fig. 1) and at high redshifts. Frequently, jet-like structures appear, though a rather loose usage is made of the word "jet". Especially at higher frequencies a compact flat spectrum core appears, presumably coincident with the nucleus of the galaxy. Fanarof and Riley (1974) introduced two classes at 178 MHz by determining the ratio of the distance between the two most intense spots on either side of the nucleus to the overall source size; in class FR II this ratio exceeds 0.5 (as in outer edge brightened doubles), while in class FR I it is less. At 178 MHz the FR II Radio Galaxies turn out to have $P > 2 \times 10^{26}$ W Hz^{-1} and the FR I $P < 2 \times 10^{26}$ W Hz^{-1}, the division being remarkably sharp. For typical steep spectrum sources ($\alpha = -0.8$) the FR I-II boundary would be at $P_{1.4\,GHz} = 4 \times 10^{25}$ W Hz^{-1} somewhat higher than our division between PRG and WRG. This is just one example of the difficulty in comparing radio source properties at different frequencies, because of important spectral index variations in the maps. In the extended sources small regions of intense emission, the "hot" spots, occur not infrequently. In 10 % of a sample of 50 such hot spots optical continuum emission has also been detected (Meisenheimer et al., 1989). Presumably, the optical emission is also synchrotron radiation. Since the lifetime of the relativistic electrons responsible for that radiation is rather short, Meisenheimer et al. argue that in situ acceleration (by a Fermi-type mechanism, i.e. between moving magnetic irregularities) is necessary. The required power could be produced by a jet from the nucleus.

Substantial linear polarization is often observed in extended radio sources ($p = 10$-30% or more) demonstrating that relatively uniformly oriented magnetic fields are present. If thermal electrons are also present, Faraday rotation of the plane of polarization will occur, over an angle ψ given by

$$\psi = 0.81 \, \lambda^2 \, (m) \int n_e \, (cm^{-3}) \, \mathbf{B} \, (\mu G) \cdot d\mathbf{s} \, (pc) \qquad \text{radians}$$

where λ is the wavelength in m, n_e the thermal electron density, $\mathbf{B}$ the magnetic field strength in 10^{-6} Gauss and $\mathbf{s}$ the path length in parsec. Varying values of ψ over the radio beam will cause depolarization. From the observation of these effects it is seen that thermal electrons ($n_e = 10^{-5}$ cm^{-3}) are present at distances of 100 kpc from the centers of large radio galaxies; closer to the center the thermal electrons produce X-ray emission (Strom and Jägers, 1988).

2. *Radio Quasars.* The radio characteristics of the Quasi-Stellar Radio Sources (QSR or RQ) resemble those of the PRG, but the optical image is dominated by a bluish (U-B < 0) unresolved ($\theta < 1''$), luminous ($M_v < -22$ or -23) nucleus with strong broad emission lines in its optical spectrum. The optical nuclei tend to be variable especially in the flat spectrum radio sources. The latter also frequently show important optical linear

polarization, while the steep spectrum sources do not. Most bright RQ have been detected as X-ray sources at a few keV. The cosmological evolution of QSR is strong.

Observations of radio quasars by VLBI (Very Long Baseline Interferometry - typically baselines of 1000 - 10,000 km) shows that in the nuclei of flat spectrum quasars frequently very compact components are present with dimensions of a few m.a.s. (10^{-3} arcsec) and that these may have apparent transverse speeds in excess of the velocity of light. Such "superluminal" velocities have been observed in more than 10 % of the 3 C quasars and also in BL Lac Objects (Table 1; more extensive lists are given by Zensus, 1989, and by Mutel, 1990). The most likely interpretation involves relativistic jets oriented at small angles to the line of sight (see for example Cohen, 1989, and the lectures by Dr Blandford).

Table 1

Superluminal Radio Sources ($H_0 = 50$ km s^{-1} Mpc^{-1}, $q_0 = 0$)

Subsequent columns give the redshift, the maximum angular velocity and the apparent value of v/c.

(a) *Quasars in the 3 C R catalogue*

	z	max μ (m.a.s. / year)	v/c
3 C 216	0.67	0.11	6
3 C 245	1.029	0.11	8
3 C 263	0.652	0.06	3
3 C 273	0.158	1.20	17
3 C 345	0.594	0.43	22
3 C 454.3	0.859	0.35	22

(b) *BL Lac Objects*

	z	max μ (m.a.s. / year)	v/c
OJ 287	0.306	0.27	7
2007 + 77	0.342	0.18	5
BL Lac	0.070	1.2	8

3. *BL Lac Objects* resemble the flat spectrum radio quasars, except that the broad optical emission lines are absent. They tend to be highly variable both at radio, optical (> 1 mag) and X-ray wavelengths with strong and variable, optical polarization. The variability time-scale at optical and X-ray wavelengths may be less than a day. The cosmological evolution of the BL Lac population appears to be weak.

4. *Optically Violent Variables* are a subclass of (mainly radio) quasars with optical characteristics like the BL Lac's, except that broad emission lines are present - though somewhat weaker than in more typical quasars. Sometimes, BL Lac's and OVV's are grouped together as "Blazars", but it is not at all clear that they form one population. Blazars tend to be somewhat less blue than other quasars.

5. *Radio Quiet Quasars* (QQ). Objects which optically resemble the QSR, but in which no radio emission was detected, have generally been taken to be QQ. In the meantime,

however, it has become clear (Kellerman et al. 1989) that most QQ do have radio emission at power levels in the range $P_{5\ GHz} = 10^{22} - 10^{24}$ W Hz^{-1}. We shall place the separation of QQ and RQ at $P_{5\ GHz} = 10^{24.7}$ W Hz^{-1}, which corresponds about to the power level separating the PRG and the WRG. Since QQ are 10-30 times more abundant than RQ, an optically or X-ray selected quasar is frequently assumed to be a QQ even in the absence of radio measurements. The X-ray emission of the QQ tends to be weaker (with respect to the optical) than that of the RQ. Several per cent of the QQ have very Broad Absorption Lines in the optical spectra, the BAL quasars.

6. *Seyfert 1 Galaxies.* The Sy 1 nuclei resemble the QQ, but their luminosity is lower. We shall take the boundary between Sy 1 and QQ at $M_v = -23$. The hydrogen lines are very wide (FWHM 1000-5000 km s^{-1}) with wings so broad that the FW0I (full width at zero intensity) ranges from 7000-20000 km s^{-1} for a typical sample (Osterbrock 1977). Most Sy 1 galaxies are radio sources with $P_{5\ GHz}$ in the range $10^{20} - 10^{23}$ W Hz^{-1}. As X-ray sources they tend to be somewhat stronger relative to the optical light than the QQ. Sy 1 nuclei are typically found in early type spiral galaxies.

The differences between Seyferts and Quasars as mentioned in the literature often reflect more the method of discovery than a well defined classification. A known galaxy in which a bright blue nucleus is discovered with broad emission lines usually is taken to be a Seyfert, while a blue stellar object with broad emission lines with a substantial redshift and without a clearly visible surrounding galaxy is frequently taken to be a QQ. Since the surface brightness of an underlying galaxy diminishes rapidly with redshift ($\Sigma \propto (1+z)^{-4}$, K-term), it is no surprise that objects with $M_v = -21$ or -22 have been called Seyferts at low ($z < 0.01$) but QQ at high ($z > 0.1$) redshifts.

7. *Seyfert 2 Galaxies.* Whereas the Sy 1 nuclei have broad wings in the permitted (and narrower forbidden) emission lines in their spectra, the Sy 2 nuclei tend to have equally wide lines for the permitted and forbidden lines without the very broad ($\pm$ 5000 km s^{-1}) wings. The precise classification of Sy 1, Sy 2 and the intermediate categories (Sy 1.5, etc.) is discussed by Osterbrock (1981). There appear to be no statistically significant differences in the radio luminosities or radio sizes of Sy 1 and Sy 2 (Edelson 1987, Ulvestad and Wilson 1989). Diameters for the "nuclear" sources range from point-like to 3 kpc with a median around 0.5 kpc; linear structures are not uncommon. Most Sy 2 are weak X-ray sources. Sy 2 and Sy 1 nuclei occur in very similar galaxies.

8. *LINERS.* Nuclei in this class are characterized by relatively strong lines of low ionization species (O I, S II, ...). In contrast to the Seyferts in which the ionization appears to be principally radiative, in the LINERS shock ionization may be important.

9. *Nuclear H II Regions.* Many galactic nuclei show narrow lines (Balmer, [O III]) characteristic of H II regions ionized by hot stars. Unless the phenomenon is very intense, this appears to be a characteristic of "normal" nuclei rather than of AGN.

10. *Star Burst Galaxies* are galaxies in which star formation takes place at a rate far higher than the average during a galactic lifetime. The phenomenon may be in evidence through optical colors and spectra (young stars) and through a large IR output. Bursts of star formation may be induced by mergers of galaxies. There is some evidence that AGN may also have sometimes high rates of star formation in the underlying galaxies.

Table 2

Qualitative Characteristics of AGN and Related Objects

		Emission Lines		Assoc. Galaxies		Evolution
		Broad	Narrow	Type	Luminosity	
Radio Galaxies ⎰	Powerf.	SW	SW	E	S	S
⎱	Weak	W	W	E	S⁻	W?
Radio Quasars		S	SW	(E)	S	S
BL Lac		O	OW	E	S⁻	W
OVV		S⁻	W			
Radio Quiet Quasars		S	SW		S⁻	S
Sy 1		SW	SW	Sa - Sbc	All	W
Sy 2		O	SW	Sa - Sbc	All	
LINERS		O	SW	E - Sbc	All	
Nuc H II		O	SW	Sb - Sc	All	
Starburst		O	S	All	All	
Strong IRAS		OS	S		S	

S : strong; W : weak; SW : strong in some, weak in others; O : absent. In the luminosity column S⁻ is somewhat less than the luminous giant ellipticals, while "all" means all luminosities with $M_v < -20$.

Table 3

Local Space Densities of Some Objects

Object		Gpc^{-3}
Spiral Galaxies	$M_v < -20$	5×10^6
	$M_v < -22$	3×10^5
	$M_v < -23$	3×10^3
Elliptical Galaxies	$M_v < -20$	1×10^6
(incl. S0)	$M_v < -22$	1×10^5
	$M_v < -23$	10^4
Rich Clusters of Galaxies		3×10^3
Radio Galaxies	$P_{1.4\ GHz} > 10^{23.5}$ W Hz^{-1}	3×10^3
	$P_{1.4\ GHz} > 10^{25}$ W Hz^{-1}	10
Radio Quasars	$P_{1.4\ GHz} > 10^{25}$ W Hz^{-1}	3
Radio Quiet Quasars	$M_v < -23$	100
	$M_v < -25$	1
Sy 1	$M_v < -20$	4×10^4
Sy 2	$M_v < -20$	1×10^5
BL Lac	$P_{1.4\ GHz} > 10^{23.5}$ W Hz^{-1}	80
Strong IRAS Galaxies	$L_{IR} > 10^{12}$ L$_o$	300

11. *Strong IRAS Galaxies.* The IRAS satellite has for the first time given a view of the sky at 100, 60, 25 and 12 microns. Some galaxies are extremely luminous in the far IR (10^{12} L_o). Much of this radiation is believed to be re-radiation from dust which is heated by an AGN or by a burst of star formation.

In Table 2 we summarize the qualitative characteristics of the various classes, and in Table 3 we present crude values for the space densities of some of the objects. The densities below 100 Gpc^{-3} are particularly uncertain because of incompleteness and cosmological evolution.

EMISSION LINE SPECTRA OF AGN

Representative examples of relative emission line intensities in different classes of AGN and in an H II nucleus are given in Table 4. All lines stronger than Hβ are listed, except [O III] λ 4959 and [N II] λ 6548 which are a fixed fraction (~ 1/3) of [O III] λ 5007, respectively [N II] λ 6583. Also a few weaker lines are included. The Fe II λ 4570 line is one of several Fe II lines seen in some spectra; in addition, a very large number of Fe II lines which cannot be resolved individually may produce a quasi-continuum in some wavelength intervals.

The spectra divide broadly in two classes: those with narrow (FWHM < 1000 km s^{-1}) permitted lines (H, He, Fe II, Mg II, C IV) and those with broad permitted lines (FWHM > 1000 km s^{-1}). The forbidden lines are generally narrow, although frequently still wide by the standards of "normal" galactic nuclei. C III] λ 1909 which has an intermediate transition probability tends to have the same width as the permitted lines.

The Narrow Line Radio Galaxies (NLRG) and the Sy 2 have very similar spectra. In the LINERS [O III] λ 5007 is weaker, while the lower ionization lines are stronger. In the H II nuclei [O III] λ 5007 may be rather strong, but the lower ionization lines then are very weak. These differences are related to the ionization mechanism: ionization by hot stars (continuum strong in a rather restricted wavelength range) tends to give a narrower range of ionization stages than ionization by a power law continuum. In the LINERS collisional ionization (shocks) may be important.

The narrow lines in the broad line objects have relative intensities which are not too different from those in the Sy 2 and NLRG. The broad line spectra have also a general similarity in most objects, except that the Fe II lines tend to be strong in the radio quiet objects (Sy 1 and QQ) and in some flat spectrum quasars, but weak in the Radio Galaxies and steep spectrum quasars. Some related differences in the H-line ratios also occur.

It therefore appears that the AGN spectra contain lines emitted in a Broad Line Region (BLR) and in a Narrow Line Region (NLR). The basic difference is due to very different densities. In the BLR no forbidden lines are detected and this implies $n_e > 10^8$ cm^{-3}, while the presence of C III] λ 1909 requires $n_e < 10^{10}$ cm^{-3} in at least part of the BLR. From the intensity ratios of the forbidden lines it is generally found that in the NLR $n_e = 10^3$-10^4 cm^{-3}. The high density in the BLR implies recombination times less than a few days. Consequently, if the ionizing radiation field changes the intensities of the broad lines should change, too. Such changes have been detected in quite a few quasars and Seyferts. For example, in the QQ/Sy 1 Fairall 9 the uv flux diminished by a factor of 33 between 1978 and 1984 and the strength of the Ly α line by a factor of 10 (Clavel et al., 1989). From the time delay between the changes in the intensities of lines and continuum

Table 4

Line Intensities in the Optical Spectra of Some Representative AGN

Subsequent columns give the wavelength, the identification and the data for a nuclear H II region, for a LINER, for a Sy 2 galaxy, for a Narrow Line Radio Galaxy, for a Broad Line Radio Galaxy, for a Sy 1, for a steep spectrum radio quasar and for a flat spectrum quasar. The intensities are given relative to the narrow component of Hβ = 1 (Hβ = 10 if no narrow component is seen) and have not been corrected for (internal) reddening. The last rows give the widths of the narrow (n) and broad lines (b) at half intensity. The forbidden lines are generally narrow; the intensities of the uv lines for NGC 7469 include both the narrow and the broad components. The uv and visual spectra have been taken at different times, and variability may have affected the corresponding intensity ratios.

λ		HII N 625	LINER N 4036	Sy 2 MARK 3	NLRG 3 C 223.1	BLRG P 1417-19	Sy 1 N 7469	QS P 2135-14	QF 3 C 273
1216	Ly α			2			34		55
1548/1551	CIV			1			22		20
1909	C III]			0.5			4		8
2796/2803	Mg II			1			8		4
3727	[O II]	1	7	2	1	1	2	2	
4570	Fe II						3		4
4861 n	Hβ	*1*	*1*	*1*	*1*	*1*	*1*	*1*	
b						7	10	27	*10*
5007	[O III]	5	1	13	11	16	6	16	0.6
6300	[O I]	0.01	2	1	0.3	0.6	0.3		
6563 n	Hα	4	5	5	4	7	5	4	
b						35	36	130	52
6583	[N II]	0.2	9	5	3	5	3		
6716/6731	[S II]	0.2	5	3	2	2	2		
FWHM (km s⁻¹) n		50	450	900	200	?	600	500	
b						1700	3000	5800	3600
Reference		a	b	c, d	e	f	d,g,h	i	i, j, k

a Phillips et al, 1983;   b Heckman et al., 1980;   c Koski, 1978;
d Wu et al., 1983;   e Cohen and Osterbrock, 1981;   f Grandi and Osterbrock, 1978;
g Osterbrock, 1977;   h Westin, 1985;   i Baldwin, 1975;
j Grandi, 1981;   k Boggess et al., 1979

sizes of a pc or less are inferred for the BLR. Since the emission per unit mass in the recombination lines is proportional to n_e, the masses required to produce the broad lines are not very large - less than $100\ M_0$ in most cases.

In the NLR the density is much lower, the recombination time much longer, and the mass much larger. No variability would be expected, and none has as yet been found. While the NLR generally has dimensions of the order of a few kpc or less, in some cases very extended emission regions (10 - 100 kpc) occur (e.g. Prieto et al., 1989) which are probably ionized by the uv radiation from the central source. The observed distribution of the ionized gas seems to indicate that the ionizing radiation is beamed in a broad cone; this is perhaps also suggested by the fact that sometimes the radiation emitted in our direction (extrapolated to the uv) appears to be inadequate for the ionization.

The ratio of Ly α to Hβ frequently has values of 2 - 5 in AGN, while recombination theory would give values ten times larger. Part of the explanation involves absorption by dust, which tends to increase with decreasing wavelength - at least for dust particle distributions as found in interstellar space in our galaxy. Such absorption may also strongly affect the estimated luminosities of the nuclei.

CONTINUA OF AGN FROM RADIO TO GAMMA-RAY FREQUENCIES

During the last decade, our knowledge of AGN spectra has much increased by the data obtained by the IRAS satellite in the far IR and by the Einstein, HEAO 1, EXOSAT and other satellites in the X-ray domain. Furthermore, the 1.3 mm data being obtained with the 30-m IRAM telescope are particularly helpful in ascertaining the precise form of the steep gradient between the radio and IR data, the IUE satellite has allowed a detailed study of the uv spectra close to the Lyman limit, and in a few cases data from gamma-ray satellites have been of interest and have given an indication that the next generation of gamma-ray instruments may detect many more AGN.

In describing the spectra of AGN we shall use νF_ν rather than F_ν, since νF_ν is a measure of the total energy emitted around the frequency ν. In fact, if $F_\nu \propto \nu^\alpha$, then for $-1 < \alpha < 1$, νF_ν is the energy contained in an interval $\Delta \log \nu = 0.4$ centered on ν to within 9 %.

The radio fluxes of the flat spectrum sources and the near IR, optical, uv, and X-ray fluxes of many AGN are variable. In some cases the X-ray variations may be very rapid with substantial variation in a matter of hours or even minutes. As a consequence, the interpretation of spectra based on observations obtained at different times may pose problems. For example, in F 9 over the six years during which the uv flux diminished by a factor of 33, the near IR flux diminished by a factor of 2 - 3 and the X-ray flux by a factor of 3 - 4 (Clavel et al., 1989), and as a consequence the spectra at different epochs show substantial differences. A further problem is that in observations with insufficient spatial resolution the underlying galaxy may make an important contribution to the optical and near IR flux, especially in relatively weak nuclei. Finally, absorption by intervening dust may have important effects on the optical and uv spectra, while soft X-rays may experience photoelectric absorption by gas between the nucleus and us.

The overall characteristics of the spectra of AGN (Figures 2-4) show a certain gross uniformity with much of the energy emitted in the IR, and frequently also in X-rays and in the uv, and with the spectral flux declining steeply or precipitously between 100 μm and the radio frequencies. In some cases - like I Zw 1 - the decline seems steeper than $F_\nu \propto \nu^{2.5}$,

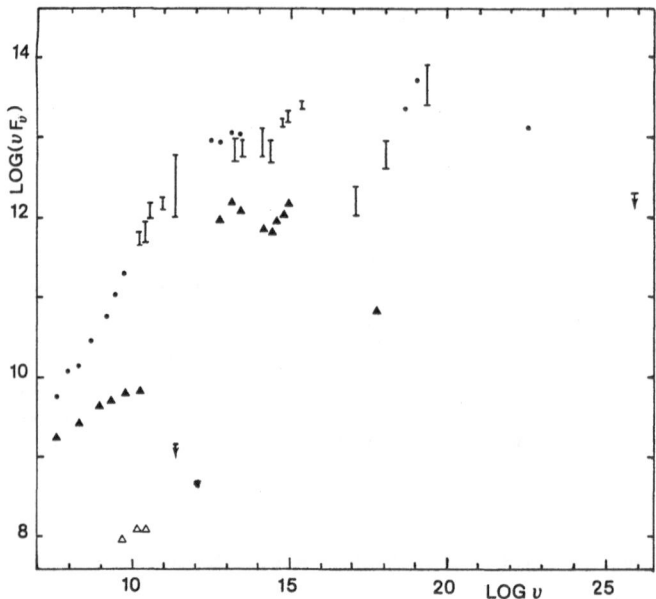

Fig. 2 : The spectra of the flat spectrum radio quasar 3 C 273 (dots and vertical lines when variability has been seen) and of the steep spectrum radio quasar 3 C 351 (filled triangles and for the nucleus only open triangles). The upper limit at 230 GHz is for the nucleus of 3 C 351, the upper limit at Log ν = 25.8 is for 3 C 273. F_ν is expressed in Jy.

which corresponds to synchrotron self-absorption for an isotropic distribution of relativistic electrons. Such steep spectra are also found in starburst galaxies (Telesco, 1988) like M 82 which do not have an active nucleus, but rather a larger central region with a high rate of star formation. Since in such objects there is direct evidence for large amounts of dust which is heated by the newly formed massive stars, it is generally thought that the far IR emission is radiated by dust at a temperature below 100 K. The similarity of the far IR spectra of starburst galaxies and of many AGN suggests that also in the latter the far IR emission is due to dust, heated either by the nucleus or by recently formed stars, or both. If the dust were radiating as a black body (which is a rather gross approximation), then combination of the Wien displacement law and the Stefan-Boltzmann law yields $L \approx 0.2\,R^2\,\lambda_{mm}^{-4}$, with λ the wavelength of maximum emission, L in erg s^{-1} and R in cm. For a modest AGN with $L_{IR} = 2 \times 10^{43}$ erg s^{-1} and with λ = 0.1 mm we already have R = 10^{20} cm = 30 pc. Hence, if the far IR emission is due to dust, the emitting region may be expected to be so large that no significant time variation can be expected. Unfortunately, the short duration of the IRAS mission does not yet allow firm conclusions.

Looking in somewhat more detail, we note first a significant difference between the "flat spectrum" (in F_ν) rapidly variable objects, like the quasar 3C 273 and the BL Lac 0J 287. In these two objects the radio spectrum fits smoothly to the IR with F_ν declining on average rather smoothly and slowly. As a consequence, there is no evidence for dust emission, and the whole radio-IR continuum may well be non-thermal. This appears to be confirmed by the evidence for variability in far IR flux of 0J 287 on timescales of the order of a few

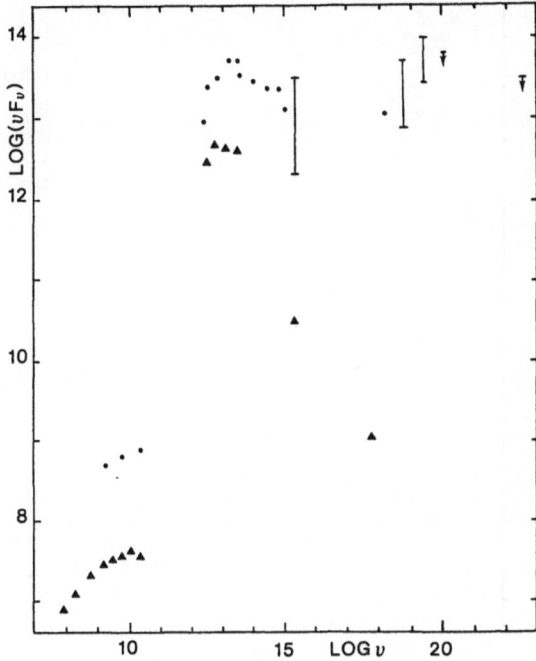

Fig. 3 : The spectra of the Sy 1 galaxy NGC 4151 (dots, vertical lines and upper limits) and of Sy 2 galaxy NGC 1068 (triangles). The vertical scale is for NGC 4151. For NGC 1068 the scale values are to be increased by 2.4.

months (Edelson and Malkan, 1987). In contrast in the steep spectrum quasars and radio galaxies, and in the Seyferts and radio-quiet quasars, the radio and IR spectra appear discontinuous with F_ν increasing between 10 GHz and 3000 GHz (100 µm) by factors of 20-1000 if only the nuclear radio flux is included, and the IR is probably radiated by dust.

McAlary and Rieke (1988) and Sanders et al. (1989) have presented evidence that this dust emission continues into the middle IR with hotter dust responsible down to 2 µm, where the dust temperature would be near 2000 K, about its evaporation temperature. Thermal radiation (free-free emission, etc.) would then account for the 0.5-2 µm part of the spectrum, while the upturn of the νF_ν spectra below 0.5 µm seen in many quasars and Sy 1 galaxies would be due to the accretion disk itself. Again, the flat spectrum quasars and BL Lac's with their frequently large optical polarization and rapid near IR and optical variability would be different, and undoubtedly their emission is mainly non-thermal.

In the X-ray region the radio quasars tend to be stronger than the radio galaxies, and the Sy 1 stronger than the Sy 2. There is, however, much variation from object to object. Typically, the X-ray νF_ν is less than that in the IR, but in the very active objects 3C 273 and NGC 4151 there is comparable or larger energy in X-rays.

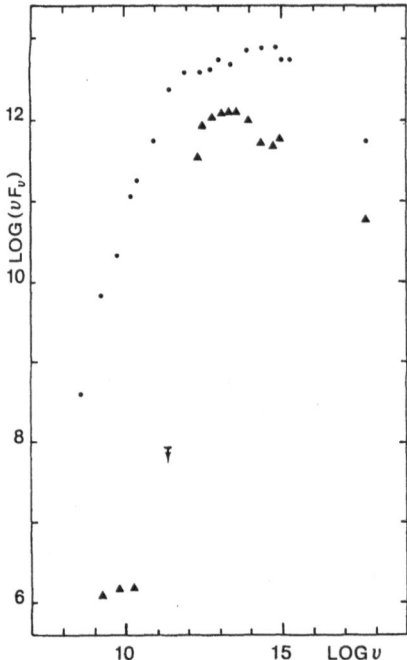

Fig. 4 : The spectra of the BL Lac Object OJ 287 (dots) and of the radio quiet quasar I Zw 1 (triangles and upper limit). The vertical scale is for OJ 287. For I Zw 1 the scale values are to be increased by 1.0.

SURVEYS FOR AGN

AGN have been discovered by the optical identification of radio or X-ray sources and by a variety of studies that included only optical observations. Since the selection effects in the three wavelength regions are different, the mix of objects found may also be different. This may also lead to different conclusions about cosmological evolution and other aspects for the same general class of objects.

Nearby AGN

Common kinds of AGN may be studied best in nearby galaxies, where much detail may be seen that is unobservable in more distant objects. Samples of nuclei for statistical studies have been drawn in particular from the NGC or the Revised Shapley Ames Catalogue (RSA, Sandage and Tammann, 1981). Véron-Cetty and Véron (1986) studied 320 nearby ($z < 0.01$) bright ($M_B < -21$) galaxies from the RSA and found nuclear emission to be common. They find among 159 Sa - Sbc spirals 7 Sy 1 and 11 Sy 2 nuclei. H II regions were found in 27 of these objects (1 in Sa-Sab) and in 52 out of 80 Sc.

We have taken all RSA galaxies, which are listed in the Véron-Cetty and Véron (1989) Catalogue (VVC) as being (possible) Seyferts and tried to somewhat conservatively establish a list of Sy 1 and Sy 2. We find 28 Sy 1 and 25 Sy 2 among the 1246 RSA galaxies. The distribution over galaxy types is given in Table 5, where it is seen that the Seyfert phenomenon is largely restricted to the early type spirals.

Table 5

Numbers and percentages of Seyfert galaxies in the Revised Shapley Ames Catalogue

Class	Sy 1	Sy 2	Total RSA	% Sy
E - S0/a	6	0	361	2
Sa - Sab	11	11	165	13
Sb - Sbc	9	11	283	7
Sc	2	3	370	1

$M_{B_T}^{o,i}$ (Sa - Sbc) -21.57 ± 0.18 $\quad$ -21.56 ± 0.21 $\qquad$ -21.33

The percentage of bars in all Sa - Sbc galaxies in the RSA is 31 %, that for the Seyferts of the same classes 36 % (15 % for Sy 1 and 54 % for Sy 2). Thus, a bar does not seem to have a major overall effect (see also the chapter "Galaxies with Active Nuclei" where the importance of weak or small bars is discussed). We note, however, that Hummel (1981) found a tendency for barred spirals to have stronger nuclear radio sources than ordinary spirals. What seems to be important, however, is group membership. Stauffer (1982) found that approximately half of the galaxies with active nuclei in his sample are in compact groups, but only 20 % of other galaxies. Kennicut and Keel (1984) found in a sample of multiple or interacting spiral galaxies a three times higher frequency of high-excitation Seyfert-like nuclei than in other spiral galaxies. The absolute magnitudes of the Sy 1 and Sy 2 *in the RSA* appear to be the same and not significantly different from those of the non-Seyfert galaxies; the small difference in Table 5 is not significant since the brighter RSA galaxies have been somewhat more extensively studied. However, some weak Seyfert nuclei may have been missed in the brightest galaxies.

The fraction of Sa - Sbc galaxies which are Seyferts has been increasing as instrumentation became available to study fainter emission lines in the galactic spectra. Future studies may well increase this fraction further. In any case, the fact that Seyferts account for 10 % of the early-type spirals shows that 10 % of all such galaxies must be Seyferts all the time or that all such galaxies are Seyferts during 10 % of the time or some situation in between these two extremes.

In some Sy 2 faint wide wings of the hydrogen lines appear in off nuclear spectra (Shields and Filippenko, 1988). Undoubtedly, these originate by scattering from a hidden Sy 1 nucleus as proposed by Miller and Antonucci (1983) to explain the observation of polarized broad emission lines in the Sy 2 galaxy NGC 1068.

The LINERS appear to be even more common than the Seyferts. Heckman (1980) found that about 1/3 of a sample of bright galaxies have LINER nuclei. Véron-Cetty and Véron (1986) found that among non Seyferts about 1/4 of E/S0 and Sb/Sbc galaxies and nearly half of the Sa/Sab galaxies have $H_\alpha < 1.2$ [N II] λ 6583, most of which are probably LINERS. Only 4 % of Sc galaxies have such spectra. By contrast they found nuclear H II regions (ionized by hot stars) in 65 % of Sc, 25 % of Sb/Sbc, 2 % of Sa/Sab and 0 % of E/S0 galaxies.

Radio galaxies also are not uncommon in the RSA. Of its 1246 galaxies, 18 are known to have $P_{5 \text{ GHz}} > 10^{23.0}$ W Hz^{-1}. Below this limit, normal spirals begin to become abundant. Among the 8 galaxies with $P_{5 \text{ GHz}}$ in the range 1-2.5 x 10^{23} W Hz^{-1} there is one Sb Sy 2, two classified respectively S pec and Sc pec, one S0 pec/Sa, one E3/S0 and three ellipticals. The ten galaxies with higher $P_{5 \text{ GHz}}$ may all be (sometimes peculiar) ellipticals (NGC 1316 is classified in the RSA as Sa pec, but Schweizer (1980) in a detailed study concludes it to be a giant elliptical of the cD variety; NGC 5128 is S0 + S pec in the RSA, but frequently considered as a merger of an E with an S). These 10 galaxies are

very luminous with $< M_{B_T}^{o,i} > = -22.4 \pm 0.2$ (-21.2 for all RSA ellipticals, excluding the dwarf ellipticals), with a dispersion of $\pm 0^m6$. Of the 10 galaxies, 6 are in groups or clusters or interacting systems. With a 151 galaxies classified E in the RSA, at least 5 % are radio galaxies with log $P_{5 \text{ GHz}} > 23.4$. Of the 35 galaxies with $M_B < -22$ and classifications E - S0/a, 7 or 20 % are radio galaxies with log $P_{5 \text{ GHz}} > 23.4$. Apparently relatively strong radio emission is common in luminous ellipticals.

In conclusion, Seyferts and radio galaxies are not at all uncommon. They make up 6 % of the brighter galaxies. Undoubtedly, an even larger percentage of these galaxies have extinct AGN. Their study in the nearest galaxies, which will be possible with the high angular resolution instruments which are becoming available, will provide valuable constraints on theories of AGN.

Radio Surveys

The 3C survey was the first high quality survey for radio sources. At a frequency of 178 MHz nearly 500 sources were discovered to a flux level of 9 Jy. A later revision with more accurate fluxes - the 3C R catalogue (Bennett, 1962) - has made the catalogue more uniform, and all sources north of Dec. -05° are listed - 328 in total. Most of these sources have been identified - a large fraction with radio galaxies and fewer with quasars, Seyferts and BL Lac. The subsequent 4C catalogue went substantially deeper. In the south the sky was surveyed by the Parkes telescope, and large numbers of sources appear in various supplements to the Australian Journal of Physics; the sensitivity reached in the later instalments was around 0.3 Jy at 2.7 GHz. Various other surveys have covered limited areas of the sky to much lower flux levels. With the VLA flux levels of 0.2 mJy at 5 GHz can be reached in about half an hour, but, of course, the total sky coverage is limited. A catalogue listing the 518 extragalactic sources >1 Jy at 5 GHz in 9.8 steradians is given by Kühr et al. (1981); it is supposed to be essentially complete.

Optical identification proceeds by inspection of a Sky Survey (Palomar in the north, ESO/SRC in the south) or by special imagery. With the increasing precision of the positions (~ 1 arcsec), positional coincidences usually suffice for identification. The most time consuming step comes next - the acquisition of optical spectra for establishing the nature of the object and for redshift determination. Even though the identifications may usually be made exclusively on the basis of positional coincidences, the follow up may be influenced by the likelihood that the object will be "interesting". For example, at the time that radio positions were less accurate and several candidate objects were found at each position, frequently uv excess objects were chosen for spectroscopy, since one had a good chance to find quasars that way. Since quasars at large redshifts have more neutral colors, the opposite could occur nowadays. As a consequence, extreme care is needed in evaluating the statistical results of a survey unless it is complete.

A large effort has been spent on identifying all the extragalactic 3C R sources. Roughly 1/4 are quasars and 3/4 radio galaxies. A few previously unidentified - or erroneously identified - sources have recently been shown to be radio galaxies with large redshifts. These include the remarkable object 3C 326.1 at $z = 1.82$ which is associated with a large Ly α emitting cloud with a luminosity of 10^{11} L_0 in Ly α (McCarthy et al., 1987a) and B2 0902+34 at $z = 3.40$ (Lilly, 1989).

The identification content of a radio survey depends on the frequency at which it is conducted. The 3C contains a high percentage of steep spectrum classical doubles. A survey at a much higher frequency will have more flat spectrum sources which are compact.

Until now we have been discussing primary radio surveys: surveys of radio sources which are subsequently identified optically. It may also be of interest to make a radio survey of objects found in surveys in other wavelength bands. Typical examples are the recent surveys of optically discovered "radio quiet" quasars. Kellermann et al. (1989) have observed all 114 objects of the Palomar Bright Quasar Survey (BQS) at 5 GHz; 84 % of all objects were detected at 0.2 mJy or above. Even at low flux levels, the higher redshift objects still may have radio emission well in excess of that in normal galaxies. Since there are about 150 sources per square degree above the 0.2 mJy limit, there should be about 1.5×10^6 in the area of the BQS. Making optical identifications of radio sources would therefore be a hard way to discover radio emission in "radio quiet" quasars.

IR Surveys

The IRAS (InfraRed Astronomical Satellite) all-sky (actually only 97 %) survey in four wavelength bands centered at 12, 25, 60 and 100 μm (Table 6) has detected of the order of 10^5 sources - a substantial fraction of which are stars. Typical error boxes for positions are 0.1 - 0.2 arcmin^2.

Table 6

The central wavelength λ_0, wavelength interval λ_1 - λ_2, and the 10 σ sensitivity for the four IRAS bands (Neugebauer, 1984)

λ_0 (μm)	λ_1 - λ_2 (μm)	10 σ (Jy)	field of view (arcmin^2)
12	8.5 - 15	0.7	3
25	19 - 30	0.65	3
60	40 - 80	0.85	7
100	83 - 120	3.0	15

Many of the identifications have been with previously known objects. For example, of a list of 186 Seyfert galaxies, 116 were detected in the IRAS survey (Miley et al., 1985). The 60 μm luminosity functions were found to be quite similar for Sy 1 and Sy 2. However, the curvature of the spectra was slightly different with the Sy 2 on average having 15 - 20 % more emission at 60 μm with respect to a power law fit between 25 μm and 100 μm than the Sy 1, though the scatter is large. Also 179 quasars were observed, with 4 σ detections in at least one of the IRAS bands for 74 (Neugebauer et al., 1986). Luminosities in the IR range up to 10^{14} L_o.

In addition to the IRAS surveys of previously known sources, some new objects were also discovered. Of particular interest was the source IRAS 13349 + 2438 which was identified with a B = 15^m1 quasar with a redshift of 0.107 corresponding to M_B = -23^{m}9. Its near IR spectrum rises exceptionally steeply towards longer wavelength. The discovery of this source, which was not in the Palomar BQS because its U-B excess is too small, shows that not all nearby quasars are included in the BQS, and consequently the local space density of quasars may have been underestimated (by 1/3, Beichman et al., 1986). A similar claim has been made by Low et al. (1988) who in an IRAS color selected sample found 12 quasars, of which 7 were said not to have been previously known. If we reclassify the quasars and Seyferts by the M_B = -23 separation, there are 11 quasars with four new objects, of which only two (including IRAS 13349 + 2438) could have been in the PG survey. With 5 PG quasars in their IRAS selected list, we again find that something like 1/3 of quasars may be missing in the BQS.

In an area of 14500 deg^2 the IRAS Bright Galaxy Survey detected 10 galaxies with IR luminosities in excess of 2.2 x 10^{12} L_o (Sanders et al., 1988b). The mean redshift of the sample is 0.055, while on average M_B = -22, not particularly high. In a two color diagram of J(1.3 μm) - H(1.6 μm) versus H - K(2.2 μm) some of these objects are relatively close to the starburst galaxies, while others are nearer to the area occupied by low redshift quasars. Some of these also show broad Hα (FW0I > 2000 km s^{-1}). Much of the IR emission appears to originate in a region less than a small number of kpc across. CO observations show the presence of large amounts of molecular gas (10^{10} -10^{11} M_o). In Arp 220 most of this is in a region less than 2 kpc in diameter. All of the objects are in strongly interacting systems, while of objects with 50 times lower luminosity only 10 % are. Sanders et al. (1988b) present an evolutionary model in which collisions between the molecular clouds in two merging galaxies lead to high rates of star formation which are followed by the formation of a merged nucleus with much gas near the center feeding an AGN. As the surrounding dust shroud is driven out by radiation pressure   ($\dot{p}$ = L_{AGN}/c suffices to accelerate 10^{10} M_o to > 100 km s^{-1} in a few times 10^7 years) the inner parts become visible, relatively more energy is radiated at shorter wavelengths and finally a conventional quasar emerges.

However, Norris (1989) shows from high resolution interferometry that whereas in several of the extremely luminous IR galaxies with Seyfert spectra a bright compact core was detected, those without Seyfert spectra never have such a core; this might indicate that these objects are actually starburst galaxies without (as yet ?) a quasar-like nucleus.

Optical Surveys

Following the discovery by Sandage (1965) that radio quiet quasars far outnumber radio quasars, special surveys have been made for QQ. Available techniques include color selection, very low dispersion spectroscopy and variability studies. Two factors determine the success of these: the spectral energy distribution in the objects being looked for, and the relative numbers of these in comparison with those of all other objects at the same apparent magnitude. A typical spectral distribution for quasars in the optical and uv results in the U-B colors as a function of redshift shown in Fig. 5. While the continuum colors in the visible and near uv are blue, the observed colors will depend much on which emission lines pass the color filters. In particular, at about $z = 2.2$ when Ly α begins to enter the B filter, the U-B colors become substantially redder. Thereafter the absorptions in the Ly α forest and in the Lyman continuum also reduce the U luminosities.

The numbers of quasars, stars and galaxies per square degree as a function of B magnitude is shown in Fig. 6. A selection technique has to be much more precise at $B = 15$ where stars outnumber quasars by a factor of more than 10^4, than at $B = 20$ where this factor is about 20.

The Palomar Bright Quasar Survey (BQS) (Schmidt and Green, 1983) resulted from the Palomar-Green Survey (PG) of stellar objects with U-B < -0.44 and *on average* $B < 16.16$. In a quarter of the sky (10,714 square degrees at galactic latitude > 30°) some 1800 objects were found of which subsequent spectroscopy showed 114 to be quasars and Seyferts, 92 of which have $M_B < -23$. As noted before, there probably are other bright nearby quasars which are less blue, while objects with $z > 2.2$ could also have been missed. In an earlier survey, Braccesi et al. (1970) had isolated 175 U-B excess objects in 37 square degrees down to magnitudes around $B = 19$. Applying various corrections, Woltjer and Setti (1982) concluded that in this field there were 0.91 quasars per square degree with U-B < -0.4 and with B < 18.27. Comparison with the BQS figures shows that the area density of objects brighter than B increases by a factor of 8 per magnitude. This immediately indicates the need for evolution: in a Euclidean static Universe this should be a factor of four per magnitude and cosmological effects would make this even less.

In a recent survey, Boyle et al. (1990) have searched a field for faint ($B \le 21$) uv-excess objects. Of 1440 objects found, 820 turned out to be halo subdwarfs in our galaxy, while 420 quasars were discovered, including 9 BAL quasars. From $B = 18 - 21^m$ at half magnitude intervals 0.7; 2.6; 5.3; 6.9; 9.4 and 11.6 quasars were found per square degree.

Multicolor techniques have been used to find quasars at larger redshifts where uv excess is no longer possible. For example, Koo and Kron (1988) found faint quasars in a U-J versus J-F diagram, with J close to the standard B ($B = J + 0.1$) and F half way between V and R. Lower redshift quasars with $z < 2.2$ are found by their uv excess, while higher redshift quasars have J-F colors bluer than stars for the same U-J. Such multicolor

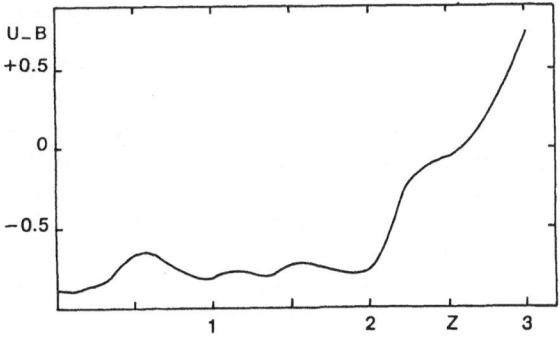

Fig. 5 : U-B colors of an average quasar as a function of redshift (after Cristiani and Vio, 1990).

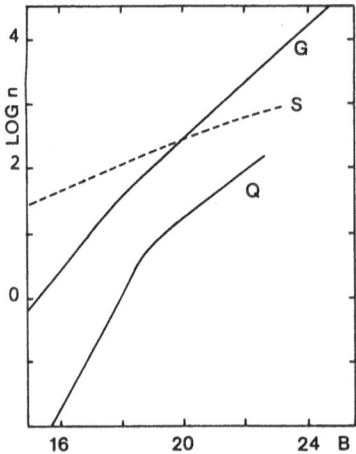

Fig. 6 : The number n per magnitude per square degree of Quasars, Stars and Galaxies as a function of B magnitude.

techniques gave the first clear indication that the relation between integral number and flux flattens off at fainter flux levels. From complete spectroscopy to J = 21.0 and partial spectroscopy for J > 21, of the objects selected in 0.58 square degrees between J = 19.5 and 22.5, Koo and Kron find in half magnitude intervals 5; 4; 12; 16; 19 and 39 objects or 9; 7; 21; 28; 33 and 67 per square degree. With about 20 % of the redshifts exceeding 2.2 the agreement with Boyle et al. (1990) is quite good.

While the multicolor techniques are useful in creating lists of candidate objects, there always remain doubts as to whether the color criteria give a complete set of quasars. Objects with different characteristics might be systematically excluded. To remedy this techniques have been developed for obtaining spectra of large numbers of objects simultaneously with sufficient spectral resolution to recognize objects with strong emission lines. Large surveys have been made with objective prisms on Schmidt telescopes, and smaller areas have been studied with grisms on large telescopes. The completeness of these photographic surveys is not always evident. The results, however, sufficed to indicate that the space density of quasars at large redshifts (z > 3) is no longer increasing and appears, in fact, to be declining (Osmer, 1982).

Recent systematic grism surveys with large CCD's at the 200-inch telescope at Palomar have led to the discovery of higher redshift quasars. In these surveys (Schmidt et al., 1987) the telescope is kept stationary and each object is followed electronically as it moves over the CCD. Typically 1-2 square degrees are covered per hour, and of the order of 0.1 or 0.2 quasars per square degree with z > 4 have been found. At present from various sources (including 3 from multicolor studies) 11 such quasars are known (Schneider et al., 1989 a,b). Some of these are surprisingly bright (one with Gunn r = 17^{m}5 and in total 8 with r ≤ 20^{m}2). The largest redshift is 4.73 and the highest luminosity M_B = -30^{m}5 (for q_o = 0). For q_o = 0.5 this would be -28^{m}6. If we were able to determine reliable M_B values from spectroscopic or other distance independent aspects of these quasars, we could obtain decisive cosmological information.

Variability of quasars poses problems, but also provides a technique for their discovery. The problems arise in surveys in which the quasars are discovered by the use of methods of low photometric accuracy and where more accurate photometry is done only later. Variable objects that have diminished to below the brightness cutoff of the survey will be eliminated, while the corresponding objects which have increased in brightness and should be included at the time the photometry is done cannot be identified. Great care is needed to correct for this.

The variability of a sample of 130 AGN has been studied by Pica and Smith (1983) on the basis of a 13 year long photographic monitoring programme. From an analysis of their data (reclassifying some objects and deleting quasars with less than 15 observations), we find the following statistics: The 18 steep spectrum RQ and 12 QQ have on average a maximum variation of 0^m7; there is no significant difference between the two. The 35 flat spectrum RQ have on average a range of 1^m2 and the subset of 10 high polarization quasars $(P > 3 \%)$ 1^m9. The 15 BL Lac have a mean maximum variation of 2^m4 - perhaps no surprise as some BL Lac have been recognized by their large variations. Since most observations in the sample of Pica and Smith have r.m.s. errors of 0^m10 -0^m15, a maximum apparent variation of the order of 0^m4 - 0^m5 for a constant source would not be surprising. Therefore the true amplitudes of the steep spectrum RQ and of the QQ are probably no more than 0^m2 - 0^m4. In none of these samples is there a significant relation between amplitude and either luminosity or redshift. Of course, all of these variability amplitudes pertain to a far from "complete" sample observed for a limited time and with very uneven coverage. In fact, longer term variability studies have found even larger amplitudes for some OVV's like 6^m7 for 3C 279 which reached B = 11^m3 (Eachus and Liller, 1975) and M_B = -31^m6. If it were an isotropic radiator this would correspond to 4×10^{14} L_o.

Hawkins (1986) has studied variable and uv-excess objects in a field repeatedly surveyed with the UK Schmidt telescope. With procedures that eliminate short period variables like RR Lyrae stars, he isolated 11 clearly variable objects between B = 17 and B = 18.5, all of which turned out to be quasars and 10 of which had uv-excess; the only exception had z = 2.94. Of the uv-excess objects (U-B < -0.4) half turned out to be quasars of which 30 % were the 11 variable objects. Several of the others (40 - 70 %) turned out to be variables with smaller amplitudes.

Trevese et al. (1989a) analyzed 9 plates taken over 11 years of the field studied previously by Koo and Kron (1988) with multicolor methods. Of the 41 known quasars in the field, 30 have been detected as variables, while only 13 other variables (of unknown class) have been found. This also gives some reassurance that in the multicolor studies not too many quasars have been missed. Subsequently, Trevese et al. (1989b) have reported that the variability appears uncorrelated with either luminosity or redshift.

The conclusion of all of this is that variability is likely to be a powerful tool for finding (largely radio quiet) quasars, without the strong redshift dependent selection effects which have plagued so many uv-excess or objective prism surveys.

While the steep spectrum RQ and the QQ generally have polarization less than 3 % (sometimes just the interstellar polarization in our galaxy), several flat spectrum objects have larger polarization (Moore and Stockman, 1984; Stockman et al., 1984). In a more recent sample of 115 flat spectrum sources Wills (1989) find about 50 % with P > 3 %, which figure rises to 70 % for objects with strong core dominance. Surveys for radio quiet polarized quasars or BL Lacs have had little success so far, but the resulting limits on their numbers are not yet very significant.

X-ray Surveys

Three relatively recent surveys have provided much of our knowledge about X-ray sources: The HEAO I survey (Wood et al. 1984) covered the whole sky. Because of the satellite orientation such surveys are generally deepest near the ecliptic poles. A much more sensitive survey at lower energies should be conducted later this year by ROSAT. The Einstein X-ray observatory made pointed imaging observations of hundreds of objects. Every such observation constitutes a mini-survey of 0.5 square degrees, the sensitivity being determined by the observation time. As a consequence, the Extended Medium Sensitivity Survey (EMSS) (Gioia et al., 1990) has a very varying sensitivity level from field to field, and care is needed when statistical studies are made. In a similar way, the EXOSAT High Galactic Latitude Survey has been made (Giommi et al., 1989). A summary of the characteristics of these surveys is given in Table 7. It is clear that ROSAT will greatly increase the X-ray detection of AGN, at least in the lower energy range.

X-ray surveys have some advantages for obtaining complete samples of quasars. With the steep integral number - flux relation of a factor of 8 per magnitude for bright quasars small fluctuations in limiting magnitude caused by interstellar absorption, variability and differing emission line strength have relatively large effects in the optical surveys. Also, while at bright optical magnitudes there are few quasars and many stars, AGN are a dominant constituent of the point sources in the X-ray sky at 2 keV.

Table 7

Recent large area X-ray surveys and the planned ROSAT survey

	HEAO A-1	EMSS	EXOSAT	ROSAT
Sky coverage (sq. deg.)	40,000	780	820	40,000
Energy range (keV)	0.25 - 25	0.3 - 3.5	0.05 - 2.0	0.1 - 2.0
Sensitivity at 2 keV (μJy)[1]	1 - 3	0.01 - 2	0.03 - 0.3	0.05
Number of sources	842	835	230	
Identified	414[2]	761	209	
Extragalactic	266	541	63	
AGN	90	429		(30,000)
BL Lac	13[2]	34	10	

[1] for a spectrum $F_\nu \propto \nu^{-1}$ with low energy cutoff at 1 keV

[2] more recently (Schwartz et al., 1989) : 550 identified of which 20 BL Lac

ENERGETICS AND TIME SCALES

Simple arguments may be made which lead to estimates of the total energies involved in radio galaxies and radio quasars. Suppose we have a spherical source with radius R uniformly filled with randomly directed magnetic fields and relativistic electrons and / or positrons. If the electrons have a differential energy spectrum

$$n(E) = k E^{-\beta}$$

the power radiated in synchrotron radiation at a frequency ν will be given by (from the equations in Salter and Brown, 1988)

$$P(\nu) = C k (\nu/L)^{-1/2(\beta-1)} B^{1/2(\beta+1)}$$

with L and C constants (the latter depending weakly on β) and B the magnetic field strength. The total energy of the electrons in the interval E_{min} to E_{max} is given by

$$E = k \int_{E_{min}}^{E_{max}} E^{1-\beta} dE = \frac{k}{2-\beta}\left(E_{max}^{2-\beta} - E_{min}^{2-\beta}\right)$$

Since the characteristic frequency ν_c around which an electron radiates is given by

$$\nu_c = L B E^2$$

we have approximately

$$E = C^{-1}\left(2-\beta\right)^{-1} L^{-1/2} \nu^{1/2(\beta-1)} P(\nu)\left[\nu_{max}^{1/2(1-\beta)} - \nu_{min}^{1/2(1-\beta)}\right] B^{-3/2}$$

where ν_{max} and ν_{min} are the frequencies corresponding to E_{max} and E_{min}. A lower limit to the electron energy is obtained by taking ν_{max} and ν_{min} equal to the maximum and minimum frequencies at which the source has been observed. Since $P(\nu)$ and the spectral index $1/2(1-\beta)$ are also observed quantities, we have

$$E = f_{\text{(observed quantities)}} B^{-3/2}$$

The total magnetic energy is given by

$$M = 1/6\ B^2 R^3$$

If B is large M is large, if B is small E is large. It is clear that $E + M$ has a minimum when the magnetic field is near B_{eq}, the value of B for which $E = M$, that is for which there is equipartition between the electron and magnetic energies. From the foregoing equations we find

$$B_{eq} \propto P^{2/7} R^{-6/7}$$

$$(E + M)_{eq} \propto P^{4/7} R^{9/7}$$

$$t_{eq} = E_{eq}\left[\int_{\nu_{min}}^{\nu_{max}} P(\nu)\ d\nu\right]^{-1} \propto P^{-3/7} R^{9/7}$$

21

where t_{eq} is the time in which the available electron energy would be radiated away. Inserting numerical values we find for the large strong radio galaxies typically $B_{eq} = 1\text{-}10\ \mu G$, $(E+M)_{eq} = 10^{59}\text{-}10^{60}$ ergs and $t_{eq} = 10^8\text{-}10^9$ years.

The time in which an electron would lose half of its energy may be written as

$$t_{1/2} = 3 \times 10^4\ B_{\perp}^{-3/2}\ \nu_c^{-1/2} \quad \text{years}$$

with $B_{\perp}$ the component of the magnetic field perpendicular to its momentum vector in Gauss. With $B_{\perp} = 10^{-6}$ Gauss and $\nu_c = 10^{10}$ s^{-1}, we have $t_{1/2} = 3 \times 10^8$ years. Hence, the very large and not extremely powerful sources could radiate a more or less constant spectrum extending up to 10 GHz for times of the order of 10^8 years without needing much energy input. This is, however, likely to be the exception. The large radio sources with "hot spots" would have much stronger fields there and $t_{1/2}$ is then much shorter. In those radio galaxies in which optical synchrotron radiation has been detected in the hot spots, the parameters are likely to be closer to $B = 10 - 100\ \mu G$ and $\nu_c = 10^{15}$ s^{-1} which corresponds to $t_{1/2} = 10^3\text{-}10^4$ years. Clearly in many radio sources a continuous input of energy is necessary - presumably coming from the AGN.

The very large radio sources must be long lived. A source like 3 C 236 even if expanding at the velocity of light would need 10^7 years to reach its present size. The actual lifetime is likely to be much longer. For this source $t_{eq} = 4 \times 10^9$ years and $B_{eq} = 1\ \mu G$, corresponding to $t_{1/2} = 10^9$ years at 1 GHz. The only way for the source to disappear faster is by expansion or by escape of the relativistic electrons. Since neither is likely to proceed at a speed much in excess of the effective sound speed [~3 B_{eq} (μG) n$^{-1/2}$ km s^{-1}], with n the proton density of the thermal gas, the time scale would be not much less than a few times 10^8 years if n is not less than 10^{-5} cm^{-3}.

Statistically it is also clear that radio galaxies are long lived and / or frequently rejuvenated. We noted already that 20 % of the giant ellipticals with $M_B < -22$ and classified E - S0/a in the RSA are at least weak radio galaxies; of the 17 with $M_B < -22.5$, two (Cen A, For A) have radio structures with extensions in excess of 0.5 Mpc. This again suggests lifetimes of the order of 10^9 years. The fact that there are not many radio sources with a break in their radio spectra due to synchrotron losses of the higher energy electrons, then also indicates that these are continuously or repeatedly replenished, presumably by the central nucleus.

Radio quasars resemble radio galaxies in their radio structures, and the energies and time scales are probably rather similar. There is no guarantee, however, that the optical emission remains constant during the lifetime of the radio source.

An interesting global argument may be made concerning the average optical quasar. From Fig. 6 it is seen that beyond B = 19 the numbers of galaxies and of quasars increase roughly by a factor of 2.5 per magnitude; hence, each magnitude interval makes about an equal contribution to the total light received. While beyond B = 22.5 we have no information on the quasar numbers, we probably do not make an error of more than a factor of 2 if we assume that the curves for galaxies and quasars remain more or less parallel. Since between B = 19 and B = 22.5 there are about 20 times as many galaxies as quasars, it would follow that the total energy radiated in B by all quasars is 0.05 times that radiated by all galaxies. If we would repeat this in the R color, we would find the ratio to be about 2 times smaller and in I perhaps 3 times. We conclude that on average galaxies emit perhaps 40 times as much light as quasars.

Let us now assume that all galaxies pass through a quasar stage and emit the same fraction of their total light as quasar light. A luminous galaxy with M_v = -23 emits during its 1.5×10^{10} year lifetime about 5×10^{62} ergs. Consequently its quasar would emit in total about 10^{61} ergs, an order of magnitude more than the typical minimum energy contained at a particular moment in a powerful radio galaxy. We shall find later some evidence that quasars tend to be associated with galaxies which have above average luminosities and consequently our value for the energy is rather a lower limit. But the basic conclusion is clear. The optical energy radiated by quasars and the energy put into relativistic electrons and magnetic fields by radio galaxies are globally of comparable magnitude.

QUASAR ABSORPTION LINES

Numerous absorption lines appear in the spectra of high redshift quasars (Fig. 7). Part of the reason for this is that the resonance lines of the abundant atoms and ions are mostly in the ultraviolet, and these are redshifted into the more easily observed parts of the spectrum. For example, Ly α absorption becomes observable from the ground ($\lambda \geq 3200$ Å) only around z = 1.6.

Since the absorption lines often are rather sharp, their wavelengths can be determined accurately. By looking for groups of lines with wavelength ratios corresponding to those for the stronger transitions of the common elements, we may identify lines with a common redshift - a line system. The criterion for the validity of the assignments to a certain line system then is the plausibility of the identification in terms of "reasonable" abundances and ionization equilibria. However, once a redshift system has been established with certainty, it becomes possible to look for weaker lines at precisely predicted places, and elements as heavy (and rare) as Zn have been detected. Three kinds of absorption lines may be distinguished:

I   Metal Line Systems in which groups of absorption lines are identified as just described. The doublets of CIV λ 1548 / λ 1551 Å and MgII λ 2796 / λ 2803 Å are particularly useful in the identification process because they are strong and close. Typical species in the metal line systems include H, various ionization stages of C, N, O, Si, S, Al and sometimes others.

II   The Ly α Forest. While most strong lines redwards of the Ly α emission line may be identified as belonging to some particular redshift system, this is not the case for most of the lines which appear very abundantly further to the blue. It is generally believed that these are Ly α absorption lines produced in intergalactic clouds with low heavy element abundances. Since the redshift of the absorbers must be less than that of the quasar, this immediately explains that such lines are not found redwards of Ly α in emission.

III   Broad Absorption Lines. Some quasars - BAL quasars - show very broad absorption features, the absorption starting blueward of strong emission lines and extending for several thousands of km s^{-1} or more. Their close connection to the emission lines indicates that they arise close to the quasar.

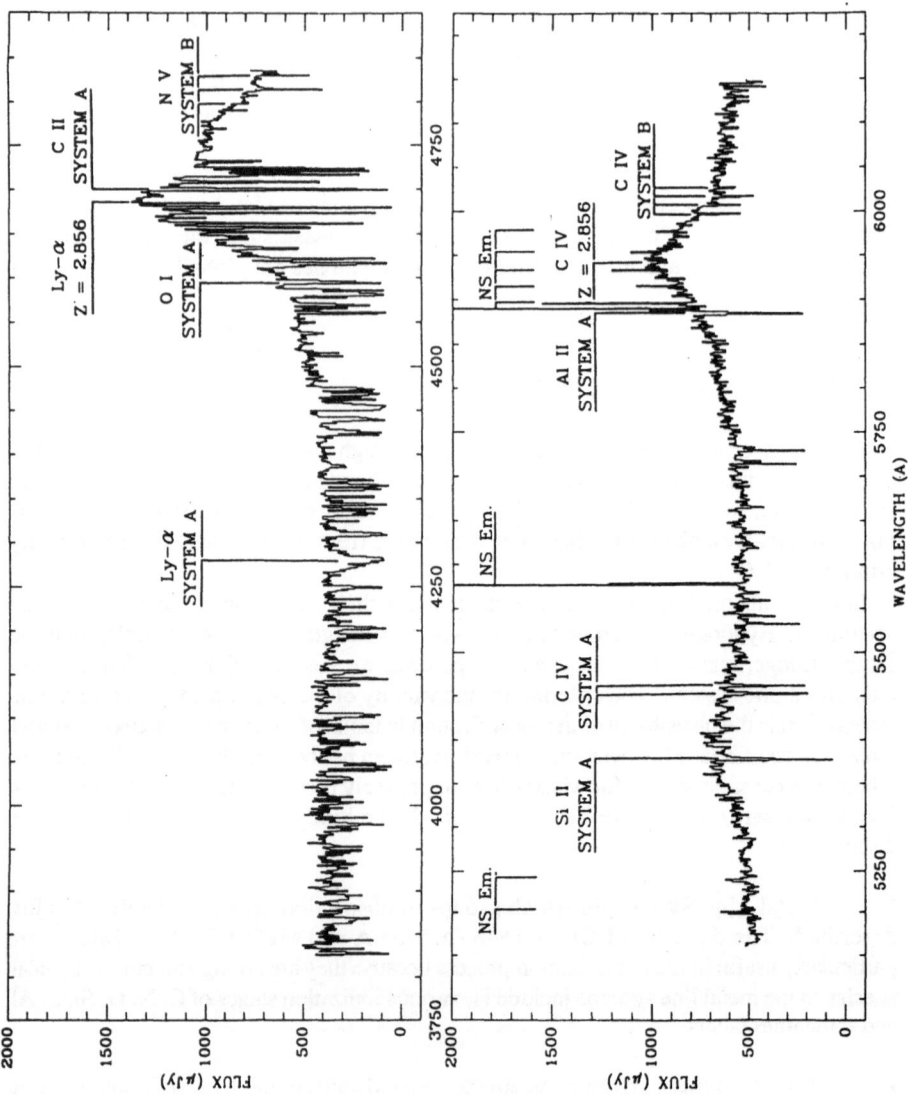

Fig. 7 : The spectrum of the quasar UM 402 (z_{em} = 2.856) obtained by Wampler (1990) with the 3.6-m ESO telescope. Several metal line absorption systems are present (system B centered at z = 2.88 composed of three sets spread over 1000 km s^{-1} and system A at z = 2.523 which is best fit by a 10 cloud model with a total velocity spread of 184 km s^{-1}; system A has generally lower ionization species than system B). The Ly α line at 4280 Å shows clear damping wings, corresponding to N_H = 10^{19} cm^{-2}. Blueward of the Ly α emission most lines belong to the Ly α forest. The narrow emission lines labeled NS Em are night sky lines which originate higher up in the earth's atmosphere.

Metal Line Systems

The redshifts range from around z_{em} to values close to zero. In principle, two explanations are possible:

(a)   absorption by matter ejected from the quasar, or
(b)   absorption by matter between the quasar and us, but unrelated to the quasar.

If the emission and absorption redshifts are z_{em} and z_{ab}, then on the ejection model the velocity of ejection V_e is given from elementary relativity by

$$V_e = c \; \frac{(1 + z_{em})^2 - (1 + z_{ab})^2}{(1 + z_{em})^2 + (1 + z_{ab})^2}$$

In one of the more extreme cases UM 675(0150-203), $z_{em} = 2.147$ and $z_{ab} = 0.389$ in the lowest redshift system, corresponding to $V_e = 0.67$ c (Sargent et al., 1988). To accelerate cool matter to such a velocity obviously poses problems. Since absorption systems are common - there is a second system in UM 675 with $V_e = 0.51$ c - we cannot assume that the matter is ejected in one very narrow jet.

Many authors have commented on the energetic problems which such a relativistic ejection poses. Perhaps this can be seen most simply in the interesting quasar pair UM 680/UM 681 (0307-195 A,B). For the two objects, Shaver and Robertson (1983) obtain emission line redshifts of $z_{em} = 2.144$ and $z_{em} = 2.122$, respectively; UM 680 has two definite absorption redshift systems at $z_{ab} = 2.1228$ and $z_{ab} = 2.0353$ (and a few more possible or probable systems), while UM 681 has definite absorption systems at $z_{ab} = 2.1220$, $z_{ab} = 2.0323$ and $z_{ab} = 1.7885$. Two of the systems are therefore coincident with velocity differences of only 80 and 300 km s^{-1}, respectively.

The difference in z_{em} corresponds (in the reference frame of the quasars) to a velocity difference $\Delta v = c \Delta z (1+z)^{-1}$ or 2100 km s^{-1}, which may indicate that the two are located in a cluster of galaxies. In a $q_o = 0$ cosmology, the transverse separation of the quasars (55″) is 750 kpc.

In an ejection model, the common system near z = 2.034 must be due to an expanding shell centered on one of the quasars and with $V_e \approx 10,000$ km s^{-1}. The radius of the shell must exceed 750 kpc. The observed equivalent widths of the Ly α absorption lines in this system are about 5.4 Å in both quasars, corresponding to $5(1+z_{ab})^{-1}$ Å at the absorber or 1.8 Å. Such absorption requires a column density of 6 x 10^{18} H-atoms cm^{-2} and therefore (neglecting ionized gas) about 1.4 x 10^{-5} g cm^{-2}. We then find for the shell a mass of 1 x 10^{45} g = 5 x 10^{11} M$_o$ and a kinetic energy of 5 x 10^{62} ergs. This is still too low since the radius of a spherical shell would have to be 4 times larger than the separation of the quasars to have a differential velocity of the absorption lines of only 300 km s^{-1}. Evidently, such figures stretch the probable by a substantial margin.

Other methods of estimating the minimum distance between quasar and absorber involve the populations of the fine structure levels of certain ions and the ionization equilibrium under the influence of the radiation field of the quasar (e.g. Wolfe et al., 1981). They confirm the high energies involved and document that radiation driven winds would be grossly inadequate (factors of 100) to generate the necessary momentum of the shells.

In view of the difficulties of the ejection models, it is now generally believed that gas clouds in intervening galaxies are responsible for the metal line absorption systems. In fact, there is increasingly direct evidence for this. The quasar PKS 2128-12 has $z_{em} = 0.500$

and Mg II absorption at z_{ab} = 0.4299. One galaxy has been found close to the quasar, and its redshift turns out to be z = 0.430 ± 0.001 (Bergeron, 1986). The line of sight to the quasar passes the galaxy at a transverse distance of 64 kpc. The quasar PKS 1327-206 (z_{em} = 1.165) has Ca II and Na I absorption at z_{ab} = 0.0179 (split into several components, the strongest at z_{ab} = 0.01750 and at z_{ab} = 0.01830). A galaxy is found at z = 0.0180 at a transverse distance of 20 kpc. Apparently, several clouds in this galaxy are responsible for the absorption (Bergeron et al., 1987). Several more such cases have been found. In a recent paper, Bergeron (1987) lists 10 cases of Mg II absorption where the associated galaxy has been found. The galaxies tend to be rather luminous, and the transverse distances range from 19-73 kpc, with a mean of 50 kpc. Earlier negative results may have been due to inadequate sensitivity. From Bergeron's study it appears that in most, though not all, cases a galaxy with the appropriate redshift may be found. Apparently, surprisingly large absorption cross sections are associated with galaxies. The galaxies causing Mg II absorption have a rather blue continuum, and 8 out of 10 have strong emission lines, suggesting that the extensive halos are associated with a rather high rate of star formation. However, such galaxies cannot be too small a fraction of all galaxies without a conflict arising between the mean cross-sectional radius R^x from statistical considerations and the mean transverse distance found in the Bergeron sample which at luminosity L^x should be about $\sqrt{2}$ times smaller. Yanni et al. (1990) have made use of the high frequency of [O II] λ 3727 emission in the absorbing galaxies to locate such galaxies by imaging through narrow band interference filters centered on λ = 3727 (1+z_{ab}). A more extreme form of this may be seen in the "associated absorption" in pairs of quasars. Cristiani and Shaver (1987) summarize the available observations for 20 quasar pairs of which six have absorption in the spectrum of the higher redshift quasar with z_{ab} approximately equal to z_{em} of the lower redshift quasar. The median transverse distance is 0.8 Mpc and the median velocity difference 700 km s^{-1}.

The large cross-sections of the absorbing galaxies are confirmed by statistical arguments. It may easily be shown that in the absence of evolution in a Friedman universe, the mean number of absorbers per unit redshift interval and with a given column density N(z) is given by

$$N(z) = c \; H_0^{-1} \; \pi \; R_0^2 \; \Phi_o(1+z) \; (1+2 \; q_o z)^{-1/2}$$

with (H_0, q_o) the cosmological parameters, Hubble constant and deceleration parameter, R_0 the radius and Φ_0 the local space density of the (spherical) absorbers. If galaxies are the absorbers, R_0 and Φ_0 would be functions of the luminosity L, and $R_0^2 \; \Phi_0$ should be integrated over L.

Taking for Φ the Schechter (1976) luminosity function

$$\Phi(L)dL = (L/L^x)^{-5/4} \; \exp(-L/L^x) \; d(L/L^x)$$

where L^x is the luminosity which corresponds to M_B^x = -20.6, and assuming $R \propto L^{0.4}$, Sargent et al. (1988) find for systems with W_o (Mg II λ 2796) > 0.3 Å : R_0^x = 95 kpc, for W_o > 0.6 Å : R_0^x = 60 kpc, and for a sample of C IV absorbers comparable, but sometimes larger, values. Such systems have column densities $N(Mg^+)$ < 10^{14} cm^{-2}, corresponding with solar abundances to $N(H)$ > 3 x 10^{18} cm^{-2}. From Tytler (1987a) we find that for spiral galaxies with L = L^x the radius at which N_H falls below this value is about 50 kpc. Taking into account also the projection factors for disk-like galaxies, Tytler concludes that the predicted number of absorbers is too small by a factor of six. Evidently, many of the numbers entering these calculations are uncertain, but the necessary cross-sections are perhaps larger than expected. However, the results of Bergeron (1987) rather conclusively prove that most absorbers are in fact associated with galaxies - either in the form of gas clouds in an extended halo or disk, or perhaps in a medium pervading small groups of galaxies.

The evolution of $N(z)$ with z has frequently been discussed with contradictory results. Inhomogeneous sampling due to different sensitivities and spectral resolutions combined with the z-dependence of the equivalent widths through $W = W_0 (1+z)$ may easily generate an erroneous z-dependence. Sargent et al. (1988), fitting a relation of the form $N(z) \propto (1+z)^s$ to the Mg II data (37 systems with $z_{ab} < 2$), obtain s = 1.45 ± 0.6, consistent with no evolution (s = 1) in $R^2\Phi$ for $q_0 = 0$. From earlier C IV data with $1.3 < z_{ab} < 3.4$, the same authors obtained s = -1.2 ± 0.7 which would imply that $R^2\Phi$ decreases at larger z values perhaps as a consequence of lower abundances of the heavy elements in young galaxies. This is confirmed by more detailed studies of some objects.

Meyer et al. (1989) found that in PKS 0528-250 zinc and sulfur are underabundant by a factor of 10 with respect to the sun in a redshift system at z = 2.81. Si, Cr, Fe and Ni are underabundant by an additional factor of 4. In interstellar clouds in our galaxy these elements have underabundances in the range of 10 - 100 which is generally ascribed to their condensation into grains, while Zn and S have more normal abundances and apparently do not participate much in grain formation. The observations of PKS 0528-250 then would seem to indicate that all heavy elements are underabundant by a factor of 10, but that only 80 % of the condensing elements are locked up in grains instead of the 99 % in our galaxy - corresponding to 8 % as much dust (per unit H) as in our galaxy.

This is confirmed by Fall et al. (1989) who found that 7 quasars with damped Ly α systems (z = 2.14 - 3.39) were redder than a control sample and concluded that this was due to dust with a dust / H ratio somewhere in the range of 0.05 - 0.25 times that in our galaxy. While much work remains to be done, ultimately much information about galactic evolution may result from absorption line studies.

The Ly α Forest

The large number of lines blueward of the Ly α emission are almost certainly mainly due to absorption by hydrogen clouds with low abundances of heavy elements. In many cases other Lyman lines may be identified. For example, in PKS 2000-330 a quasar with z_{em} = 3.78 a total of 280 absorption lines were found between 4190 Å and 5860 Å (corresponding to λ_0 = 880 Å - 1230 Å) by Hunstead et al. (1986). Four metal line systems and 45 Lyman systems were identified - including 16 with 3 or more Ly lines. The fact that in several such systems 3 unblended lines can be identified gives confidence that the basic identification of the Ly α forest is correct.

The evolution of the Ly α systems has been studied repeatedly during the last decade. Again, much care is needed in the selection of homogeneous samples and in the elimination of Ly α absorption belonging to metal line systems, Ly β, etc. Also the possibility that H_2 lines could be important still has not yet been fully excluded. The most remarkable result to have come out of these studies (Carswell et al., 1982; Murdoch et al., 1986; Tytler, 1987b) has been that whereas statistically when several quasars are combined the comoving number density of systems appears to increase with z, the opposite is the case in each individual quasar (the "inverse effect"). In fact, Tytler (1987b) finds that if one writes for the number of absorbers an expression involving both z_{ab} and z_{em} as follows

$$N(z_{ab}, z_{em}) \propto (1+z_{ab})^s (1+z_{em})^t$$

a best fit yields s = -1.1 and t = 3.4. In an individual quasar z_{em} is obviously a constant, while in a sample of quasars z_{ab} and z_{em} are statistically coupled since the Ly α forest is observed for 1216 Å $> \lambda_o > 912$ Å only, since in many quasars the continuum beyond the Ly limit is too faint.

The above expression may also be written as

$$N(z_{ab}, \lambda_r) \propto (1+z_{ab})^{s+t} \lambda_r^{-t}$$

where λ_r is the wavelength at which the absorbed radiation is emitted by the quasar. The explanation then would be that on the one hand the number of absorbers increases with redshift, but that it also becomes larger as λ_r diminishes. This could be due to the size of the continuum emitting region of the quasar becoming smaller with diminishing λ_r, provided the absorbing clouds have comparable dimensions. Tytler (1987b) mentions that hypothetically if the "blue bump" in the quasar spectra is due to thermal emission from an accretion disk with temperature increasing inwards, this would be the case. The Ly α clouds then would have to be very small, indeed (10^{16} cm). But this possibility is unlikely in view of the observation by Foltz et al. (1984) of correlated Ly α absorption in the gravitationally lensed quasar pair 2345+007 AB. They find 13 sets of Ly α lines common to both objects and perhaps 3 Ly α lines in only one component. Since the transverse distance of the light paths to the two components at the absorber ranges from 5 - 25 kpc, the cloud sizes would most likely be of this order or perhaps larger. This also makes it unlikely that the pc size scale of the Ly α emitting region could lead to a significant reduction in the number of Ly α components for λ_r around 1216 Å.

Alternatively, it has been proposed that clouds near the quasar might have higher ionization and consequently lower Ly α absorption due to the ionizing radiation of the quasar itself. This would reduce the number of Ly α absorbers for λ_r near 1216 Å. While Murdoch et al. (1986) conclude that this effect might well be quantitatively sufficient, Tytler (1987) reaches the opposite conclusion. A larger sample of high resolution observations which would more accurately delineate the $N(z_{ab}, \lambda_r)$ relation could settle this matter. The issue is important, because the magnitude of the effect depends on the average ionizing radiation field at large z which is still quite uncertain.

The Ly α lines have different equivalent widths, and consequently different column densities N_H are responsible. The relation between w and N_H is given by the "curve of growth", which depends on the velocity distribution of the absorbing atoms. We have

$$w = \int_0^\infty (1 - e^{-\tau_\lambda}) \, d\lambda$$

with $\tau_\lambda = C \, \psi(\lambda) \, N_H$, where $C = \pi e^2 \, m_e^{-1} \, c^{-2} \lambda_o \, f$ with λ_o pertaining to the line center, f the oscillator strength and $\psi(\lambda)$ the normalized profile of absorption coefficient due to both intrinsic and Doppler effects. Taking the Doppler profile for simplicity as rectangular with width a km s^{-1}, we have for low and intermediate N_H values:

$$w=\left[1-e^{-CN_Hc/(a\lambda_o)}\right]\frac{a}{c}\lambda_o$$

For unsaturated lines this yields $w = C N_H$, for saturated lines $w = (a/c) \lambda_o$. With a Gaussian profile the latter becomes proportional to a term containing $\ln N_H$. At still higher densities the line is broadened by the damping wings, in which case we have

$$\psi=\frac{\Delta}{\pi}\left[(\lambda - \lambda_o)^2+\frac{1}{4}\Delta^2\right]^{-1}$$

Since the intrinsic width Δ is small compared to the Doppler width in the cases of interest we may neglect the Δ^2 term in ψ and obtain

$$w=\left(\frac{C\Delta}{\pi}\right)^{\frac{1}{2}} N_H^{\frac{1}{2}} \int_{-\infty}^{+\infty} \left(1-e^{-u^2}\right)du$$

For Ly α we then have

unsaturated : $w = 5.4 \times 10^{-15} N_H$ Å

saturated : $w = 1216\ a/c$ Å

damping : $w = (N_H/1.9 \times 10^{18})^{1/2}$ Å

and the transitions are about where the subsequent expressions are equal. If we take $a = 40$ km s^{-1}, the saturated part of the curve of growth has $w = 0.16$ Å, the unsaturated part of the curve of growth terminates at $N_H = 3 \times 10^{13}$ cm^{-2}, while the damping part begins at $N_H = 5 \times 10^{16}$ cm^{-2}. If the velocity distribution is Gaussian with dispersion σ the saturated part of the curve of growth becomes

$$w = 5200\ (\sigma/c)\ [\log(1.3 \times 10^{12}\ N/\sigma)]^{1/2}$$

with σ in km s^{-1}. With $\sigma = 20$ km s^{-1}, the damping part begins around $N_H = 10^{18}$ cm^{-2}. In the absence of precise information about the velocity field, it is difficult to obtain accurate results for saturated lines.

Converting the w_o distribution into an N_H distribution, Tytler (1987a) concludes that the latter is given by

$$f(N_H) \propto N_H^{-1.5}$$

where $f(N_H)$ is the number of systems per unit interval in N_H. Per unit interval in $\log N_H$ we would find $N f (N_H) \propto N_H^{-0.5}$. This result is obtained by including all Ly α lines, including those from the metal line systems. From the fact that there is no special feature in this distribution, Tytler concludes that all the systems must have the same origin - presumably in intergalactic clouds. This conclusion has been contradicted, however, by Bechthold (1987) who finds that a two power law fit is much to be preferred. More high quality data are apparently needed to settle the issue.

The Nature of the Ly α Clouds

It is generally believed that the Ly α lines are due to intergalactic clouds (Arons, 1972). Sargent et al. (1980) considered the evidence for this to be:

The large number density of Ly α lines would require very large cross-sections (R = 0.4 Mpc) if galaxies were responsible.

The Ly α lines are not clustered like the metal line systems - the latter supposedly associated with galaxies.

The heavy element abundances in the Ly α clouds are low.

If this point of view is accepted, there are three different possibilities for the nature of the clouds.

(a)   The clouds may be pressure confined by an intergalactic medium.

(b)   The clouds may be in self-gravitating equilibrium.

(c)   The clouds may be more dynamic structures resulting from shock waves in the intergalactic medium.

Let us consider a Ly α line with $N_H = 10^{14}$ cm^{-2}. Since the cloud is optically thin in the Lyman continuum, photo ionization will proceed uniformly through the cloud and over a wide range of parameters a hydrogen cloud will be highly ionized and heated to 30,000 K by the (uncertain) radiation field produced collectively by quasars, young galaxies, etc.

If, from a correlation between Ly α lines in a gravitationally lensed quasar (Foltz et al., 1984) we take the cloud radius to be 30 kpc, we have $n_H = 5 \times 10^{-10}$ cm^{-3}. Writing the ionization equilibrium as

$$n_H \int_0^\infty J(v)\,\sigma_H(v)\,dv = \alpha(T)\,n_e^2$$

with $J(v)$ the omnidirectional radiation flux, $\sigma_H(v)$ the continuous absorption coefficient of hydrogen ($\sigma_H = 0$ for $\lambda > 912$ Å and 6.3×10^{-18} $(\lambda/912)^3$ for $\lambda < 912$ Å) and α the recombination coefficient ($\alpha = 4 \times 10^{-10}$ T$^{-3/4}$ for T > 5000 K), we obtain for plausible values of $J(v)$: $n_H = 0.1$ n_e^2 and therefore $n_e = 7 \times 10^{-5}$ cm^{-3}, corresponding to a cloud mass of 2×10^8 M$_\odot$. If the cloud is in pressure equilibrium with the surrounding intergalactic medium, $n_e T$ should be the same inside and outside the cloud and intergalactic parameters $n_e = 10^{-5}$ cm^{-3}, T = 2×10^5 K or $n_e = 10^{-6}$ cm^{-3}, T = 2×10^6 K could be appropriate. Higher intergalactic temperatures might lead to evaporation of the clouds (Ostriker and Ikeuchi, 1983), although small scale magnetic fields could counteract this.

Alternatively, we may consider the cloud to be self-gravitating. With the parameters chosen, the gravitational energy is about 30 times smaller than the thermal energy. Since the gravitational energy per unit mass (for fixed N_H) is proportional to $R^{3/2}$, a 10 times larger radious would bring a balance, but the mass would become $6 \times 10^{10} M_0$, and the total mass of all Ly α clouds would exceed that of all galaxies.

If, however, the condition for self-gravity is met by introducing dark matter, a more reasonable situation may be brought about. In fact, Ikeuchi and Norman (1987) have developed a model of galaxy formation involving both cold dark matter (10^{-4} eV particles) and hot dark matter (30 eV) and find that numerous dwarf galaxy like structures form with masses below $10^{10} M_0$ and that the expected density of the objects suitable for the Ly α absorption is reasonable. As time goes on, many such objects would collapse and undergo star formation to become dwarf galaxies. Consequently, an evolutionary effect would be expected with fewer Ly α systems at low z. Some of the dwarf galaxies might produce metal line absorption systems.

Explosive events in young galaxies might send shock waves through the intergalactic medium, and these might lead to compressed regions with Ly α absorption (Chernomordik and Ozernoy, 1983).

BAL Quasars

While it is difficult to believe that the absorption lines with a redshift very different from z_{em} could be due to ejection, there is, nevertheless, much evidence that ejection at more modest velocities does take place. IUE observations of the nearby Seyfert galaxy NGC 4151 have revealed a rich absorption spectrum with resonance lines of H, C II, C IV, N V, O I, Si II (Bromage et al., 1985). Absorption extends from slightly above the systematic velocity to 1000 km s^{-1} below it, indicating that outflow takes place.

Larger velocities are found in the BAL quasars. Broad absorption troughs are found blueward of some emission lines which extend for several thousands of km s^{-1}, in some cases up to 0.1 c or more. Much structure may be present in the absorption troughs. Such absorption is mainly found in high ionization resonance lines (N V, C IV), while it is weak or absent in Mg II or Ly α. In typical cases of resonant line scattering, as is likely to apply to N V, the number of photons is conserved and the loss of photons from the continuum source (or the occulted emission line region) must result in an equal amount of emission line photons appearing (but at different frequencies due to the outflow) in case the continuum source is fully covered. From the fact that the emission is much less, it is generally concluded that the covering factor is no more than 0.2 (e.g. Turnshek et al., 1988), and that therefore in only 20 % of the cases the BAL phenomenon would be observable. With the BAL quasars relatively common (5-10 % of quasars), it would follow that a large fraction of all quasars would have such outflows. However, Surdej and Hutsémekers (1987) and Braun and Milgrom (1990) argue that in flows with large velocity variations the emission to absorption ratio arguments may not be convincing. Various differences in emission line characteristics between BAL and other quasars have been found, but it is unclear if these relate to aspect angles or to intrinsic differences. Also the BAL quasars appear to have weak radio emission, but the statistical significance of this is still unclear (Turnshek, 1984). For the object UM 232 Bregman (1984) finds no absorption of X-rays at 1 keV (in the quasar frame) and concludes that the column density of absorbing gas is less than 10^{22} cm^{-2}. Since line emission coming from a pc size region is absorbed by the BAL gas, it would follow that $n < 10^3$ cm^{-3}, much less dense than the BLR clouds, unless the gas is in thin sheets.

GALAXIES WITH ACTIVE NUCLEI

Quasar Galaxies

We have already discussed some aspects of the galaxies with Seyfert nuclei. These were sufficiently near to study the galaxies in detail. The situation is much more difficult in the case of brighter quasars. Because of their relative rarity, their distances are on average much larger, while the luminous nucleus tends to blot out the galaxy. In addition, the K-corrections tend to weaken the galaxy with respect to the quasar when observations are made in B or V colors. As a consequence, in most cases it is difficult to determine with confidence even the most general aspects of the galaxies, for example whether they are ellipticals or spirals. In a typical photometric cross-section of a quasar image the central part is entirely dominated by the unresolved nucleus, while at large distances small errors in the determination of the sky background have big effects. As a consequence, we can only determine the surface brightness in an intermediate annulus.

Typical elliptical galaxies have an intensity profile following a "de Vaucouleurs law" or an "$r^{1/4}$ law" with the surface brightness varying as

$$\Sigma(r) \propto \exp\left[-7.65 \ (r/r_e)^{1/4}\right]$$

with r_e the radius containing half the total light. Typical spiral galaxies contain an "exponential disk" in which

$$\Sigma(r) \propto \exp(-r/r_o)$$

with r_o a scale length. However, spiral galaxies also have a bulge with an $r^{1/4}$ profile, which is relatively strong in Sa and weak in Sc galaxies. Unfortunately, the classification of galaxy types as usually done by inspection of photographic plates and the classification by the photometric profile - usually determined from CCD images - have only an imperfect correlation. Since in addition the most visible differences between the $r^{1/4}$ law and the exponential disk occur rather close to the center, the nature of most quasar galaxies is open to doubt and results are frequently obtained *assuming* a galaxy type.

The $r^{1/4}$ law allows a rather good determination of the integrated brightness of a galaxy when only limited information in an annulus is available. For example, it may be seen in Figure 8 that if the light is measured in an annulus with radii r^* and $2r^*$, the correction to the full brightness is a rather flat function of r^*/r_e over a range of a factor of ten in r^*/r_e. The advantage of this is that the luminosity of elliptical galaxies underlying quasars may be rather well determined. Unfortunately, the determination of r_e is usually not very reliable since only $r_e^{1/4}$ is measured and since small errors in sky background have rather large effects. To a lesser degree the determination of the integrated brightness of an exponential disk may also be made rather well in the absence of a precise determination of r_o.

Radio Quasars

Since spiral galaxies rarely, if ever, have strong radio emission, while in ellipticals this is not uncommon, we shall treat all galaxies associated with radio quasars ($P_{1.4\,GHz} > 10^{25}$ WHz^{-1}) as ellipticals. Recent samples of radio quasars used in imaging

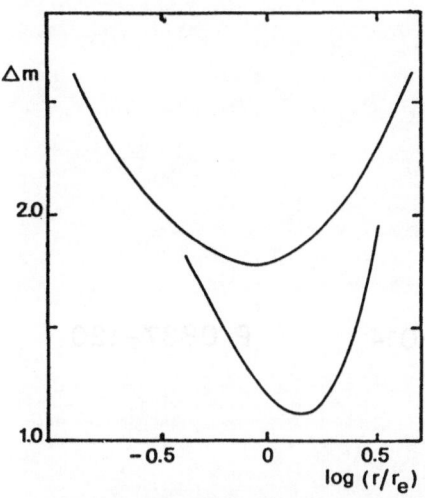

Fig. 8 : The difference in magnitude between the light in an annulus with radii r and 2r and that in the whole galaxy model as a function of r/r_e or r/r_0 for models of elliptical ($r^{1/4}$ law) galaxies, respectively exponential disks.

studies include those of Malkan (1984), Smith et al. (1986) and Romanishin and Hintzen (1989) who all used rather similar reduction techniques. In addition, Véron-Cetty and Woltjer (1990a) studied a statistically "complete" sample of 20 objects extracted from the Véron-Cetty - Véron (1989) catalogue, which itself, of course, is not at all complete. For the underlying galaxies (Fig. 9) they found (extrapolated to infinite radius) $<M_V> = -23.26 \pm 0.12$ with a dispersion of $\pm 0^m50$, while the other samples (including only quasars with $z < 0.5$) contained 29 objects with $<M_V> = -23.40 \pm 0.14$ with a dispersion of $\pm 0^m67$. The VW sample is less likely to have been affected by emission lines, and since the results were based on imaging at a rather long wavelength (i band 7200 - 8500 Å), the separation of nucleus and galaxy should have been slightly more easily accomplished than in samples imaged in bluer colors.

It is interesting to compare these results with those obtained for powerful radio galaxies. Sandage's (1972) photoelectric data transformed to $q_o = 0$ and "infinite" radius yield $M_V = -23.28$ with a dispersion of $\pm 0^m49$, while those of Smith and Heckman (1989) obtained from CCD photometry would correspond to $M_V = -23.34$ with a dispersion of $\pm 0^m65$. These data therefore suggest that radio galaxies and radio quasar galaxies may well be objects of the same kind. This would be consistent with theories in which quasars and radio galaxies are identical objects with variability or aspect dependent effects determining the observability of the quasar nucleus.

Only fragmentary data are available on the colors of the radio quasar galaxies. These suggest that they are somewhat bluer than typical giant ellipticals, perhaps as a result of recent star formation. The spectroscopic data of Boroson et al. (1985) point in the same direction. Also Miller (1981) observed a number of radio quasars spectroscopically and showed that the absence of the Mg I absorption feature in the integrated light is inconsistent with the assumption that the underlying galaxy is a gE with an old population.

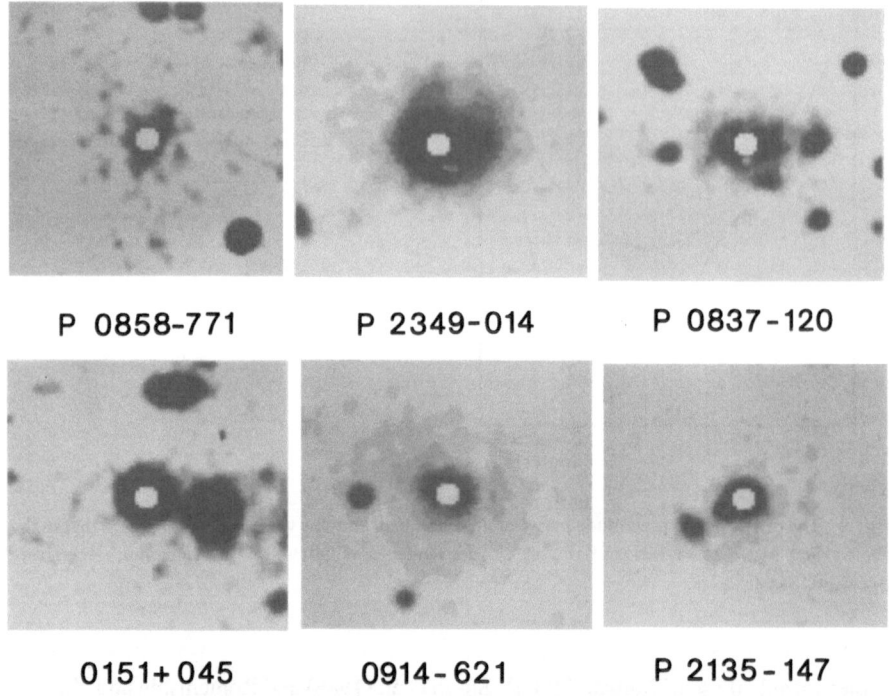

P 0858-771   P 2349-014   P 0837-120

0151+045   0914-621   P 2135-147

Fig. 9 : Some quasar galaxies (Véron-Cetty and Woltjer, 1990a). PKS 0837-120 (3 C 206) is located in a cluster, confirmed by the redshifts of two other galaxies in the field. PKS 2135-147 has two companion galaxies, with Stockton's compact galaxy at 2″ from the quasar accounting for the protrusion towards the more distant galaxy. PKS 0858-771 and especially PKS 2349-014 exhibit clear asymmetries. Near the radio quiet quasar 0151+045 (PHL 1226, z = 0.404) a galaxy is located at 6″ (with z = 0.159) and the spectrum of the quasar shows Mg II absorption at z = 0.160 (Bergeron et al., 1988). The radio quiet quasar 0914-621 was first found as an IRAS source (Strauss et al., 1988). The central area of the images, where the subtraction of the nucleus is most uncertain, has been left white. Each image is about 28″ x 28″ in size.

Radio Quiet Quasars

Since typical Seyferts are "radio quiet" spiral galaxies, it is usually assumed that QQ are also spirals. Little direct evidence is available to support this point of view. In the comparison of the properties of the RQ and QQ galaxies much care is needed to ensure that comparable samples are obtained of each kind - samples with the same redshift distribution, the same absolute quasar luminosities, etc. For relatively luminous quasars Véron-Cetty and Woltjer (1990a) find the QQ-galaxies to be 0^m6 fainter than the RQ-galaxies if both are ellipticals, and about 1^m1 if the QQ-galaxies are spirals (Fig. 10). This would be compatible with a model in which the QQ are ellipticals with the same relationship between radio emission and galactic absolute magnitude, as was found for radio galaxies by Auriemma et al. (1977). Alternatively, it could be that the QQ are the continuation of the Sy 1 sequence and that only very luminous spirals can have quasar like nuclear luminosities. For the nearby objects F 9 (QQ/Sy 1) amd 0914-621 (QQ) Véron-Cetty and Woltjer (1990b) find photometric profiles which accurately follow an $r^{1/4}$ law, and which consequently may well be ellipticals.

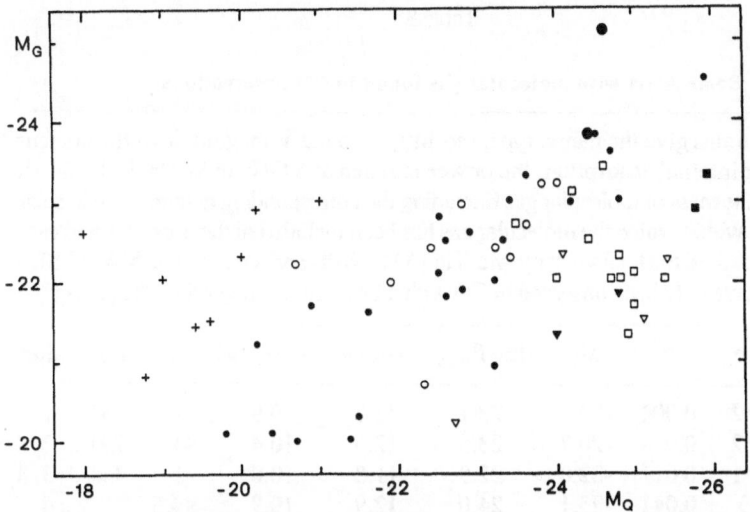

Fig. 10 : The relation between the absolute visual magnitude of the nucleus M_Q and that of the underlying galaxy M_G (on the assumption of spiral galaxies) for various samples of quasars (Véron-Cetty and Woltjer, 1990a). Inverted triangles are upper limits. While among the Seyferts there is not much of a relation between M_Q and M_G, among the quasars ($M_Q < -23$) few low luminosity galaxies are found while some are among the most luminous galaxies known.

Other Properties of Active Galaxies

To power the AGN and to provide the material for the observed outflow from these, an ample supply of gas is needed to the region very close to the nucleus. Recent observations are beginning to show that, in fact, much gas is frequently present in the central region and that asymmetries in the mass distribution may assist the movement of the gas to the center. At the same time this gas may partly condense into stars, thereby giving rise to mixed AGN-starburst objects. A few examples of gas rich, IR luminous AGN are given in Table 8. The optical data are taken from the VV catalogue (Véron-Cetty and Véron, 1989). The molecular gas is inferred from measurements in the 2.6 mm CO line, with the conversion of CO intensity to H_2 mass based on data for molecular clouds in our own galaxy and with the molecular gas mass also including the corresponding amount of helium. Of course, it is not certain that the conversion factor from CO to H_2 is entirely appropriate for these AGN.

If all the gas were used to generate the luminosity of these objects by falling onto a black hole and releasing $0.1\ mc^2$, the amount of gas detected would suffice for 10^{11} years. However, it is likely that much of the gas will form stars, because the area densities are very high, frequently $> 1000\ M_o\ pc^{-2}$. In fact, several galaxies without known AGN have very similar characteristics, and it appears probable that much of the IR radiation is powered by recently formed massive stars. The detection of much molecular gas in some radio galaxies - presumably associated with E galaxies - is especially interesting. In 3C 84 (NGC 1275) this is supposed to be matter coming in from the outside in a cooling flow, while in other cases an encounter with a gas rich system may well play a role.

Table 8

Some AGN with molecular gas found in CO observations

Subsequent columns give the name, type, redshift, absolute V magnitude of the nucleus uncorrected for internal absorption, the power radiated at 5 GHz in $W Hz^{-1}$, the far IR luminosity and the mass of molecular gas (including the corresponding helium) both in solar units, the radius within which the molecular gas has been included or the upper limit in case of unresolved sources, the total velocity width in CO (FW0I) and a reference. MARK 1014 and 4 C 12.50 have only been observed in CO with a beam much larger than the galaxy.

Galaxy	type	z	M_V	Log $P_{5 GHz}$	Log L_{IR}	Log(H_2+He)	R	ΔV	Ref
NGC 1068	Sy 2	0.004	-20.1	23.1	11.7	9.9	4	250	1
Arp 220	Sy 2	0.018	-20.7	23.5	12.5	10.4	<1.1	750	2, 8
NGC 7469	Sy 1	0.017	-22.0	22.9	11.8	10.0	2	350	3, 8
MARK 231	QQ	0.041	-23.1	24.0	12.9	10.9	<4.5		4
MARK 1014	QQ	0.163	-24.2	24.0	12.8	11.1		400	5
NGC 1275	RG	0.018		25.7	11.7	10.3	<3.5	400	6
4 C 12.50	RG	0.122		26.3	12.6	11.3		950	7

References: 1. Myers and Scoville, 1987; 2. Scoville et al., 1986; 3. Sanders et al., 1988a; 4. Scoville et al., 1989; 5. Sanders et al. 1988d; 6. Lazareff et al., 1989; 7. Mirabel et al., 1989; 8. Sanders and Mirabel, 1985.

Elliptical galaxies without AGN may also contain more gas than has been believed on the basis of 21-cm observations. NGC 4272 with $<1.1 \times 10^7 M_0$ of H I turned out to have $2 \times 10^7 M_0$ of molecular gas (Huchtmeier et al. 1988), while the blue E0 galaxy NGC 3928 has $1 \times 10^9 M_0$ of H_2 within 2 kpc from the center, about 1.5 times as much as the total H I which mainly is situated further out (Gordon, 1990). While it is true that most CO observations have been done in selected systems with a high far infrared luminosity, nevertheless the data presented in this section show that there may not be too much of a problem in finding gas that can accrete on the central black hole.

Even if there is enough gas, it is not necessarily easy to make this flow to the center. As long as angular momentum is conserved this is virtually impossible, and therefore it is necessary to have sufficiently strong deviations from axial symmetry in the galactic gravitational field - at least if no very strong magnetic fields are present. Such fields may well play a role very close to the center, but it is doubtful that they are strong enough at 1 kpc distance from it.

Two kinds of asymmetry are actually observed. First of all, in barred spirals angular momentum is certainly not conserved along flow lines and complex in- and out-flow patterns may occur. We noted earlier that the percentage of galaxies classified as SB is not significantly higher among Seyferts than among all spirals in the RSA catalogue. However, it may well be that bars in the inner parts of Seyferts occur more frequently than the standard classification indicates. For example, in the prototypal Sy 2 NGC 1068 (Sb) Scoville et al. (1988) discovered a bar in observations in the K-band (2.2 μm), which is not apparent at visual wavelengths. Subsequent imaging studies in K by Thronson et al. (1989) show this structure very clearly. The bar has a length of about 30″ (3.5 kpc if the distance is 24 Mpc) and the main CO concentration is in a ring just outside it, indicating a close connection. It is not known at present if such hidden bars are found in other Seyferts. The bar in NGC 1068 has a strong rounder concentration of light in the inner 1/3 of the bar. It may be relevant that in a recent study Pfenniger and Norman (1990) found that it is exactly such a combination which fosters a rapid inflow.

Another type of asymmetry results from interactions between galaxies and the consequent mergers. The interaction may be noticeable by various tails coming out of the galaxies, while the merging process is more specifically characterized by double nuclei in one perturbed galaxy. Such configurations are common in powerful radio galaxies (Smith and Heckman, 1989) and in the powerful IR emitters (Sanders et al., 1988c, Melnick and Mirabel, 1990). While the nuclei spiral in towards each other by dynamical friction, the gravitational field will be much perturbed leading to possible gas flows to the center which may result in star formation, formation of an AGN or both.

While the studies of quasar galaxies, Seyferts and radio galaxies, which we have discussed, generally refer to redshifts of at most a few tenths, recently some radio galaxies have been discovered at substantially larger distances (mainly z = 1-2, but up to z = 3.4). These tend to have bizarre shapes (see the plates accompanying Le Fèvre et al., 1988) with strong condensations. In the case of 3C 326.1 (z = 1.82) a very large (100 kpc) relatively smooth Ly α emitting cloud (Fig. 11) is found in association with the galaxy (McCarthy et al., 1987a). This cloud which emits 10^{11} L_0 in Ly α appears also to emit weakly in lines of C II and C III and is therefore presumably not composed of primordial material. The line width (FWHM) is about 1000 km s^{-1}. About 10^6 massive O stars would be required to maintain its ionization.

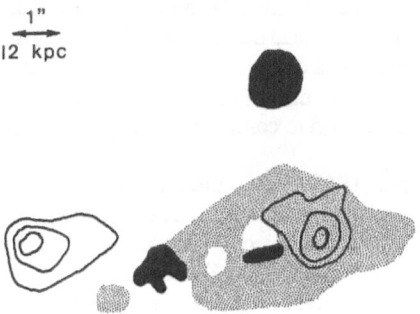

Fig. 11: Sketch of 3 C 326.1 (z = 1.82). Indicated are the isophotes of the radio source and the stronger parts of the Ly α emission (dotted area), both from McCarthy et al. (1987a). The concentrations of continuum emission in black are from Le Fèvre et al. (1988). Note the parallel orientation of the axes of the radio emission and of the Ly α cloud and associated continuum emission.

One of the remarkable findings about these objects has been that the optical continuum and the radio structure are aligned (Chambers et al., 1987, McCarthy et al., 1987b). Evidence has been presented (Chambers and McCarthy, 1990) that the optical spectra of 3 C 239 and 3 C 256 (z ~ 1.8) show absorption features which correspond to those seen in the uv spectra of O and B stars. This would suggest that the stellar distribution is related to the radio structure, perhaps due to the radio source having somehow induced star formation. Chambers and McCarthy find that the stellar population corresponding to an age of several times 10^8 years and diminishing star formation rate gives a good fit to the spectral data.

Recently, di Serego Alighieri et al. (1990) have measured suprisingly large linear polarization in the uv (rest frame) continuum of 3 C 277.2 (z = 0.77, p = 21 %) and 3 C 368 (z = 1.13, p = 8 %), with the E vectors roughly (within 25°) perpendicular to the radio axis. While a contribution from synchrotron radiation cannot be excluded, these authors prefer a model in which beamed radiation from the nucleus is scattered by dust or perhaps by electrons. Evidently, more spectro-polarimetric data on the high redshift galaxies are needed before their nature and age can be ascertained.

LUMINOSITY FUNCTIONS

A survey for a certain class of objects frequently involves different wavelength ranges with different sensitivity limits. For example, a survey for quasars may be done in X-rays, but the identification and the redshift determination require optical data. A sample is said to be "complete" if in an area of the sky all the objects within certain limits are included: for example, all quasars with B-magnitudes brighter than 20 within 10° from the south Galactic pole. In principle, such a sample is obtained by inspecting all the objects in that area with B < 20 and retaining only those objects which are quasars as seen from their spectra.

The situation becomes more complex if an X-ray sample of quasars is to be "complete". We may, of course, find all X-ray sources down to a certain flux level and have a complete sample of X-ray sources. However, when then we proceed to identify these sources, not all need to have an optical counterpart within the limit of our optical observation. All we can do is to construct the complete sample of all quasars in an area of the sky which have an X-ray flux above a certain limit and an optical flux above some other limit. If then may be the limits are different in different parts of the sky and we wish to combine the samples further problems arise.

To deal with this type of problems, Schmidt's (1968) V/V_m method may be used. Suppose that in a Newtonian non-expanding universe we survey within a solid angle Ω a set of objects with an apparent luminosity limit l_0. Let a certain object Q with luminosity L be found at a distance R. The apparent luminosity $l = L / (4 \pi R^2)$. The volume V nearer than Q is $1/3 \, \Omega R^3$. The maximum volume V_m in which the object could have been found is $1/3 \, \Omega R_m^3$ where R_m is given by $l_0 = L / (4 \pi R_m^2)$.

Suppose that the class of objects we look at is uniformly distributed in space. The chance of finding an object in a volume V is then proportional to V, and consequently the average value of V/V_m should be 0.5. Suppose we have overestimated the completeness of our sample - for example, by l_0 being in reality larger than we thought; in this case we find fewer objects in the more distant parts of V_m, and we will have $<V/V_m>$ smaller than 0.5. Suppose there is evolution in our volume, in the sense that the density is larger at large distances; the more distant parts of V_m will be more heavily populated, and we will find $<V/V_m>$ larger than 0.5. Finally, suppose like in our X-ray selected sample we have two limits to consider; all we have to do is to compare the volumes accessible for each of these and the smaller one is the V_m.

In the real Universe we have to use the expressions appropriate to the chosen cosmological model to connect L and I and to determine V and V_m, and we have to take into account the effects of the redshift, which effectively means the K-term. The fact that Schmidt found $<V/V_m> = 0.7$ for a complete sample of optically identified radio quasars constituted proof that the quasar population has substantial evolution.

A similar methodology may be used to determine the luminosity function from observations of a complete sample of objects. For each object i we determine the $V_{m,i}$. We consider all objects we have found in a luminosity interval dL around L. The contribution to the density of each detected object is $V_{m,i,L}^{-1}$ and the total density $\Phi(L)$ in that interval is given by

$$\Phi(L) = \sum_i V_{m,i,L}^{-1}$$

This is for a uniform distribution through the largest volume $V_{m,i,L}$. Binning at the same time by redshift interval, we may obtain $\Phi(L,z)$ - that is the evolution of the luminosity function. For the special problems which arise when V_m has angular dependence or when samples with different limits are to be combined, see Felten (1976).

When only a small number of objects is available, the determination of $\Phi(L,z)$ poses problems. Frequently, it turns out to be necessary to assume an evolutionary law with one or two parameters, and then to obtain the parameters which lead to the best fit. One way of doing this is to use "density weighted volumes" V' (Schmidt, 1968). Writing

$$\Phi(L,z) = \Phi(L,o)\, \rho(L,z)$$

with ρ the evolution function, we have

$$V' = \int_o^z \rho(L, z)(dV/dz)\, dz$$

Since in V' the distribution is uniform, we should have $<V/V'> = 0.5$. In practice, one then takes the simplest smooth $\rho(L,z)$ which is not contradicted by observation and which satisfies this condition. While this procedure leads to results, it is not necessarily unique, and one has the impression that much unnecessary controversy has occurred by the use of different $\rho(L,z)$ functions. The situation would be different if physical arguments could be used to a priori select a particular type of function, but our models of AGN are still too primitive to do so in a convincing manner. As a consequence, $\rho(L,z)$ is more a type of interpolation function, and its results for parts of the L,z plane where no data exist should be treated with caution. In his first attempt at a determination of ρ, Schmidt (1968) observed that for small z we have

$$V = A^3 (1 + A)^{-3}$$

independent of q_o (Sandage 1961), where the "luminosity distance" A is given by

$$A = q_o^{-1} z + q_o^{-2} (q_o - 1)[(1 + 2 q_o z)^{1/2} - 1]$$

$[q_0 = 0 : A = z(1 + 1/2 z)]$, and therefore decided for his quasar sample to try

$$\rho = (1 + A)^n$$

which would give a smooth behavior locally. Choosing $n = 5$ yielded $\langle V'/V'_m \rangle = 0.5$. As more data became available, it became clear that the evolution was luminosity dependent, and Schmidt and Green (1983) took

$$\rho(M_B,z) = \exp[k(M_0 - M_B)\,\tau(z)] \qquad M_B < M_0$$

$$\rho(M_B,z) = 1 \qquad M_B > M_0$$

where M_B is the blue absolute magnitude in the reference frame of the quasar, $\tau(z)$ the light travel time as a fraction of the age of the Universe, and k and M_0 two constants to be determined from observation. The expression for τ is $\tau = 1 - (1 + z)^{-1}$ for $q_0 = 0$ and $\tau = 1 - (1 + z)^{-3/2}$ for $q_0 = 0.5$. For q_0 ranging between 0.1 and 0.5 values of M_0 ranged between -23 and -20 and the corresponding k values between 4 and 2. At high luminosities the evolution is very strong. A $(k = 4, M_0 = -23)$ model gives at $\tau(z) = 0.8$ for $M_B = -29$, an increase of 2×10^8 compared to local values, very consistent with the observation that no such quasars have been found locally!

The first transit surveys (Schmidt et al., 1986), while agreeing with these models for $z < 2.9$, found far fewer objects at larger redshifts than predicted by the model (0 found for 30 - 60 predicted in 7.8 square degrees down to magnitudes around 20) confirming a rather strong cutoff around $z = 3$ (Osmer, 1982). Subsequently, however, a slightly deeper survey covering 7 square degrees effectively yielded 9 quasars with $z = 3.0 - 3.8$ (Schmidt et al., 1987), thereby indicating a more gradual decline of the comoving density beyond $z = 3$. This was confirmed by the subsequent discovery of several quasars with $z > 4$. Hazard et al. (1986) found six quasars in about 60 $\deg^2$ with $z = 3.3 - 3.8$ at very bright magnitudes $(R = 17 - 18)$. This seems to indicate that the luminosity function at high redshift is rather flat and that at the highest luminosities the maximum co-moving density is reached at a higher redshift (see also Warren et al., 1988).

In the mean time, Boyle et al. (1988) had firmed up the evolution of uv-excess quasars. They found that a two power law fit $\rho(>L) \propto L^{-\alpha}$ with $\alpha = 3.9$ (for $q_0 = 0$) at high luminosities and $\alpha = 1.75$ at low luminosities gave a good fit. For the switch-over point they obtain $M_B^* = -23.2$ at $z = 0$. Exponential evolution did not fit their data very well, but a power law fit $L_B^x \propto (1+z)^{3.4}$ did. However, predicting the number of quasars for $z = 2.5 - 3.0$, they obtained $2.0/\deg^2$ for $B = 18.5 - 19.5$ about a factor of 2-3 higher than found by the objective prism surveys, perhaps again indicating a reduction near $z = 3$.

The main conclusion from these studies appears to be that we have a fair idea about the evolution of quasars with $z < 2.2$, but that at larger z much work remains to be done. The various formulae that have been proposed for $\rho(M_B, z)$ are interpolation formulae without predictive value, and the precise evolution at larger redshifts can only come from observation. In Fig. 12 we show a global picture of the quasar luminosity function (density per unit co-moving volume and per magnitude as a function of absolute magnitude for $H_0 = 50$ km s^{-1} Mpc^{-1} and $q_0 = 0$) at various redshifts. For $q_0 = 0.5$ the diagram contracts with at $z = 4.3$, the absolute magnitudes 1^m8 fainter and the densities 12 times higher. The curves for $z > 3$ are still extremely uncertain.

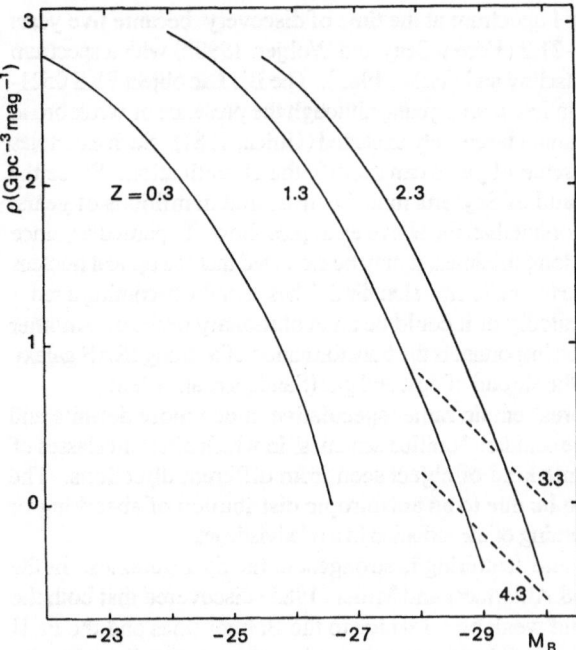

Fig. 12 : The quasar luminosity function at different redshifts. The curves for z = 3.3 and z = 4.3 are still extremely uncertain.

In the literature appears frequently the distinction between density evolution and luminosity evolution. In Fig. 12 the first corresponds to a translation of the z = 0 curve vertically upwards, the second to a translation to the right. In the first, $\Phi(L,z)$ would be obtained from $\Phi(L,o)$ by multiplication with a function of z. In luminosity evolution it would be L that is multiplied by a function of z. Since it hardly seems likely that the quasars which are seen near z = 0 have already existed at much larger redshifts, it is not clear that the luminosity evolution has a physical meaning. Statistically, evidently luminosity dependent density evolution can reproduce any observed luminosity function.

UNIFIED MODELS

At the beginning of these lectures we have discussed eleven classes of AGN and related objects. We shall now consider the question to which extent these classes are really distinct and different. Could it be that the same object seen pole on and from its equatorial plane would be placed in two different classes ? Could it be that variability causes an object to belong to different classes at different times ? Or does one class of object evolve into another class ?

Variability on a time scale of years has changed the classification of several objects. For example NGC 3516 - one of the 11 objects first described by Seyfert (1943) - has been changed from a Sy 1 to something close to a Sy 2 (Andrillat and Souffrin, 1968). Fairall

9, a quasar (M_V = -24) with a Sy 1 spectrum at the time of discovery, became five years later a Seyfert galaxy with M_V = -21.2 (Véron-Cetty and Woltjer, 1990b) with a spectrum approaching that of a Sy 2 (Kollatschny and Fricke, 1985). The BL Lac object PKS 0521-36 turned into a Sy 1 like object in less than 6 years, although the presence of weak broad emission lines at the earlier date cannot be entirely excluded (Ulrich, 1981). Such examples show that variability on a time scale of years can modify the classification. Since the lifetime of large radio galaxies and of Seyferts must be measured in millions of years, variability may well be more important than the above examples show. In particular, since the large radio sources have such long lifetimes, it may be expected that the optical nucleus of a radio quasar could fade before the radio emission diminishes, thereby becoming a radio galaxy. This could happen repeatedly or it could be an evolutionary process. Another evolutionary process which may be important is the transformation of a strong IRAS galaxy into a quasar by the expulsion of the shroud of dust and gas (Sanders et al., 1988b).

While the evolutionary pictures remain rather speculative, much more definite and testable predictions result from the standard "unified schemes" in which different classes of AGN are supposed to be the same kind of object seen from different directions. The directionality is usually taken to be due to an anisotropic distribution of absorbing or scattering matter or to intrinsic beaming of the radiation in a relativistic jet.

The evidence for absorption and scattering is strongest in the Sy 2 galaxies. In the prototype Sy 2 galaxy NGC 1068 Antonucci and Miller (1985) discovered that both the nonstellar nuclear continuum and weak broad wings to the Balmer lines and the Fe II emission are linearly polarized (p ~ 16%) with the polarization plane perpendicular to the symmetry axis of the nuclear radio morphology. Since such strong polarization in the emission lines must indicate scattering, it follows that there must be a hidden nucleus, some of the light of which is scattered to us. As the nucleus apparently has broad emission lines (the narrow lines have low polarization), it would present itself as a Sy 1 nucleus if seen directly. The simplest geometry which would produce the observed pattern would involve an absorbing ring around the nucleus in its equatorial plane (the "absorbing torus") with above it some free electrons to scatter the light of the nucleus; alternatively, the electrons might constitute the hot inner surface of the torus seen along a line of sight which makes a modest angle with the line of sight. Scattering by dust would not reproduce the wavelength independent polarization seen in NGC 1068, but could perhaps also make a contribution in other objects. Goodrich (1989) discusses several objects which have varied between Sy 1 and Sy 1.8 - 1.9 (NGC 2622, NGC 7603, MARK 1018) and shows that changes in extinction may play a dominant role, as originally suggested by Tohline and Osterbrock (1976); large polarization of broad Hα is also observed. It emerges from these studies that many more galaxies have a Sy 1 nucleus than are observed to have one. Whether this means that all Sy 2 galaxies have such a nucleus is still unproven.

Analogous effects may sometimes be seen in radio galaxies (Antonucci, 1984), where scattered light from a quasar nucleus could be involved. More recently, Barthel (1989) has proposed that quasars and radio galaxies have the same relation as Sy 1 and Sy 2, with the central quasar in the radio galaxies obscured by the absorbing torus. As confirming evidence Barthel cites the fact that the relative distribution of radio diameters of the two classes is similar, but with the radio galaxies on average 3 times larger than the quasars. If the opening angle of the torus as seen from the quasar nucleus is about 45°, this is exactly what would be expected if the orientation of the axes is random. The relative numbers of quasars and radio galaxies (1:3 in the 3 C catalogue) would also roughly fit. In addition, this would explain that even the largest quasars may show superluminal motion

($v_{apparent}/c > 1$) because a jet in the polar direction would never be seen under too large an angle.

If we take these unified models very strictly, with the obscuring torus having a fixed geometry, two consequences follow: All properties of the corresponding classes which do not depend on orientation should be statistically identical and all differences should result from absorption in the torus. With regard to the first point, the associated galaxies should have the same luminosity function (as is the case for radio quasars and radio galaxies, Véron-Cetty and Woltjer, 1990a), the narrow emission lines formed outside the torus the same intensities, etc. The second point is particularly relevant to the X-ray emission. Since radio galaxies and Sy 2 are on average weaker emitters at a few keV than quasars and Sy 1, it follows that the torus should be able to block X-rays as well, which requires a high column density ($N_H \approx 10^{23}$-10^{24} cm^{-2} with solar abundances).

If, however, we allow the properties of the torus to change as a function of other parameters of the galaxy (gas content, star formation, etc.), the predictive power of the scheme is much diminished. The limiting case would be one in which in some kinds of galaxies the nucleus is obscured in all directions and in others not at all - which would be equivalent to saying that for example Sy 1 and Sy 2 are intrinsically different. While this is an extreme case, it shows that it is very well possible that orientation dependent effects are important, but that the relative proportions of Sy 1 and Sy 2 would be different for galaxies with somewhat different properties as a result of correlated changes in the torus; this, in turn, could lead to statistical differences in the isotropic properties of Sy 1 and Sy 2. For example, if the IR emission at 60 µm were isotropic, then the observations which show that almost all of the Sy 2 have a substantial flux excess at 60 µm, while only 1/3 of the Sy 1 do (Edelson et al., 1987), could be interpreted as showing that 2/3 of the Sy 1 have not much dust and no absorbing torus, while only the remaining 1/3 of the Sy 1 should be "unified" with the Sy 2.

The physical conditions in the absorbing torus have been considered by Krolik and Begelman (1986, 1988) and by Krolik and Lepp (1989). The torus would be composed of dense molecular clouds with column densities near 10^{24} cm^{-2}, which are relatively warm (10^3 K) and dense (10^7 cm^{-3}) and confined by the pressure of the hot gas in the inner region. The clouds undergo many collisions, and from time to time a cloud has low enough angular momentum to accrete on the black hole. With the inner edge of the torus at 1 pc, the necessary influx of matter is obtained to yield $L = 10^{44}$ erg sec^{-1}. Ablation of the clouds by the radiation from the central object produces a hot wind which (after some cooling by adiabatic expansion) can scatter the light from the nucleus.

An entirely different kind of unification involves relativistic jets. Many radio galaxies have jet like features in their radio emission which may be one-sided or two-sided. Some of the one-sided jets when observed with sufficient dynamic range (the ability to see faint features near very much stronger ones) become two-sided but with intensity ratios of 10^3-10^4. The usual interpretation is that this is due to Doppler effects in relativistically moving matter in which the approaching jet is much amplified and the receding one weakened. This picture has received confirmation in some cases from the observation that the receding part of the radio source has the larger Faraday rotation and consequently is presumably behind the galaxy. Also the superluminal motions and the absence of excessive inverse Compton radiation require relativistic effects.

The relativistic effects introduce a strong anisotropy and, as a consequence, the aspect of a source with a jet will depend very much on the angle between the jet axis and the line of sight. When looking into the jet at a small angle a greatly enhanced intensity and rapid time variations could be expected. Blandford and Rees (1978) proposed that this is what gives rise to the BL Lac objects. Since the jets radiate into a relatively small solid angle, there should then be many more identical objects - the "parent population" - in which the jets are seen at a larger angle and, therefore, much less conspicuous. Browne (1983) noted that the only reasonable and sufficiently numerous parent population is composed of relatively weak radio galaxies. This picture is quantitatively consistent with what is known about the luminosity of the diffuse radio emission around BL Lacs and about the absolute magnitudes of the associated galaxies, as well as qualitatively with the apparent absence of emission lines and of strong cosmological evolution in the BL Lac population (Woltjer, 1989). When an extended radio galaxy with a relativistic jet is observed in a direction close to the jet axis, the extended structure should still be visible. Since this radiation should be more or less isotropic, its intensity should be the same in a BL Lac Object as in its parent radio galaxy. Antonucci and Ulvestad (1985) have measured the flux at 1.5 GHz of the extended component of a number of BL Lac's. For 9 objects with $z < 0.1$ the radiated power is in the range $10^{23.5 - 25}$ W Hz^{-1} - corresponding to the power radiated by weak radio galaxies. With of the order of 80 BL Lac Gpc^{-3} locally and 3000 corresponding radio galaxies Gpc^{-3} the total solid angle of the jets would correspond to that of a cone of 26° for each jet if all radio galaxies have two jets. Frequently, this angle is identified with the relativistic beaming angle which then would yield v/c for the matter in the jet, but it is also possible that v/c is larger and that the jet is intrinsically conical. In the former case the luminosity function tends to be very flat at low luminosities (Urry, 1989) which appears to be consistent with the limited available observational evidence on X-ray selected BL Lacs (Maccacaro et al., 1989).

Powerful radio galaxies have an average optical absolute magnitude which is independent of the radio power. However, when the radio power at 1.4 GHz falls below 10^{25} W Hz^{-1}, it diminishes on average by 0^{m}4 per factor of 10 reduction in $P_{1.4 GHz}$ (Auriemma et al., 1977). We therefore would expect the BL Lac galaxies to be somewhat less luminous ($\approx$ 0^{m}4) than the PRG, and this appears to be what is observed.

A potential difficulty in these proposals is the requirement that a good fraction of the weak radio galaxies should have relativistic jets. Since the observed jets in such galaxies tend to be rather symmetrical, this is not obvious. In fact, in a recent study Bicknell et al. (1990) have modeled these jets and conclude that the velocities range from $10^3 - 10^4$ km s^{-1}.

The phenomena in the OVV quasars show a strong similarity to those in the BL Lac, and it is generally believed that similar relativistic effects play a role. The precise relation, however, remains in doubt as does the possible continuity between the two classes of objects.

Where the numerous "radio quiet" quasars fit into these schemes is still unclear. Are they simply the continuation of the Sy 1 sequence ? If so, how come that their evolution and luminosity function are so similar to those of the radio quasars ? Or are they related to the weak radio galaxies, or to the strong IRAS galaxies ? More in general the very basic question as to why all radio galaxies are associated with ellipticals and the Seyferts with spirals, remains unanswered. Speculatively one might perhaps think that there is a basic difference in the magnetic configuration. To have a jet, confinement is needed, and the most promising mechanism appears to be based on magnetic confinement (e.g. Heyvaerts and Norman, 1989). But if the field is weak, there is no focussing into a jet and a lower velocity outflow in a much larger cone could be expected. Are the BAL quasars perhaps an

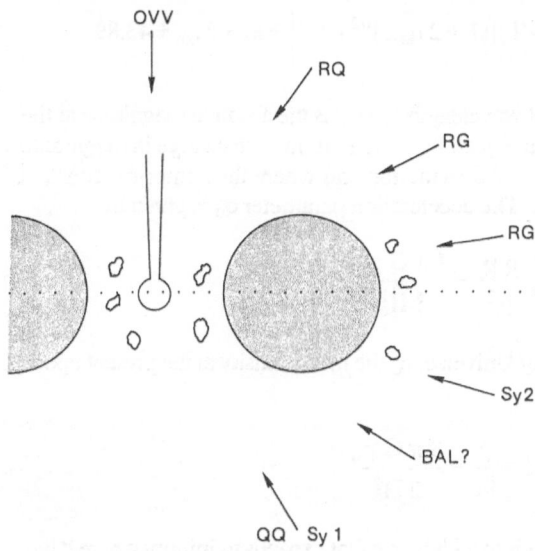

Fig. 13 : A unified model for AGN. An absorbing torus with a radius of a few pc surrounds the radiating nucleus. Inside the torus are the fast moving clouds of the Broad Line Region and outside the slower clouds of the Narrow Line Region. In the upper half of the figure is the case where relativistic jets are generated in the nucleus; in the lower half such jets are absent. Looking (nearly) into the jet the rapid variability of OVV's and BL Lac Objects would be seen; when the broad line clouds would be absent or the Doppler boosting very strong, no broad emission lines would be visible. At a larger angle from the jet axis a radio quasar would result. When the central source is obscured, but some of the broad line clouds still in view, the appearance is that of a radio galaxy with broad lines and at still larger angles a radio galaxy with only narrow lines would remain.

When, as in the lower half of the figure, no jet (or only a very weak one) is present radio quiet quasars and Sy 1 would result if the nucleus is in view and Sy 2 if not. Even though the collimated jet is missing a fast wind from the nucleus could be expected. Perhaps when this wind ablates the torus or a broad line cloud the Broad Absorption Line quasars would result. While this figure shows how much of the phenomenology of AGN may be produced in a simple model, the real situation is likely to be more complex with variability of the nucleus and intrinsic differences in the distribution of the gas clouds also playing a role.

example of this ? It certainly would explain the absence of the BAL phenomenon among radio quasars. In Fig. 13 a possible unified picture of the various classes of AGN is illustrated.

AGN AS COSMOLOGICAL PROBES

The simplest way of gaining information about the type of cosmology that applies to our Universe would be to observe a set of objects of known luminosity and to determine their redshift and apparent brightness. Standard cosmological theory for a pressure free Universe ($\Lambda = 0$) then gives (Mattig, 1958; Sandage, 1961)

45

$$m_\lambda = 5 \log \left[\frac{1}{q_0^2} \{ q_0 \, z + (q_0 - 1)[(1 + 2 \, q_0 z)^{1/2} - 1]\} \right] + k_\lambda + M_{\lambda 0} + 43.89$$

where m_λ is the observed magnitude at wavelength λ, $M_{\lambda 0}$ is the absolute magnitude at the same λ in the frame of reference of the object, k_λ is the K-term - the change in magnitude due to the redshift displacing the spectral distribution and where the numerical constant corresponds to $H_0 = 50$ km s^{-1} Mpc^{-1}. The deceleration parameter q_0 is given by

$$q_0 = - \frac{\ddot{R} R}{\dot{R}^2} = \frac{4 \pi G \rho_0}{3 H_0^2} = \frac{\Omega}{2}$$

with R the scale factor of the expanding Universe, ρ_0 the mean density at the present epoch and Ω the density parameter

$$\Omega = \frac{\rho}{\rho_c} = \frac{8 \pi G \rho_0}{3 H_0^2}$$

where ρ_c, the critical density, corresponds to a Universe that expands to infinity where it has zero velocity. For $\rho < \rho_c$ the Universe expands forever, for $\rho > \rho_c$ contraction follows expansion. Current inflationary cosmologies would lead to $\Omega \approx 1$. The luminous matter in the Universe would by itself give only $\Omega \approx 0.01$, and if also the dark matter believed to be associated with galaxies and clusters of galaxies is included, perhaps $\Omega \approx 0.1$. With some dark matter distributed more uniformly through the Universe, we could have $\Omega = 1$.

The basic problem in making use of the m-M relation to determine q_0 is that at small z, where we have some understanding of galactic evolution and therefore of M(z) for galaxies, the dependence on q_0 is not very strong, while at large z we do not understand enough about quasar evolution to determine their M(z) with any degree of confidence. However, the difference in m-M can become very large. For the two cases $q_0 = 0$ and $q_0 = 1/2$ we have

$$m - M - k - 43.89 = \begin{cases} 5 \log \{z(1 + 1/2 \, z)\} & q_0 = 0 \\ 5 \log [2 \{(1 + z) - (1 + z)^{1/2}\}] & q_0 = 1/2 \end{cases}$$

For z = 0.5, m-M is 0^m28 larger for $q_0 = 0$ than for $q_0 = 1/2$, for z = 1 : 0^m54, for z = 2 : 0^m99 and for z = 4 : 1^m69.

Baldwin (1977) discerned a relation between the absolute luminosity of a quasar and the equivalent width w_0 of the C IV emission doublet λ 1548/1551. Osmer et al. (1988) more recently obtained the relation (for $q_0 = 0.5$)

$$w_0(C \, IV) = 65 \, (10^{-30} \, L_{1450})^{-0.28} \, \text{Å}$$

with L_{1450} the continuum luminosity per frequency unit at $\lambda_0 = 1450$ Å. Since the more luminous quasars tend to have larger z, the relation in practice depends on q_0. In principle, however, we may obtain both the Baldwin relation and the q_0 value from the condition that the relation be independent of z (Wampler et al., 1984). These authors show that with the data in their samples with $q_0 = 0$ the quasars with $1.9 < z < 2.6$ are systematically brighter for the same $w_0(C \, IV)$ than those of lower redshifts, while for $q_0 = 1$ this effect largely disappears. However, since the physical foundation of the L-w relation is still not

46

very evident and since the Seyferts do not seem to follow the relation, it remains uncertain whether there could not be some evolution in the relation. The quasars with $z > 4$ are not inconsistent with the relation between $w_0(C\ IV)$ and L_{1450} given above (if $q_0 = 0.5$ is adopted) but their scatter is too large to give further confirmation (Schneider et al., 1989 b).

The next classical test in cosmology involves the angular diameters. Here we have

$$\theta'' = 0.034\ L_{kpc}\ q_0^2\ (1+z)^2\ [q_0\ z + (q_0 - 1)\ \{(1 + 2\ q_0\ z)^{1/2} - 1\ \}]^{-1}$$

and therefore

$$\theta'' = 0.034\ L_{kpc}\ \begin{cases} (1+z)^2\ (1 + 1/2\ z)^{-1}\ z^{-1} & q_0 = 0 \\ 1/2\ (1+z)^2\ [1 + z - (1+z)^{1/2}]^{-1} & q_0 = 1/2 \end{cases}$$

For $z = 0.5$ the angular diameter θ is 1.14 times smaller for $q_0 = 0$ than for $q_0 = 1/2$, for $z = 1 : 1.27$, for $z = 2 : 1.58$ and for $z = 4 : 2.17$. The differences are therefore quite large. However, the only metric diameters L we may use at the moment are the diameters of the radio sources of steep spectrum radio quasars. Since these sizes are determined by the interaction of the energetic beams from the quasar with the intergalactic or circumgalactic medium, considerable evolution may be expected. In fact, a relation like $\theta \propto z^{-1}$ would seem to give a better fit, which then would say that we live in a strictly Euclidean Universe or more plausibly that the maximum diameters of the radio sources evolve - being smaller at $z = 3$ by a factor of 6 ($q_0 = 0$) or 10 ($q_0 = 0.5$) compared to local values. Such evolution may perhaps also be seen from the shapes of the sources which become more contorted at large redshifts (Barthel et al., 1988).

The third classical cosmological test - number counts as a function of redshift - is clearly unsuitable because of the strong evolution of the AGN population.

In principle, gravitational lenses offer excellent possibilities for determining the cosmological model. When a variable source is lensed, the path lengths to the different images are slightly different and, as a consequence, there will be time shifts between the light curves of these images. As shown by Refsdal (1966) and by Kayser and Refsdal (1983), this time difference may be written (for $\Lambda = 0$)

$$\Delta t = H_0^{-1}\ T\ (z_s,\ z_d,\ q_0)\ f\ (obs)$$

with H_0 the Hubble constant, T the "cosmological correction function" ($T = 1$ at small redshifts) and f(obs) a function of observable quantities - the geometry of the images and the deflector, the mass distribution in the defelector and the redshifts of the source z_s and of the deflector z_d. In principle, by observing lensed sources with different z_s and z_d we may determine both H_0 and q_0. As an example with $z_s = 4$ and $z_d = 2$ we would have $T(q_0 = 0) = 1.5$ and $T(q_0 = 0.5) = 0.9$. While this might seem a satisfactorily large difference, the mass distribution of the lens is difficult to determine with sufficient accuracy and, as noted by Kayser and Refsdal (1983), inhomogeneities in the Universe may introduce further complications. In fact, they find that for a Universe in which all matter is clumped into galaxies (empty tube model) with $z_s = 4$ and $z_d = 2 : T(q_0 = 0.5) = 2.0$. Hence, the effect of extreme clumpiness is larger than that of the change from $q_0 = 0$ to $q_0 = 0.5$. In the "empty tube model" such a clumpy Universe is modeled by taking a homogeneous Universe and emptying one tube along the light path (Zeldovitch, 1964). In this tube also the classical cosmological tests come out rather different. However, as noted by Weinberg (1976), energy conservation requires that averaged over all light paths (those that do not pass near matter and those that do) the classical relations for a homogeneous Universe remain valid. What is gained along some paths by focussing, must be lost along others. Of course, for one individual light path the effects may be substantial and one has to insure that the selection effects in a sample do not bias the statistics.

The X-ray Background

An isotropic diffuse extragalactic background has been detected at radio and X-ray wavelengths. At longer wavelengths this background is difficult to separate from the radio emission in our galaxy. Various estimates of the brightness temperature (the temperature a black body would have in order to have the same brightness at a particular frequency) yield values in the range of 10-40 K at 178 MHz. Integrating over the source counts one predicts (with some extrapolation) comparable values (Longair, 1978). Hence, it seems likely that unresolved faint sources account for the observed background and that there need not be a truly diffuse component. The energetically more important (0.3 eV cm^{-3}) 2.7 K radiation field (around $\lambda = 1$ mm) is generally believed to be really diffuse, and attempts to explain it as a superposition of discrete sources have generally been unsuccessful. The recent measurements with the COBE satellite (Mather et al., 1990) have demonstrated that the spectrum perfectly resembles that of a black body, and it seems excluded that sources at different redshifts could ever mimic this spectrum with the required precision.

The status of the X-ray background (XRB) is much less clear. The observed 3-50 keV spectrum is well fitted by the spectrum of optically thin Bremsstrahlung with a temperature of about 40 keV (Marshall et al., 1980), and this would suggest that a hot gas in the Universe could produce the XRB. It is clear, however, that sources must make an important contribution. Clusters of galaxies are estimated to contribute at 2 keV somewhat less than 10 % of the (extrapolated) XRB and much less above 10 keV. AGN undoubtedly make a larger contribution: Quasars should account for 8-15 % and non evolving Seyferts 29 % (Schmidt and Green, 1986). With evolution the Seyfert contribution could be larger. At 2 keV the total source contribution therefore can hardly be less than 50 %. Since the Seyferts have an average spectral index (2-10 keV) of -0.7 (Turner and Pounds, 1989) and since the more limited data on quasars are consistent with such a value (Inoue, 1989), the contribution of these sources at 20 keV should be half as large at 20 keV as at 2 keV. If this contribution is subtracted from the observed XRB, the remaining spectrum is no longer thermal. It seems, therefore, that the good fit of the total XRB to a Bremsstrahlung spectrum must be an accident. At the same time, it is clear that the whole 2-20 keV background cannot be made up of Seyferts and quasars since a power law fit to the XRB spectrum over this interval would give $\alpha = -0.4$, much flatter than the AGN spectra. Setti and Woltjer (1989) have suggested that the absorbing torus in Sy 2 and radio galaxies (see the unified models) could be responsible for a class of X-ray sources which would be absorbed up to 10 keV. These would only contribute to the XRB at the higher energies and thereby flatten the spectrum of the total AGN contribution. Alternatively, other sources could be important, like for example low metallicity starburst galaxies (Griffiths and Padovani, 1990). In any case for the moment there is no evidence for any truly diffuse X-ray emission from a hot intergalactic gas. As our evaluation of the source contribution improves, it should become possible to set more meaningful limits on the diffuse component.

At higher energies too little information is available for an evaluation of the source contribution to the hard X- and gamma-ray backgrounds. If the claimed COS B detection of 3C 273 above 100 MeV were real, a significant contribution seems probable.

The X-ray background may also be used to confirm the average energy requirements for quasars which we have evaluated before on the basis of an uncertain estimate that quasars generate 0.05 times as much light in B as galaxies. From the analysis of his deep galaxy counts Tyson (1988) concludes that at 4500 Å galaxies produce a background equal to 6.8×10^{-10} erg cm^{-2} s^{-1} sr^{-1} Å^{-1}. With an average $\nu^{-0.7}$ optical spectrum for quasars we then find that between 3000 and 10 000 Å the optical background light due to quasars is 2×10^{-7} erg cm^{-2} s^{-1} sr^{-1}, of which around one third is accounted for by the observed quasars with B < 22.5. The X-ray background between 2 and 20 keV equals about 1×10^{-7} erg cm^{-2} s^{-1} sr^{-1}, and if AGN make up half of this, the energy radiated in this range is already a quarter of the optical estimate. We conclude that the observed brightness of the X-ray background confirms, but does not greatly change, the energy requirements for AGN.

CONCLUSION

AGN are very common; more than 10 % of E-Sbc galaxies with M_B < -20 are Seyfert or radio galaxies.

There are essentially three different types of AGN:
(1) The BL Lac objects and OVV quasars, associated with elliptical galaxies and presumably involving relativistic jets seen at small angles.
(2) The radio galaxies and radio quasars associated with elliptical galaxies.
(3) The Seyferts and Radio Quiet Quasars associated with spiral galaxies at the lower luminosities and possibly at least in part with ellipticals at higher luminosity.

There are of the order of 10^7 quasars and perhaps only 10^4 BL Lac Objects in the Universe. The population of high luminosity Quasars and Powerful Radio Galaxies has undergone strong evolution - with a maximum density more than a hundred times the local value around z = 3. Less luminous sources evolve less.

We do not know yet whether quasars exist with z > 5. Observations near 1 μm are required to find these. If none are found, this may say something about the formation epoch of galaxies or it may indicate that the Universe becomes opaque; in the latter case, X-ray observations perhaps combined with IR data should allow their detection.

In the AGN there is evidence for both inflow and outflow of gas, and also for star formation. As a consequence, much more gas is needed in the inflow than that required for the energetics of the nucleus $\left(\sim 0.1 \ Mc^2 \right)$. In fact, observation shows that much gas is actually present. In many cases this gas appears to have its origin in mergers in which one of the participants is a gas rich system, but the gas content in E and S0 galaxies may also have been underestimated, and more distant encounters may also activate a quiescent galaxy. Deviations from axial symmetry appear to be essential in bringing the gas closer to the center, where magnetic effects also may play a role.

Finally, we note that in most of our discussion we have neglected the possibility that gravitational lensing could play an important part. Lensing by large mass concentrations (galaxies or larger) may produce multiple images and amplification in some directions. Solar mass type objects (microlensing) may give strong amplification, but the cross-sections are relatively small. Some investigators believe that gravitational lensing is common and generates part of the morphology of AGN. Proposals which have been made include:

(a)   that the morphology of high redshift galaxies may be strongly affected by lensing (Le Fèvre et al., 1988);

(b)   that BL Lac objects and/or OVV quasars may be lensed objects. In particular, Ostriker and Vietri (1990) have proposed that some BL Lac's are lensed OVV's;

(c)   that BAL quasars may owe their characteristics to micro-lensing (Angonin et al., 1990);

(d)   that the luminosity functions of quasars may be very much flattened by gravitational lensing (Schneider, 1987);

There is little doubt that some quasars with multiple images are, in fact, lensed objects. However, whether lensing is sufficiently common for the above proposals to be justified is still unclear. A priori calculations of the frequency of lensing depend strongly on the numbers of compact objects in the Universe (Canizares, 1982), and the answer will have to depend on further observational studies. For the moment, it is our impression that gravitational lensing, while sometimes providing valuable diagnostics (e.g. as to the sizes of absorbing clouds) is unlikely to fundamentally alter our picture of AGN.

REFERENCES

Andrillat, Y., Souffrin, S.: 1968, *Astrophys. Lett.* 1, 111
Angonin, M.C., Remy, M., Surdej, J., Vanderriest, C.: 1990, *Astron. Astrophys.* 233, L5
Antonucci, R.R.: 1984, *Astrophys. J.* 278, 499
Antonucci, R.R., Ulvestad, J.S.: 1985, *Astrophys. J.* 294, 158
Antonucci, R.R., Miller, J.S.: 1985, *Astrophys. J.* 297, 621
Arons, M.: 1972, *Astrophys. J.* 172, 553
Auriemma, C., Perola, G.C., Ekers, R., Fanti, R., Lari, C., Jaffe, W.J., Ulrich, M.H.: 1977, *Astron. Astrophys.* 57, 41
Baldwin, J.A.: 1975, *Astrophys. J.* 201, 26
Baldwin, J.A.: 1977, *Astrophys. J.* 214, 679
Barthel, P.D., Schilizzi, R.T., Miley, G.K., Jägers, W.J., Strom, R.G.: 1985, *Astron. Astrophys.* 148, 243
Barthel, P.D., Miley, G.K., Schilizzi, R.T., Lonsdale, C.J.: 1988, *Astron. Astrophys. Suppl.* 73, 515
Barthel, P.D.: 1989, *Astrophys. J.* 336, 606

Bechtold, J.: 1987, *High Redshift and Primeval Galaxies*, eds. J. Bergeron, D. Kunth, B. Rocca-Volmerange, J. Tran Thanh Van, Editions Frontières, p. 397

Beichman, C.A., Soifer, B.T., Helou, G., Chester, T.J., Neugebauer, G., Gillett, F.C., Low, F.J.: 1986, *Astrophys. J.* 308, L1

Bennett, A.S.: 1962, *Mem. Roy. Astron. Soc.* 68, 163

Bergeron, J.: 1986, *Astron. Astrophys.* 155, L8

Bergeron, J.: 1987, *High Redshift and Primeval Galaxies*, eds. J. Bergeron, D. Kunth, B. Rocca-Volmerange, J. Tran Thanh Van, Editions Frontières, p. 371

Bergeron, J., Boulade, O., Kunth, D., Tytler, D., Boksenberg, A., Vigroux, L.: 1988, *Astron. Astrophys.* 191, 1

Bergeron, J., D'Odorico, S., Kunth, D.: 1987, *Astron. Astrophys.* 180, 1

Bicknell, G.V., de Ruiter, H.R., Fanti, R., Morganti, R., Parma, P.: 1990, *Astrophys. J.* 354, 98

Blandford, R.D., Rees, M.J.: 1978, Pittsburgh Conf. on *BL Lac Objects*, ed. A.M. Wolfe, University of Pittsburgh, p. 328

Boggess, A., Daltabuit, E., Torres-Peimbert, S., Estabrook, F.B., Wahlquist, H.D., Lane, A.L., Green, R., Oke, J.B., Schmidt, M., Zimmerman, B., Morton, D.C., Roeder, R.C.: 1979, *Astrophys. J.* 230, L131

Boroson, T.A., Persson, S.E., Oke, J.B.: 1985, *Astrophys. J.* 293, 120

Boyle, B.J., Fong, R., Shanks, T.: 1988, *Monthly Notices Roy. Astron. Soc.* 231, 897

Boyle, B.J., Fong, R., Shanks, T., Peterson, B.A.: 1990, *Monthly Notices Roy. Astron. Soc.* 243, 1

Braccesi, A., Formiggini, L., Gandolfi, E.: 1970, *Astron. Astrophys.* 5, 204 (Erratum *Astron. Astrophys.* 23, 159)

Braun, E., Milgrom, M.: 1990, *Astrophys. J.* 349, L35

Bregman, J.N.: 1984, *Astrophys. J.* 276, 423

Bromage, G.E.et al.: 1985, *Monthly Notices Roy. Astron. Soc.* 215, 1

Browne, I.W.: 1983, *Monthly Notices Roy. Astron. Soc.* 204, 23P

Canizares, C.R.: 1982, *Astrophys. J.* 263, 508

Carswell, R.F., Wheelan, J.A., Smith, M.G., Boksenberg, A., Tytler, D.: 1982, *Monthly Notices Roy. Astron. Soc.* 198, 91

Chambers, K.C., Miley, G.K., van Breugel, W.: 1987, *Nature* 329, 604

Chambers, K.C., McCarthy, P.J.: 1990, *Astrophys. J.* 354, L9

Chernomordik, V.V., Ozernoy, L.M. : 1983, *Nature* 303, 153

Clavel, J., Wamsteker, W., Glass, I.S.: 1989, *Astrophys. J.* 337, 236

Cohen, M.H.: 1989, *BL Lac Objects*, in Lecture Notes in Physics, Vol. 334: Proc, Como, eds. L. Maraschi, T. Maccacaro, M.-H. Ulrich, Springer, Berlin Heidelberg New York, p. 13

Cohen, R.D., Osterbrock, D.E.: 1981, *Astrophys. J.* 243, 81

Cristiani, S., Shaver, P.A.: 1987, *High Redshift and Primeval Galaxies*, eds. J. Bergeron, D. Kunth, B. Rocca-Volmerange, J. Tran Thanh Van, Editions Frontières, p. 383

Cristiani, S., Vio, R.: 1990, *Astron. Astrophys.* 227, 385

di Serego Alighieri, A., Fosbury, R.A.E., Quinn, P.J., Tadhunter, C.N.: 1989, *Nature* 341, 307

Eachus, L.J., Liller, W.: 1975, *Astrophys. J.* 200, L61

Edelson, R.A.: 1987, *Astrophys. J.* 313, 651

Edelson, R.A., Malkan, M.A., Ricke, G.H.: 1987, *Astrophys. J.* 321, 233

Edelson, R.A., Malkan, M.A.: 1987, *Astrophys. J.* 323, 516

Ekers, R.D., Goss, W.M., Wellington, K.J., Bosma, A., Smith, R.M., Schweizer, F.: 1983, *Astron. Astrophys.* 127, 361

Fall, S.M., Pei, Y.C., McMahon, R.G.: 1989, *Astrophys. J.* 341, L5

Fanaroff, B., Riley, J.: 1974, *Monthly Notices Roy. Astron. Soc.* 167, 31P

Felten, J.E.: 1976, *Astrophys. J.* 207, 700

Filippenko, A.V., Sargent, W.L.: 1988, *Astrophys. J.* 324, 134

Foltz, C.B., Weyman, R.J., Röser, H.J., Chaffee, F.H.: 1984, *Astrophys. J.* 281, L1

Gioia, I.M., Maccacaro, T., Schild, R.E., Wolter, A., Stocke, J.T., Morris, S.L., Henry, J.P.: 1990, *Astrophys. J. Suppl.* 72, 567

Giommi, P. et al.: 1989, *BL Lac Objects*, in Lecture Notes in Physics, Vol. 334: Proc, Como, eds. L. Maraschi, T. Maccacaro, M.-H. Ulrich, Springer, Berlin Heidelberg New York, p. 231

Goodrich, R.W.: 1989, *Astrophys. J.* 340, 190

Gordon, M.A.: 1990, *Astrophys. J.* 350, L29

Grandi, S.A.: 1981, *Astrophys. J.* 251, 451

Grandi, S.A., Osterbrock, D.E.: 1978, *Astrophys. J.* 220, 783

Griffiths, R.E., Padovani, P.: 1990, *Astrophys. J.* (in press)

Hawkins, M.R.: 1986, *Monthly Notices Roy. Astron. Soc.* 219, 417

Hazard, C., McMahon, R., Sargent, W.L.: 1986, *Nature* 322, 38

Heckman, T.M.: 1980, *Astron. Astrophys.* 87, 152

Heckman, T.M., Balick, B., Crane, P.C.: 1980, *Astron. Astrophys. Suppl.* 40, 295

Heyvaerts, J., Norman, C.: 1989, *Astrophys. J.* 347, 1005

Huchtmeier, W.K., Bregman, J.N., Hogg, D.E., Roberts, M.S.: 1988, *Astron. Astrophys.* 198, L17

Hummel, E.: 1981, *Astron. Astrophys.* 93, 91

Hunstead, R.W., Murdoch, H.S., Peterson, B.A., Blades, J.C., Jauncey, D.L., Wright, A.E., Pettini, M., Savage, A.: 1986, *Astrophys. J.* 305, 496

Ikeuchi, S., Norman, C.A.: 1987, *Astrophys. J.* 312, 485

Inoue, H.: 1989, in *Big Bang, Active Galactic Nuclei and Supernovae*, Universal Ac. Press, Tokyo, p. 301

Kayser, R., Refsdal, S.: 1983, *Astron. Astrophys.* 128, 156

Kellermann, K.I., Sramek, R., Schmidt, M., Shaffer, D.B., Green, R.: 1989, *Astron. J.* 98, 1195

Kennicut, R.C., Keel, W.C.: 1984, *Astrophys. J.* 279, L5

Kollatschny, W., Fricke, K.J.: 1985, *Astron. Astrophys.* 146, L11

Koo, D.C., Kron, R.G.: 1988, *Astrophys. J.* 325, 92

Koski, A.T.: 1978, *Astrophys. J.* 223, 56

Krolik, J.H., Begelman, M.C.: 1986, *Astrophys. J.* 308, L55

Krolik, J.H., Begelman, M.C.: 1988, *Astrophys. J.* 329, 702

Krolik, J.H., Lepp, S.: 1989, *Astrophys. J.* 347, 179

Kühr, H., Witzel, A., Pauliny Toth, I.I., Nauber, U.: 1981, *Astron. Astrophys. Suppl.* 45, 367

Lazareff, B., Castets, A., Kim, D.-W., Jura, M.: 1989, *Astrophys. J.* 336, L13

Le Fèvre, O., Hammer, F., Jones, J.: 1988, *Astrophys. J.* 331, L73

Lilly, S.J.: 1989, *Astrophys. J.* 340, 77

Longair, M.S.: 1978, in *Observational Cosmology*, eds. A. Maeder, L. Martinet, G. Tammann, Geneva Observatory, p. 142

Low, F.J., Huchra, J.P., Kleinmann, S.G., Cutri, R.M.: 1988, *Astrophys. J.* 327, L41

Maccacaro, T., Gioia, I.M., Schild, R.E., Wolter, A., Morris, S.L., Stocke, J.T.: 1989, *BL Lac Objects*, in Lecture Notes in Physics, Vol. 334: Proc, Como, eds. L. Maraschi, T. Maccacaro, M.-H. Ulrich, Springer, Berlin Heidelberg New York, p. 222

Malkan, M.A.: 1984, *Astrophys. J.* 287, 555

Marshall, F.E., Boldt, E.A., Holt, S.S., Miller, R.B., Mushotzky, R.F., Rose, L.A., Rothschild, R.E., Serlemitsos, P.J.: 1980, *Astrophys. J.* 235, 4

Mather, J.C. et al.: 1990, *Astrophys. J.* 354, L37

Mattig, W.: 1958, *Astron. Nachrichten* 284, 109

McAlary, C.W., Rieke, G.H.: 1988, *Astrophys. J.* 333, 1

McCarthy, P.J., Spinrad, H., Djorgovski, S., Strauss, M.A., van Breugel, W., Liebert, J.: 1987a, *Astrophys. J.* 319, L39

McCarthy, P.J., van Breugel, W., Spinrad, H., Djorgovski, S.: 1987b, *Astrophys. J.* 321, L29

Meisenheimer, K., Röser, H.-J., Hiltner, P.R., Yates, M.G., Longair, M.S., Chini, R., Perley, R.A.: 1989, *Astron. Astrophys.* 219, 63

Melnick, J., Mirabel, I.F.: 1990, *Astron. Astrophys.* 231, L19

Meyer, D.M., Welty, D.E., York, D.G.: 1989, *Astrophys. J.* 343, L37

Miley, G.K., Neugebauer, G., Soifer, B.T.: 1985, *Astrophys. J.* 293, L11

Miller, J.S.: 1981, *Publ. Astron. Soc. Pacific* 93, 681

Miller, J.S., Antonucci, R.R.: 1983, *Astrophys. J.* 271, L7

Mirabel, I.F., Sanders, D.B., Kazès, I.: 1989, *Astrophys. J.* 340, L9

Moore, R.L., Stockman, H.S.: 1984, *Astrophys. J.* 279, 465

Morganti, R., Fanti, C., Fanti, R., Parma, P., de Ruiter, H.R.: 1987, *Astron. Astrophys.* 183, 203

Murdoch, H.S., Hunstead, R.W., Pettini, M., Blades, J.C.: 1986, *Astrophys. J.* 309, 19

Mutel, R.L.: 1990 (Preprint)

Myers, S.T., Scoville, N.Z.: 1987, *Astrophys. J.* 312, L39

Neugebauer, G. et al.: 1984, *Astrophys. J.* 278, L1

Neugebauer, G., Miley, G.K., Soifer, B.T., Clegg, P.E.: 1986, *Astrophys. J.* 308, 815

Norris, R.P.: 1989, IAUSymp. 134, 379

Osmer, P.S., Porter, A.C., Green, R.F.: 1988, *Bull. Am. Astron. Soc.* 20, 968

Osmer, P.S.: 1982, *Astrophys. J.* 253, 28

Osterbrock, D.E.: 1977, *Astrophys. J.* 215, 733

Osterbrock, D.E.: 1981, *Astrophys. J.* 249, 462

Ostriker, J.P., Ikeuchi, S.: 1983, *Astrophys. J.* 268, L63

Ostriker, J.P., Vietri, M.: 1990, *Nature* 344, 45

Peimbert, M., Torres-Peimbert, S.: 1981, *Astrophys. J.* 245, 845

Pfenniger, D., Norman, C.: 1990, *Astrophys. J.* (to be published)

Phillips, M.M., Charles, P.A., Baldwin, J.A.: 1983, *Astrophys. J.* 266, 485

Pica, A.J., Smith, A.G.: 1983, *Astrophys. J.* 272, 11

Prieto, A., di Serego Alighieri, S., Fosbury, R.A.: 1989, *Extranuclear Activity in Galaxies*, ESO Conf. Proc. 32, 31

Refsdal, S.: 1966, *Monthly Notices Roy. Astron. Soc.* 132, 101

Romanishin, W., Hintzen, P.: 1989, *Astrophys. J.* 341, 41

Salter, C.J., Brown, R.L.: 1988, *Galactic and Extragalactic Radio Astronomy*, ed. G.L. Verschuur and K.I. Kellermann, Springer-Verlag, p. 6

Sandage, A.: 1961, *Astrophys. J.* 133, 355

Sandage, A.: 1965, *Astrophys. J.* 141, 1560

Sandage, A.: 1972, *Astrophys. J.* 178, 25

Sandage, A., Tammann, G.A.: 1981, *A Revised Shapley-Ames Catalog of Bright Galaxies*, Carnegie Inst. Washington, Publication 635

Sanders, D.B., Mirabel, I.F.: 1985, *Astrophys. J.* 298, L31

Sanders, D.B., Scoville, N.Z., Sargent, A.I., Soifer, B.T.: 1988a, *Astrophys. J.* 324, L55

Sanders, D.B., Soifer, B.T., Elias, J.H., Madore, B.F., Matthews, K., Neugebauer, G., Scoville, N.Z.: 1988b, *Astrophys. J.* 325, 74

Sanders, D.B., Soifer, B.T., Elias, J.H., Neugebauer, G., Matthews, K.: 1988c, *Astrophys. J.* 328, L35

Sanders, D.B., Scoville, N.Z., Soifer, B.T.: 1988d, *Astrophys. J.* 335, L1

Sanders, D.B., Phinney, E.S., Neugebauer, G., Soifer, B.T., Matthews, K.: 1989, *Astrophys. J.* 347, 29

Sargent, W.L., Young, P.J., Boksenberg, A., Tytler, D.: 1980, *Astrophys. J. Suppl.* 42, 41

Sargent, W.L., Steidel, C.C., Boksenberg, A.: 1988, *Astrophys. J.* 334, 22

Saripalli, L., Gopal-Krishna, Reich, W., Kühr, H.: 1986, *Astron. Astrophys.* 170, 20

Schechter, P.: 1976, *Astrophys. J.* 203, 297

Schmidt, M.: 1968, *Astrophys. J.* 151, 393

Schmidt, M., Green, R.F.: 1983, *Astrophys. J.* 269, 352

Schmidt, M., Green, R.F.: 1986, *Astrophys. J.* 305, 68

Schmidt, M., Schneider, D.P., Gunn, J.E.: 1986, *Astrophys. J.* 310, 518

Schmidt, M., Schneider, D.P., Gunn, J.E.: 1987, *Astrophys. J.* 316, L1

Schneider, D.P., Schmidt, M., Gunn, J.E.: 1989a, *Astron. J.* 98, 1507

Schneider, D.P., Schmidt, M., Gunn, J.E.: 1989b, *Astron. J.* 98, 1951

Schneider, P.: 1987, *Astron. Astrophys.* 183, 189

Schwartz, D.A., Brissenden, R.J.V., Tuohy, I.R., Feigelson, E.D., Hertz, P.L., Remillard, R.A.: 1989, *BL Lac Objects*, in Lecture Notes in Physics, Vol. 334: Proc, Como, eds. L. Maraschi, T. Maccacaro, M.-H. Ulrich, Springer, Berlin Heidelberg New York, p. 209

Schweizer, F.: 1980, *Astrophys. J.* 237, 303

Scoville, N.Z., Sanders, D.B., Sargent, A.I., Soifer, B.T., Scott, S.L., Lo, K.Y.: 1986, *Astrophys. J.* 311, L47

Scoville, N.Z., Matthews, K., Carico, D.P., Sanders, D.B.: 1988, *Astrophys. J.* 327, L61

Scoville, N.Z., Sanders, D.B., Sargent, A.I., Soifer, B.T., Tinney, C.G.: 1989, *Astrophys. J.* 345, L25

Setti, G., Woltjer, L.: 1989, *Astron. Astrophys.* 224, L21

Seyfert, C.K.: 1943, *Astrophys. J.* 97, 28

Shaver, P.A., Robertson, J.G.: 1983, *Astrophys. J.* 268, L57

Shields, J.C., Filippenko, A.V.: 1988, *Astrophys. J.* 332, L55

Shuder, J.M., Osterbrock, D.E.: 1981, *Astrophys. J.* 250, 55

Smith, E.P., Heckman, T.M., Bothun, G.D., Romanishin, W.R., Balick, B.: 1986, *Astrophys. J.* 306, 64

Smith, E.P., Heckman, T.M.: 1989, *Astrophys. J.* 341, 658

Stauffer, J.R.: 1982, *Astrophys. J.* 262, 66

Stockman, H.S., Moore, R.L., Angel, J.R.: 1984, *Astrophys. J.* 279, 485

Strauss, M.A., Kirhakos, S.D., Yahil, A.: 1988, *Astrophys. J.* 332, L45

Strom, R.G., Jägers, W.J.: 1988, *Astron. Astrophys.* 194, 79

Surdej, J., Hutsémekers, D.: 1987, *Astron. Astrophys.* 177, 42

Telesco, C.M.: 1988, *Ann. Rev. Astron. Astrophys.*, 26, 343

Thronson, H.A. Jr., Hereld, M., Majewski, S., Greenhouse, M., Johnson, P., Spillar, E., Woodward, C.E., Harper, D.A., Rauscher, B.J.: 1989, *Astrophys. J.* 343, 158

Tohline, J.E., Osterbrock, D.E.: 1976, *Astrophys. J.* 210, L117

Trevese, D., Pitella, G., Kron, R.G., Koo, D.C., Bershady, M.: 1989a, *Astron. J.* 98, 108

Trevese, D., Koo, D.C., Kron, R.G.: 1989b, IAUSymp. 134, 47

Turner, T.J., Pounds, K.A.: 1989, *Monthly Notices Roy. Astron. Soc.* 240, 833

Turnshek, D.A., Foltz, C.B., Grillmair, C.J., Weyman, R.J.: 1988, *Astrophys. J.* 325, 651

Turnshek, D.A.: 1984, *Astrophys. J.* 280, 51

Tyson, J.A.: 1988, *Astron. J.* 96, 1

Tytler, D.: 1987a, *Astrophys. J.* 321, 49

Tytler, D.: 1987b, *Astrophys. J.* 321, 69

Ulrich, M.H.: 1981, *Astron. Astrophys.* 103, L1

Ulvestad, J.S., Wilson, A.S.: 1989, *Astrophys. J.* 343, 659

Urry, C.M.: 1989, *BL Lac Objects*, in Lecture Notes in Physics, Vol. 334: Proc, Como, eds. L. Maraschi, T. Maccacaro, M.-H. Ulrich, Springer, Berlin Heidelberg New York, p. 435

Véron-Cetty, M.P., Véron, P.: 1986, *Astron. Astrophys. Suppl.* **66**, 335

Véron-Cetty, M.P., Véron, P.: 1989, *A Catalogue of Quasars and Active Nuclei* (4th ed.), ESO Sci. Rep. 7

Véron-Cetty, M.P., Woltjer, L.: 1990a, *Astron. Astrophys.* **236**, 69

Véron-Cetty, M.P., Woltjer, L.: 1990b, *Astron. Astrophys.* (to be published)

Wampler, E.J., Gaskell, C.M., Burke, W.L., Baldwin, J.A.: 1984, *Astrophys. J.* **276**, 403

Wampler, E.J.: 1990 (to be published)

Warren, S.J., Hewett, P.C., Osmer, P.S.: 1988, *Astron. Soc. Pacific Conf. Series* **2**, 96

Weinberg, S.: 1976, *Astrophys. J.* **208**, L1

Westin, B.A.: 1985, *Astron. Astrophys.* **151**, 137

Wills, B.J.: 1989, *BL Lac Objects*, in Lecture Notes in Physics, Vol. 334: Proc, Como, eds. L. Maraschi, T. Maccacaro, M.-H. Ulrich, Springer, Berlin Heidelberg New York, p. 109

Wolfe, A.M., Briggs, F.H., Jauncey, D.L.: 1981, *Astrophys. J.* **248**, 460

Woltjer, L.: 1989, *BL Lac Objects*, in Lecture Notes in Physics, Vol. 334: Proc, Como, eds. L. Maraschi, T. Maccacaro, M.-H. Ulrich, Springer, Berlin Heidelberg New York, p. 460

Woltjer, L., Setti, G.: 1982, *Astrophysical Cosmology*, eds. H.A. Brück, G.V. Coyne, M.S. Longair, Pont. Ac. Scient. Scripta Varia 48, 293

Wood, K.S., Meekins, J.F., Yentis, D.J., Smathers, H.W., McNutt, D.P., Bleach, R.D., Byram, E.T., Chubb, T.A., Friedman, H., Meidav, M.: 1984, *Astrophys. J. Suppl.* **56**, 507

Wu, C.C., Boggess, A., Gull, T.R.: 1983, *Astrophys. J.* **266**, 28

Yanni, B., York, D.G., Williams, T.B.: 1990, *Astrophys. J.* **351**, 377

Zeldovitch, Y.B.: 1964, *Sov. Astron.* **8**, 13

Zensus, J.A.: 1989, *BL Lac Objects*, in Lecture Notes in Physics, Vol. 334: Proc, Como, eds. L. Maraschi, T. Maccacaro, M.-H. Ulrich, Springer, Berlin Heidelberg New York, p. 3

AGN Emission Lines

Hagai Netzer

With 40 Figures

1 The AGN Family

Active galaxies are distinguished from other galaxies in that they show indications of having energy output not related to ordinary stellar processes. The activity is centered in a small nuclear region and associated with strong emission lines. The nuclei of such galaxies are named Active Galactic Nuclei (hereafter AGNs). This group of objects includes bright quasars, with luminosity exceeding 10^{47} erg s^{-1}, as well as faint LINERs emitting no more than 10^{41} erg s^{-1}.

For the purpose of this work an object will be classified as an AGN if at least one of the following is observed:

a: Compact nuclear region, brighter than the corresponding region in galaxies of similar Hubble type.
b: Nonstellar nuclear continuum emission.
c: Nuclear emission lines indicating excitation by a nonstellar continuum.
d: Variable continuum and/or emission lines.

AGNs are further classified into subgroups, according to their luminosity and spectrum. The following main groups are usually identified:

1.1 Quasars and Seyfert 1 Galaxies

These are the most luminous AGNs, showing all the above characteristics. They are easily recognized by their strong, broad permitted emission lines and their nonstellar continuum. The line widths, if interpreted as being due to Doppler motion, are typically 3000-5000 km s^{-1}, with extreme cases of up to 40,000 km s^{-1} Full Width at Zero Intensity (FWZI). Seyfert 1s show also strong narrow (400-1000 km s^{-1}) permitted and forbidden lines. This is not so common in bright quasars, where, in many cases, the narrow lines are weak or absent. Examples of a $z = 2.051$ quasar spectrum and a Seyfert 1 galaxy spectrum are shown in Fig.1.

The shape of the nonstellar continuum in quasars and Seyfert 1 galaxies is different at different energies. Over a limited frequency range it can be described as a power-law in frequency,

$$F_\nu = C\nu^{-\gamma}, \tag{1}$$

where F_ν is the observed flux in erg s^{-1} cm^{-2} Hz^{-1}. The spectral index, γ, is about 1-1.2 between 0.8-1.5μm, 0.3-0.7 between 0.2-0.6 μm and 0.7 at hard X-ray energies (2-10 keV). Observations of high redshift quasars show considerable steepening of the continuum ($\gamma \geq 2$) below 1000Å. This is likely to be due, at least in part, to the large number of neutral hydrogen absorption systems on the line of sight. There is some indirect evidence that the intrinsic continuum at those frequencies is much harder, at least up to 3 or 4 Ryd. (chapter 10). The nonstellar continuum must have a sharp decline in the far ultraviolet, the part not accessible to direct observations. This is inferred from the comparison of the ultraviolet and X-ray data. A useful parameter, which has been used to describe the relative flux in those wavelength bands, is γ_{ox}, which is the spectral index needed to connect the two continuum points at 2500Å and 2 keV. Its value ranges from about 1.2, for several Seyfert 1 galaxies, to 1.8 in some bright quasars. There are indications that radio-loud AGNs have smaller γ_{ox} compared with the radio-quiet ones, i.e. they emit more 2 keV energy per given optical luminosity.

Some powerful radio galaxies show a typical Seyfert type 1 spectrum. These are sometimes classified as a separate group under the name of Broad Line Radio Galaxies (BLRGs). There are some spectral differences that justify, perhaps, this type of classification, mainly to do with the line profiles. Here they are simply recognized as radio-loud Seyfert 1s, much like the distinction between radio-loud and radio-quiet quasars.

1.2 Seyfert 2 Galaxies

Seyfert 2 galaxies are less luminous than Seyfert 1 galaxies and quasars and do not show broad emission lines. Their narrow lines are similar in width and excitation to the narrow lines of Seyfert 1s. The energy distribution of the nonstellar continuum is not very well known. It is confused with the stellar background at optical and near infrared wavelengths, and too weak to be accurately observed in the ultraviolet, with pre-HST instruments. There are some indications that the X-ray to optical luminosity ratio is similar to the one observed in the more luminous AGNs. There is also clear evidence that several Seyfert 2 galaxies are obscured Seyfert 1s (chapter 11). A typical Seyfert 2 spectrum is shown in Fig.1.

1.3 Low Ionization Nuclear Emission Line Regions (LINERs)

These are the least luminous AGNs. The nonstellar continuum luminosity of most LINERs is so small, compared with the stellar continuum, that there are not yet good observations of its shape. The strongest emission lines are of low ionization species and are somewhat narrower (200-400 $km\ s^{-1}$) than the narrow lines of Seyfert galaxies. LINERs are recognized by their low excitation narrow lines but some of them show weak, broad emission lines. This links them to the more luminous members of the AGN family and is discussed, along with their other properties, in chapter 11.

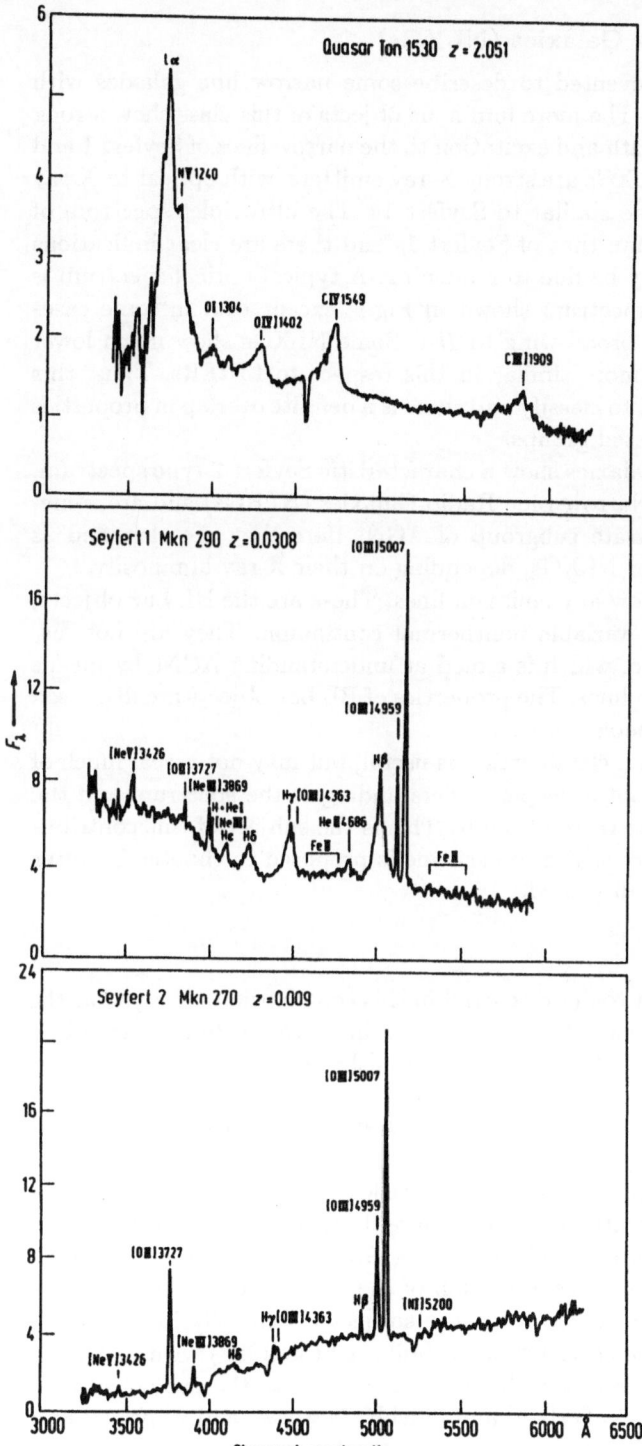

Fig. 1. Spectra of three AGNs

1.4 Narrow Line X-ray Galaxies (NLXGs)

The name NLXGs was invented to describe some narrow line galaxies with unusual X-ray properties. The more luminous objects of this class show strong narrow lines, similar in width and excitation to the narrow lines of Seyfert 1 and Seyfert 2 galaxies. The NLXGs are strong X-ray emitters, with optical-to-X-ray continuum luminosity ratio similar to Seyfert 1s. The ultraviolet spectrum of most NLXGs is weaker than that of Seyfert 1s and there are clear indications (chapter 11) that this may be due to reddening. A typical optical spectrum is similar to the Seyfert 2 spectrum shown in Fig.1, except that in some cases there exists a weak, very broad wing to $H\alpha$. Some NLXGs show much lower excitation lines, and are more similar in this respect to LINERs. Thus, this group of objects is difficult to classify and there is a definite overlap in properties with the previously discussed groups.

Some powerful radio galaxies show a characteristic Seyfert 2 type spectrum. These have been named Narrow Line Radio Galaxies (NLRGs) and are sometimes classified as a separate subgroup of AGN. Here they are classified as either Seyfert 2 galaxies or NLXGs, depending on their X-ray luminosity.

Some AGNs do not show any emission lines. These are the BL Lac objects, identified by their highly variable nonthermal continuum. They are not discussed in this contribution, which is aimed at understanding AGNs by means of analysing their emission lines. The properties of BL Lac objects are discussed by other authors in this book.

The above observational classification is useful, but may not reveal much of the nature of AGNs without a deeper understanding of the spectrum and the information that can be extracted from it. This is the subject of this contribution. A more meaningful classification scheme is proposed in chapter 11, after all these issues have been discussed.

1.5 Bibliography

References to the different topics discussed in this contribution are given at the end of the relevant chapters. This is far from being a complete list, since it is impossible to review the hundreds of papers published each year on the subject. The emphasis is on papers discussing in detail some of the topics addressed in this contribution and papers that show typical examples of the observational material.

There are several comprehensive review articles on the subject that contain much of the material presented here. The more detailed ones are by Davidson and Netzer (1979), Osterbrock and Mathews (1986) and Lawrence (1987), as well as a few chapters in the second edition of the book by Osterbrock (1989). There is also a very complete book on the subject ("Astrophysics of Active Galaxies and Quasi-Stellar Objects", J. S. Miller editor, 1985) with many excellent reviews. The articles in that book most relevant to the present discussion are by Keel, Osterbrock, Ferland and Shields, Mathews and Capriotti and MacAlpine. Recent review papers on some of the topics can be found in the proceedings of the IAU Symposium no. 134 (1989).

For AGN classification see Osterbrock (1984, 1989), Lawrence (1987) and references therein.

2 Primary Observations

Much of our understanding of AGNs is due to their emission lines. There are three types of primary observations, and three corresponding ways of analysis, that can be used. They are related to the line intensities, the line variability and the line profiles. Another experimental result relates the gas distribution, and optical depth, to the observed ultraviolet continuum. The observations are described in this chapter; the analysis and the theory in chapters 3-9.

2.1 Line Intensity

Emission line intensities and emission line ratios supply information on *the physical conditions* in the line emitting gas. The electron density and temperature, the degree of ionization and excitation and the chemical composition, can all be deduced from line ratio analysis. There are some 20 broad lines, and a similar number of narrow lines, that can be measured in a single AGN, and the amount of information conveyed by the many line ratios is very large. Fig.2, which is a composite spectrum of a large number of quasars, shows many of these lines and Table 1 contains a representative line intensity list for the main AGN groups discussed in chapter 1.

A characteristic feature of all broad line objects is the presence, in their spectrum, of both high and low excitation lines. Examples are $OVI\lambda1035$ and $MgII\lambda2798$, both very strong in many quasars. The low excitation lines indi-

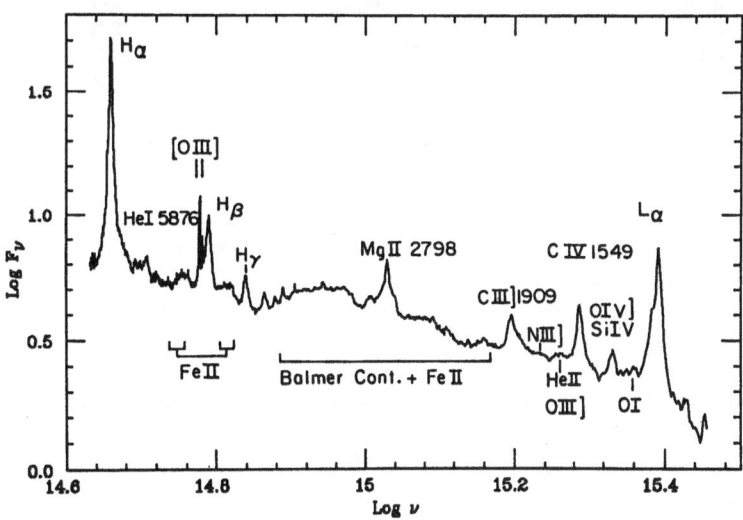

Fig. 2. A composite spectrum, obtained from the addition of spectra of quasars with different redshifts (courtesy of J. Baldwin)

Table 1. Composite AGN Spectra* (line intensity relative to $H\beta$)

Line	Quasars and Seyfert 1s (broad + narrow)	Seyfert 2 and NLXGs (narrow)	LINERs (narrow)
$CIII\lambda977$	<1		
$OVI\lambda1035$	3		
$L\alpha$	8-15	30-70	
$NV\lambda1240$	3		
$OI\lambda1304$	0.5		
$CII\lambda1336$	0.3		
$SiIV, OIV]\lambda1400$	1.3		
$NIV]\lambda1486$	0.7		
$CIV\lambda1549$	5-8	5-20	
$HeII\lambda1640$	0.6		
$OIII]\lambda1663$	0.5		
$NIII]\lambda1750$	0.4		
$CIII]\lambda1909$	2-4	2-8	
FeII (2200-2800Å)	5-10		
$MgII\lambda2798$	3	1-3	
$[NeV]\lambda3426$	0.2	1	
$[OII]\lambda3727$	0.3	1-4	2-5
$[NeIII]\lambda3869$	0.5	1.5	0.3
$HeII\lambda4686$	0.1		
$H\beta$	1	1	1
$[OIII]\lambda5007$	0.1-1	8-15	1-2
FeII(4500-5400Å)	1-3		
$HeI\lambda5876$	0.1-0.2	0.15	0.1
$[FeVII]\lambda6087$		0.1	
$[OI]\lambda6300$	0.05	0.6	1
$[FeX]\lambda6374$		0.01-0.08	
$H\alpha$	4-6	2.8-3.3	2.8-3
$[NII]\lambda6583$	0.1-0.3	0.6-1.5	2-4
$[SII]\lambda\lambda6716, 6731$	0.2	1	2-3
$CaII\lambda8498 - 8662$	0-0.3		
$[SIII]\lambda\lambda9069, 9532$			0.3
P_α	0.4		
L($H\alpha$) ($erg\ s^{-1}$)	10^{42-46}	10^{40-42}	10^{38-41}
EW($H\beta$)	100 Å	5-30 Å	1-10 Å

* Reddening corrections have been applied to the narrow lines. The $H\beta$ equivalent width is relative to the continuum within a 2-3 arcsec aperture. EW($H\beta$) relative to the nonstellar continuum can be much larger in Seyfert 2s and LINERs.

cate regions of low ionization and suggest that at least part of the gas is neutral and optically thick to the Lyman continuum radiation. This can be put to a simple observational test, as discussed below. The high excitation lines indicate highly ionized material. A fundamental issue, that ought to be addressed, is whether the high and low excitation lines originate in the same part of the emission line region.

An interesting property of AGNs is the great similarity of line ratios in objects of very different luminosities. Thus, the composite spectrum shown in Table 1 is indeed representative of the spectrum of many individual objects. This must indicate that the physical conditions in the line emitting gas are similar in bright and faint objects. As a result, much of the following analysis is aimed at the understanding of this canonical spectrum, rather than the observations of a particular object. All this, and more, is the subject of chapters 4-6.

2.2 Line Variability

Given enough time, all broad-line AGNs show continuum variability. Spectrophotometric observations, obtained in the last decade, show variable broad emission lines in most, perhaps all Seyfert 1 galaxies. Some quasars show variable emission lines too but there are very few, if any, systematic observations of this kind. Fig.3 shows the variable lines and continuum of NGC 5548, a typical Seyfert 1 galaxy.

The observed line and continuum variability in Seyfert galaxies are clearly correlated. The emission lines respond to the continuum variability after a certain lag (which seems to be longer in more luminous objects). Assuming that the observed line variability is driven by changes in the continuum luminosity, the emission line light curve depends on the gas distribution in the nucleus and the continuum light curve. The study of the correlated line and continuum variability is thus an important tool for investigating *the gas distribution* in the nucleus. This is the subject of chapter 8.

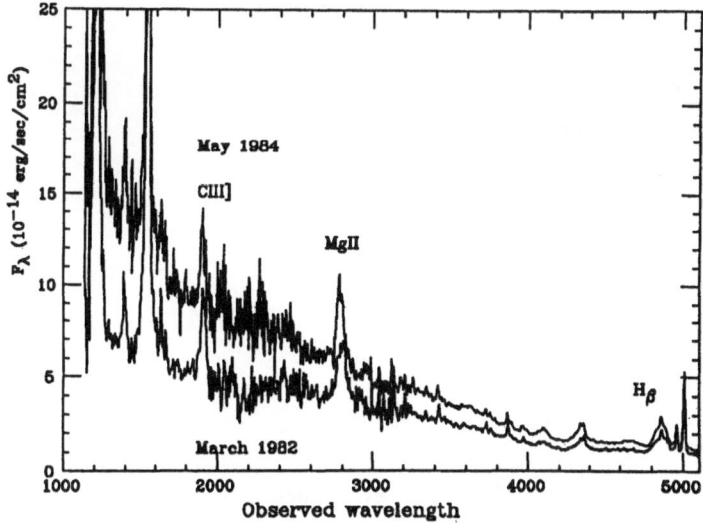

Fig. 3. The spectrum of the Seyfert 1 galaxy NGC 5548 at two epochs seperated by 26 months, showing the large variation that occured in the line and continuum luminosities (data from Wamsteker et al 1990)

Except perhaps for one case, there is no evidence for narrow line variability in AGNs. The gas emitting the narrow lines is therefore more extended and its distribution must be studied by other means.

2.3 Line Profiles

There are two distinct classes of line profiles, narrow (200-1000 $km\ s^{-1}$) and broad (1500-10,000 $km\ s^{-1}$), that seem to come from two distinct emission line regions. They are present, in different proportion, in different AGNs. Even the narrow emission lines are too broad to be interpreted as due to pure thermal motion. The gas producing the lines must be moving at high speed and a given wavelength in a profile is associated with a given projected velocity. Thus studying the line profiles is a way for understanding *the gas motion* in the nucleus.

Different AGNs can differ a lot in their line profiles. In some objects the broad emission lines of different ions have similar profiles, in others the profiles are quite different. Some broad line profiles are symmetric and others are not. The narrow emission line profiles of many objects show a clear blue asymmetry which seems to depend on the gas density and/or its level of ionization. In some objects the lines have smooth profiles, in others there is a clear evidence for a multi-component structure. Even the line center redshift is not always the same for all lines and in some high luminosity objects the broad high ionization lines are blue-shifted with respect to the broad low ionization lines. Several examples of line profiles are shown in Figs.4 and 5 and others are discussed later. All this information must be considered when modeling the **gas motion and dynamics** in AGNs, which is the subject of chapter 9.

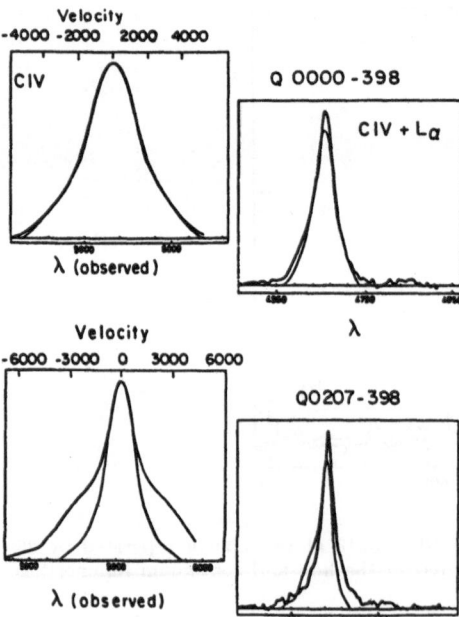

Fig. 4. Broad emission line profiles in two quasars. Left: $CIV\lambda1549$ with its reflection superposed. Right: Superposition of $CIV\lambda1549$ upon $L\alpha$. Note the very asymmetric $CIV\lambda1549$ profile in Q0207-398. (courtesy of B. Wilkes)

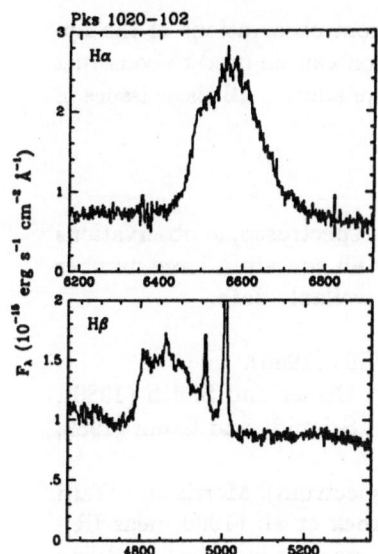

Fig. 5. $H\alpha$ and $H\beta$ profiles in a low redshift quasar (courtesy of G. Stirpe)

2.4 Lyman Continuum Optical Depth

An additional, important piece of information about the gas distribution and optical depth comes from studying the wavelength region around the Lyman limit, at 912Å. This wavelength range has been observed only in high luminosity AGNs and most of these objects do not show any significant Lyman jump in absorption or in emission. A conservative limit on the Lyman discontinuity (the relative change in the continuum level at the Lyman limit), in most of the observed cases, is about 0.1. If the line emitting gas completely surrounds the continuum source this must mean $\tau(912\text{Å}) \ll 1$. Alternatively, the gas may be clumped into small clouds, covering only a small fraction of the contiuum source. In this case, individual clouds can have large Lyman optical depth without violating the observational limit. The strong observed broad lines of MgII and FeII tend to support the second hypothesis, at least for the broad line gas. Whatever the case may be, it seems that only a small fraction of the total available continuum energy is absorbed by the line emitting gas. More observations are needed to confirm this finding in low luminosity AGNs. This issue is of primary importance for the understanding of AGNs and is further discussed in chapters 6 and 10.

An additional, important piece of information about the relative size of the clouds and the source of continuum emission comes from X-ray studies. It is found that low luminosity AGNs tend to show larger X-ray opacity and more material on the line of sight. Very luminous objects do not show any X-ray absorption. This may or may not be related to the emission line clouds discussed later on.

To summarize, the study of AGN line intensities, variability and profiles reveals information about the physical conditions in the line emitting gas, the gas distribution and dynamics. The optical depth of the gas and the amount of continuum energy absorbed by it, is deduced from observations of the Lyman and the soft X-ray continuum. This information can be used to construct theoretical models for AGNs. The following chapters address all these issues in greater detail.

2.5 Bibliography

Line intensity: Out of the hundreds of papers with spectroscopic observations of AGNs only a few will be mentioned here. They all contain a large number of observations and provide good examples of the available data.

Low redshift quasars: Baldwin (1975)

Intermediate redshift quasars: Wills, Netzer and Wills (1985).

High redshift quasars: Baldwin and Netzer (1978), Osmer and Smith (1980), Sargent, Steidel and Boksenberg (1989), Schneider, Schmidt and Gunn (1989, $z > 4$ quasars).

Seyfert 1 galaxies: Osterbrock (1977, 1984, optical spectrum), Morris and Ward (1988, near IR), Persson (1988, near IR), Osterbrock et al. (1990, near IR), Wu et al (1983, ultraviolet spectra), Cohen (1983, narrow lines in broad line objects).

Seyfert 2 galaxies: Koski (1978, optical lines) and Ferland and Osterbrock (1986, ultraviolet lines).

LINERs: Keel (1983), Filippenko and Sargent (1985) and Bonatto et al. (1989).

Molecules in AGNs: Kawara et al. (1990).

Line variability: A comprehensive review of all observations until 1988 is given by Peterson (1988). More recent emission line variability is described in Maoz et al (1991, detailed optical study of NGC4151); Clavel et al (1990, best ultraviolet variability data of a type 1 Seyfert, NGC 5548); Peterson et al. (1991, best optical variability data of a type 1 Seyfert, NGC 5548); Zheng (1988, low redshift quasasr), Perez et al. (1988, OVVs) and Gondhalekhar (1990, IUE observed quasars). For "vanishing" broad lines see Penston and Perez (1984). Narrow line variability is discussed in Clavel and Wamsteker (1987).

Line profiles: The following is a partial list of references, covering many of the important aspects:

Line profiles in bright AGNs, including profile comparison: Baldwin and Netzer (1978), Wilkes and Carswell (1982), Wilkes (1986), Mathews and Wampler (1985).

Broad line profiles in faint AGNs: Osterbrock and Shuder (1982), De Robertis (1985), Crenshaw (1986), Stirpe et al. (1989), Wamsteker et al. (1990).

Profiles changes: Stirpe et al. (1989), Zheng and O'Brien (1990) and references therein.

Narrow line profiles: Whittle (1985a, 1985b), De Robertis and Osterbrock (1984, 1986), Busko and Steiner (1989).

Relative redshift of broad emission lines: Gaskell (1982), Wilkes (1986), Espey et al (1989), Sulentic (1989) and Corbin (1990).

Disk-type line profiles: Osterbrock, Koski and Phillips (1986), Perez et al. (1987), Halpern and Filippenko (1988, a spectacular example of a double hump profile in Arp 102B). See also Miley and Miller (1979) for the highly disturbed profiles of radio galaxies. A double-hump is observed in the *difference spectrum* of variable emission line objects, see Alloin et al. 1988), Stirpe et al. (1989) and Goodrich (1989).

Lyman continuum absorption: See Smith et al (1981), Antonucci et al. (1989) and Reichert et al. (1985).

3 Theoretical Models

3.1 The BLR and the NLR

The first and obvious consequence of the spectral observations is the division into broad and narrow emission lines that seem to come from two distinct parts of the nucleus. The first is named the "Broad Line Region" (hereafter BLR). Its typical size, as deduced from the broad line variability, is 10-100 light-days in Seyfert 1 galaxies, and up to a few light years in bright quasars. The electron density in the BLR is at least $10^8 \, cm^{-3}$, as judged from the absence of strong, broad forbidden lines, and the typical gas velocity is 3000-10,000 $km \, s^{-1}$. The second region is the "Narrow Line Region" (hereafter NLR). Here the typical density is $10^3 - 10^6 \, cm^{-3}$, and the gas velocity 300-1000 $km \, s^{-1}$. The NLR must be much larger than the BLR, since no clear variation of the narrow emission lines is observed in objects undergoing large continuum variations. The NLR is resolved by ground-based observations in several nearby Seyferts, showing dimensions of 100-300 pc. There are good theoretical reasons to believe that the NLR in bright quasars, in those cases where it is observed, is much larger than that, perhaps a few kpc in diameter.

The crude division into BLR and NLR works very well when modelling AGNs. In particular, there is no clear evidence, so far, for a transition region between the two, with intermediate dimension, density and velocity.

The general picture adopted here, and in many review papers on the subject, is that of a small continuum source, around a massive black hole, surrounded by a much larger emission line region. Support for the small dimension of the continuum source comes from the short time scale variation of the optical, ultraviolet and X-ray continuum. The situation may be different at longer wavelengths, where the observed continuum radiation can originate in a region comparable in size to the BLR. Most of the discussion in the following chapters is independent of this extra complication related to the dimension of the millimeter-infrared continuum source. The parts that are likely to depend on it are specifically explained.

3.2 Photoionization Models

Photoionization is the most likely source of excitation for the emission line gas in AGNs. The intense, nonstellar continuum is directly observed in many objects, in some cases well beyond the Lyman edge, and there is nothing to prevent this radiation from interacting with the surrounding material. Direct evidence for this is the correlated line and continuum variations in many Seyfert 1 galaxies, where the line luminosity changes are clearly a response to the continuum luminosity changes.

Some observed line ratios give further support to the photoionization assumption. For example, the $CIII\lambda977/CIII]\lambda1909$ ratio is a good BLR temperature indicator for densities below about $10^{10}\,cm^{-3}$. The observed line ratio in several quasars (an upper limit usually), indicates an electron temperature below 25,000K in the C^{+2} zone. This is well below the temperature required to ionize carbon by collisions to C^{+2}, and very typical of the temperature in a photoionized nebula.

An important concept in this kind of modelling is that of a "cloud", introduced to distinguish small individual entities in the emission line region. As explained earlier, the broad line gas is optically thick to the ionizing radiation, as deduced from the presence of strong MgII and FeII lines. On the other hand, in many objects the low and the high ionization lines have similar profiles, indicating that the line ratios are about the same throughout the emission line region. Thus a reasonable assumption is that there are many individual optically thick clouds, each one producing all the observed lines in roughly the same proportion. Later on we show that the number of clouds is very large and they occupy only a tiny part of the volume of the emission line region. An alternative picture that has been considered involves a spherical shell around the ionizing source, producing high excitation lines from its inner part and low excitation lines from its outer part. The total continuum optical depth of the shell can be large but it is hard to imagine a dynamical situation in which the high and the low excitation lines would have similar profiles. Accepting the cloud picture, we are led to the conclusion, based on the very smooth line profiles, that the number of clouds must be very large. There is no direct way to obtain information about the shape and dimension of individual clouds and some assumptions about it must be made when constructing the model.

3.3 Other Models

Other models have been proposed to explain the observed emission lines. Shock-wave excitation is the most likely alternative. This cannot provide a full explanation of the broad line spectrum since the role of photoionization has been clearly established by the line variability observations. The situation is somewhat different in the NLR and some shock-wave models are moderately successful in explaining the narrow line spectrum. This is of greater interest in faint AGNs, in particular in LINERs. Recently, there have been several attempts to combine shock-wave and photoionization into a composite model. This is discussed in chapter 11.

A completely different, perhaps parallel view of activity and line excitation, is found in the so called "Warmers" model of AGN. In this model, the nuclear activity is due to a burst of star formation. The so called "nonstellar continuum" is in fact the spectrum of a young metal rich cluster containing a few, extremely hot, Wolf-Rayet stars. The broad emission lines and the observed variability are due, in this scenario, to supernovae explosions in the nuclear cluster. In this picture there is no need to assume a massive central object to provide the observed continuum. The model is getting to the stage that detailed calculations of the line and continuum spectra are becoming available. It has some very interesting features, as well as difficulties and drawbacks. It will not be discussed here any further.

Fast particles and radio wave heating have been mentioned, in several papers, as possible additional sources of heating and excitation. In particular, radio-loud AGNs are thought to host such an additional energy source. There is too little work, so far, to address this idea in any detail. There are some obvious drawbacks, similar to the ones associated with any model where much of the ionization is due to collisions. There are uncertainties to do with the unknown magnetic fields and the remarkably small observed differences between the emission line spectrum of radio-quiet and radio-loud objects.

The discussion in the following chapters concentrates on a particular AGN model in which the emission line region is made out of a large number of clouds that are photoionized by a central continuum source. The principles involved in this kind of modelling, and the physical processes affecting the spectrum of individual clouds, are discussed in chapter 4. Chapter 5 addresses the question of how to combine the flux emitted by many individual clouds into a composite theoretical spectrum and chapter 6 contains a detailed comparison with AGN observations.

3.4 Bibliography

References on AGN photoionization models are given in chapters 4 and 5. References on shock-heated AGN clouds are given in chapters 9 and 11. "Warmers" are discussed in a series of papers by Terlevitch, Melnick and collaborators, see Terlevich and Melnick (1988) and Terlevich (1988) for references. Some papers on the role of fast particles are Ferland and Mushotzky (1984, emphasis on FeII emission) and Viegas-Aldrovandi and Gruenvald (1988, NLR models with relativistic electrons).

4 Photoionization Models for Isolated Clouds

4.1 Photoionization equilibrium

Consider an isolated cloud exposed to a monodirectional flux of ionizing photons. Let L_ν be the monochromatic luminosity of the central source, per unit frequency, and r the cloud-center distance. Consider a point at a depth s into the cloud, where the optical depth is τ_ν. Assume also that all ions are in their ground level and the ionization field is not very intense. The rate of photoionization events per unit volume, of an ion X^i with a number density N_{X^i}, is

$$ N_{X^i} \int_{\nu_0}^{\infty} \frac{L_\nu e^{-\tau_\nu} \sigma_\nu(X^i)}{4\pi r^2 h\nu} d\nu \; , \tag{2} $$

where ν_0 is the threshold ionization frequency and $\sigma_\nu(X^i)$ the photoionization cross section. The rate of the inverse process, radiative recombination, is

$$\alpha(X^{i+1})N_{X^{i+1}}N_e ,\qquad (3)$$

where N_e is the electron density and $\alpha(X^{i+1})$ is a recombination coefficient which includes recombinations to all levels.

If L_ν is not varying in time or the recombination time, $t_{rec} = 1/\alpha(X^{i+1})N_e$, is short enough, the ionization fraction $N_{X^{i+1}}/N_{X^i}$ can be solved for by equating (2) and (3). The overall degree of ionization is found by solving such equations for all successive stages of ionizations, with the additional requirement on the total abundance of X,

$$N_{X^1} + N_{X^2} + ... = N_X .\qquad (4)$$

Evidently, the degree of ionization depends on the ratio of the ionizing photon flux to the gas density. It is convenient therefore to introduce an *ionization parameter* that specifies this ratio. The ionization parameter for hydrogen, at the illuminated face of the cloud, is designated U and is given by

$$U = \frac{ionizing\ photon\ flux}{cN_e} = \frac{\int_{\nu_0}^\infty L_\nu d\nu/h\nu}{4\pi r^2 cN_e} ,\qquad (5)$$

where c is the speed of light which is introduced to make U dimensionless. [1] As explained below, typical values for U in AGN clouds are $10^{-3} - 1$.

In principle there is a different ionization parameter for each element and at each point in the cloud. In practice, U as defined for hydrogen at the front of the cloud, is very useful and characterizes the overall level of ionization very well. For example, in quasars' BLR clouds,

$$\frac{N_{H^+}}{N_{H^0}} \simeq 10^{5.3}U \quad and \quad \frac{N_{He^{++}}}{N_{He^+}} \simeq 10^{3.2}U .\qquad (6)$$

This approximate relation is already enough to estimate the size of the broad line clouds. The photoionization cross section for hydrogen $(\sigma_\nu(H^0))$ at 1 Rydberg is about 6.3×10^{-18} cm^2. Using the estimated degree of ionization from (6) with $U = 0.1$, for a density of $N_e = N_H = 10^{10} cm^{-3}$, we find that an optical depth of 10 at the Lyman limit corresponds to a physical thickness of ionized hydrogen of about 3×10^{12} cm. The column density of ionized gas is about 3×10^{22} cm^{-2}. The amount of neutral material is not well known (see below) but a rough estimate for its column density is some 10 times the column of ionized gas. If clouds are roughly spherical, and they cover about 10% of the source (chapter 2), we need more than 10^7 clouds in a bright AGN whose BLR radius is of the order of 0.1 pc. (chapter 8). The associated filling factor (the

---

[1]　　The term *ionization parameter* has been used loosely in the literature. In some papers it is not divided by c, which leave it in velocity units. Some authors define it as $L_\nu/4\pi r^2 hN_e$ at a chosen frequency (usually at 1 Ryd.) and in other cases (chapter 9) it is the ratio of radiation pressure to gas pressure (i.e. divided by the temperature).

fraction of the total volume filled with such clouds) is tiny, of the order of 10^{-12}! The number of clouds is smaller for less luminous AGNs, especially if they are non-spherical, or have much larger neutral zones. Note also the small Compton depth of such clouds, which is the main reason for neglecting electron scattering in most of the following discussion.

The ionization parameter is also a measure of the way the ionization structure is changing, and it is easy to show that the thickness of ionization fronts is inversely proportional to U. Thus a large ionization parameter corresponds to sharp transition between successive stages of ionization and small U results in thick regions of gradual change in the level of ionization.

The steady state ionization structure is obtained by solving the ionization and recombination equations for all elements and at all points in the cloud. Ionization from all levels must be considered as well as corrections due to induced processes. For example, the photoionization rate per unit volume from the hydrogen level i is

$$n_i \int_{\nu_i}^{\infty} \frac{L_\nu e^{-\tau_\nu} \sigma_\nu(i)}{4\pi r^2 h\nu} [1 - \frac{\exp(-h\nu/kT_e)}{b_i}] d\nu \,, \tag{7}$$

where the second term on the right is the correction due to induced recombination and b_i is the departure coefficient for the level. The continuum optical depth includes a similar correction factor, thus

$$d\tau_\nu = \sigma_\nu(i) n_i [1 - \exp(-h\nu/kT_e)/b_i] ds \,. \tag{8}$$

Stimulated processes are important in regions of intense radiation fields and they are usually more important for high energy levels. The innermost part of the BLR is one place where they should not be neglected.

The photoionization of heavy ions by high energy X-ray photons is somewhat more complicated to treat. Such photons can eject a K-shell electron and the following ionic readjustment can often cause a removal of an additional electron. This "Auger effect" couples the ion X^i to X^{i+2}. The process is important in regions of the cloud where much of the softer ionizing radiation has already been exhausted and only the small cross-section, high energy photons can penetrate. The removal of L-shell electrons ($2s$, $2p$) is important too at high photon energies. Generally speaking, in a solar composition material with $T_e \sim 10^4 \, K$, the main sources of opacity between 1 and 20 Ryd are H^0, He^0 and He^{+1}. The metal opacity is more important at higher energies; carbon, nitrogen and oxygen up to $\sim 50 \, Ryd$ and iron etc. beyond that. Much of the X-ray opacity is due to K-shell absorption, which is not very sensitive to the exact degree of ionization. Thus the opacity at those frequencies reflects mainly the chemical composition of the gas and not its degree of ionization.

Photoionization from excited levels is, in some cases, of great importance. An example is the ionization of O^{++} from the two metastable levels, $^1S_0$ and $^1D_2$. These levels are in their Boltzmann population in AGN high density clouds, and photoionization out of them has a comparable rate to the ground level photoionization. The ground level ionization threshold of O^{++} is just short of the He^+ ionization edge, at 54.4 eV. The gas opacity at all energies higher than 54.4 eV is almost entirely controlled by He^+, whose abundance is very high, and the O^{++} ground level opacity plays almost no role. The O^{++} metastable levels ionization is about 5 eV lower than the ground level ionization. This extends the O^{++} photoionization cross section into frequencies where it may become a dominant source of opacity. It can influence, in some cases, the ionization of C^{++} (threshold at 47.9 eV) which has a large effect on the local cooling and line emission. Other important examples of this type are ionization from excited states of N^0 and Mg^+.

The ionization structure in AGN clouds is very different from that of HII regions and planetary nebulae. The clouds can be thick enough to be highly ionized (N^{+4}, O^{+5}) at their illuminated face and almost completely neutral at the back. The large flux of X-ray photons maintains a low degree of ionization ($\sim 10\%$) over a large part of the cloud, more than 90% of its thickness in some cases. Such extended low ionization regions are thought to be the origin of the strong FeII and MgII lines. The ionization structure in the highly ionized part depends on the value of U. Large ($\sim 0.1 - 1$) U results in sharply defined ionization fronts. Smaller U enables several stages of ionization to co-exist over large parts of the cloud.

4.2 Thermal Equilibrium

The equilibrium temperature of the gas is the result of heating, by absorption of the central source radiation, and cooling via several atomic processes. The temperature referred to is the kinetic temperature of the charged particles, or the *electron temperature*, T_e, which is well defined for all particles with a Maxwellian distribution of kinetic energies. It should not be confused with the *radiation temperature* (or color temperature), T_{rad}, characterizing the radiation field or the *excitation temperature*, T_{ex}, describing the populations of the atomic levels. Conditions in gaseous nebulae are far from thermodynamic equilibrium and T_e is usually different from both T_{rad} and T_{ex}. However, in some AGN clouds the density and optical depth are large enough so that for some ions $T_e=T_{ex}$, i.e. the level populations are given by the Boltzmann excitation equation.

The most important heating-cooling processes in AGNs clouds are:

4.2.1 Bound-free heating-cooling. Consider the absorption of a photon with energy $h\nu$ by the ion X^i in a level whose threshold ionization frequency is ν_0. The initial kinetic energy of the freed electron is $(h\nu - h\nu_0)$, and this energy is quickly spread, by elastic collisions, among the charged particles. [2] The heating rate per unit volume due to this ionization is

$$N_{X^i} \int_{\nu_0}^{\infty} (h\nu - h\nu_0) \frac{L_\nu e^{-\tau_\nu} \sigma_\nu(X^i)}{4\pi r^2 h\nu} d\nu \ . \tag{9}$$

---

[2] see however the note below about secondary electrons.

Summation over all levels of all elements gives the total bound-free heating. A correction due to stimulated recombination, as in equation (7), must be included in some cases.

The average energy of a recombining electron is close to kT_e and the total energy loss due to spontaneous radiative recombination is obtained by summing over expressions of the form

$$\alpha_T(X^{i+1})N_e N_{X^{i+1}} kT_e \, , \tag{10}$$

for all ions and all levels. Here α_T is an *energy averaged recombination coefficient* which is somewhat different from the coefficient α used in equation (3).

4.2.2 Free-free heating-cooling. The free-free heating rate per unit volume, due to the ion X^{+Z}, is

$$\int_0^\infty \frac{L_\nu e^{-\tau_\nu(ff)} \sigma_\nu(ff)}{4\pi r^2} d\nu \, , \tag{11}$$

where

$$\sigma_\nu(ff) = 3.69 \times 10^8 g_{ff}(\nu, T_e)\nu^{-3}T_e^{-1/2}N_e N_{X+z} Z^2[1 - \exp(-h\nu/kT_e)] \, . \tag{12}$$

In this equation $g_{ff}(\nu, T_e)$ is the thermal average of the Gaunt factor and allowance is made for stimulated emission.

Free-free absorption is a significant heating source for the gas in cases of intense low frequency radiation and large columns of ionized gas. Most low density nebulae are optically thin to free-free absorption, but some BLR clouds, with densities exceeding 10^{10} cm^{-3} and large column densities, may become opaque to this radiation, especially at low frequencies. For example, a unit free-free optical depth at a wavelength of 30μm, for $N_e = 10^{10}$ cm^{-3} and $T_e = 10^4 K$, is obtained for a column density of $\simeq 10^{22}$ cm^{-2}.

Free-free cooling is the result of bremsstrahlung events converting some kinetic energy into radiation via electron-ion Coulomb collisions. The rate, per unit volume, is given to a good approximation by

$$1.42 \times 10^{-27} Z^2 T_e^{1/2} g_{ff} N_e N_{X+z} \, . \tag{13}$$

The large abundance of hydrogen and helium ensures that their contribution to free-free cooling is the most important one.

A modification of the heating-cooling rate is required in cases of significant free-free optical depth, since some of the radiation is re-absorbed and heating by the diffuse free-free radiation was not included. An approximate way to introduce the correction is to multiply the cooling rate by the factor $\exp(-h\nu_{cut}/kT_e)$, where ν_{cut} is the depth dependent frequency, where the gas becomes optically thin to free-free absorption.

4.2.3 Collisional excitation and de-excitation heating-cooling. Inelastic collisions of free electrons with ions, followed by a radiative decay, convert kinetic energy into excitation energy and contribute to the cooling of the gas. Collisional de-excitation returns energy to the electron gas and is thus a heating process. It is convenient to discuss the net cooling, which is the cooling minus heating, per unit volume and time.

Consider the two level system, i and j ($j > i$), with statistical weights g_i and g_j respectively. The levels are coupled by an optically thin line, of energy E_{ij} and a radiative transition rate A_{ji}. The collisional excitation rate between the levels is

$$C_{ij} = \frac{8.629 \times 10^{-6} N_e \Omega_{ij}}{g_i T_e^{1/2}} \exp(-E_{ij}/kT_e) \,, \tag{14}$$

and the collisional de-excitation rate

$$C_{ji} = \frac{g_i}{g_j} \exp(E_{ij}/kT_e) C_{ij} \,, \tag{15}$$

where Ω_{ij} is the effective, temperature averaged, collision strength. In the absence of other populating mechanisms, the relative population of the levels is

$$\frac{N_j}{N_i} = \frac{C_{ij}}{(C_{ji} + A_{ji})} \,. \tag{16}$$

A useful concept is the so called "critical density", which is obtained for each transition by solving for the electron density for which $C_{ji} = A_{ji}$. Collisional de-excitation can be neglected for densities much smaller than this critical density, while the Boltzmann excitation equation can replace (16) for densities much above it.

The net cooling is the energy emitted by the atoms per unit volume and time

$$E_{ij} A_{ji} N_j \,. \tag{17}$$

In the limit of low density this is reduced to

$$E_{ij} N_i C_{ij} \,. \tag{18}$$

Thus, in this limit, the net cooling is proportional to N_e^2. In the high density limit $N_j \propto N_i$ and the net cooling is proportional to the gas density. Since the net cooling is basically the line emission, this is also the density dependence of the emergent line flux.

The formalism used here is easily generalized to a multi-level system and to the case where other atomic processes contribute to the level populations.

The steady state electron temperature is obtained by solving the simple energy conservation equation

$$\Sigma Heating = \Sigma Cooling \,. \tag{19}$$

This requires the full solution of the statistical equilibrium equations at all points in the cloud. The ionization and thermal solutions are of course coupled and iterative methods must be used to solve them, simultaneously.

4.3 Additional Heating and Ionization Processes

The following are additional processes that are likely to be important in AGN clouds.

4.3.1 Dielectronic recombination. Radiative recombination is the term used to describe the interaction of the ion X^{i+1} with a free electron, leading to the ion X^i in a ground or excited state *below* the ionization limit. Recombination can also proceed via autoionization states, *above* the ionization limit, in which case it is called dielectronic recombination. At high temperatures the free electrons cover an infinite number of autoionization states. Dielectronic recombination, in this case, is fast but density dependent. At low temperatures individual autoionization states are the main contributors to the recombination rate. The process is important in AGN clouds and can also result in emission lines of ion X^i. Such lines are weak but are of great importance in determining the chemical composition of the gas.

4.3.2 Collisional ionization and three-body recombination. The rates of collisional ionization and its inverse process, three-body recombination, depend on N_e^2, T_e, and the statistical weight of the level in question. Such processes are not very important in photoionized galactic nebulae, where the density is low and most ions are in their ground state. This is not the case in the broad line clouds, where many atoms are in excited states, due to the high density and large optical depth. For example, in the BLR clouds, the $n \geq 10$ levels of hydrogen, helium and some metals are collisionally coupled to the continuum and their populations are completely controlled by such processes. The $n < 10$ levels of hydrogen are affected too.

The treatment of these processes is rather tricky. A very large number of levels must be considered, and many unknown cross sections must be guessed. As a general rule, three-body recombination is faster than radiative recombination, for many ions, at $N_e \geq 10^{11}\ cm^{-3}$. The process is therefore important in the very dense parts of the BLR.

4.3.3 Compton heating-cooling and ionization. Compton scattering by bound and free electrons can be an important ionization and heating source for the gas, because of the intense X-ray radiation in AGNs. For nonrelativistic electrons, the energy transfer per scattering is approximately

$$\frac{h\nu}{m_e c^2}(4kT_e - h\nu)\,, \tag{20}$$

thus photons with energies greater than about 2.6 keV can ionize hydrogen from its ground level. Heavier elements can be ionized too, at higher energies, with an effective cross section which is proportional to the number of bound electrons. In realistic situations, the process is important only for mostly neutral gas, and hydrogen ionization is the only important source of free electrons because of the small abundance of the heavier elements.

The Compton heating rate is

$$\frac{N_e}{m_e c^2} \int_{\nu_0}^{\infty} \frac{L_\nu \sigma_h h\nu}{4\pi r^2} d\nu \ . \tag{21}$$

Compton cooling ("inverse Compton") occures when a low energy photon, ($h\nu < 4kT_e$) is scattered by an electron. The cooling rate is

$$\frac{4kT_e N_e}{m_e c^2} \int_{\nu_0}^{\infty} \frac{L_\nu \sigma_c}{4\pi r^2} d\nu \ . \tag{22}$$

In these expressions σ_h and σ_c are heating and cooling effective cross sections that agree with the Thomson cross section at low photon energy and with the Klein-Nishina formula at $h\nu \sim m_e c^2$. The correction due to stimulated processes, that can be important in intense radiation fields, was not included in these equations.

Compton heating and cooling is most important for the (hypothetical) hot gas suggested as a confining medium for AGN clouds (chapter 9).

4.3.4 Secondary electrons. Photoionization by high energy photons results in the ejection of an energetic electron. Because of the smaller Coulomb-scattering cross section at high energies, the mean free path of such electrons is long and several other reactions can occur before thermalization takes place. Most important are collisional ionization (leading to more secondary electrons) and collisional excitation. The result is more ionizations and less heating per incident high energy photon. The process is well known in the interstellar medium.

Secondary electrons related processes are very important in the partly neutral zone of AGN broad line clouds. The fast electrons are produced mainly by photoionization of He^0 and some metals, and they loose much of their energy by collisional excitation and collisional ionization of hydrogen. Calculations giving the number of extra ionizations, and the effective heating due to such electrons, are available. Care must be taken in calculating the cooling rate since excitation to high levels may be followed by collisional de-excitation, i.e. additional heating. There is also some extra line emission due to this process.

4.3.5 Charge exchange. These are reactions of the type

$$X + Y^+ \rightleftharpoons X^+ + Y \ , \tag{23}$$

where X and Y are two ions. The most important cases at $T_e \sim 10^4\,K$, involve neutral hydrogen. Well known examples are charge exchage reactions of H^0 with O^+ and N^+, that control the ionization of oxygen and nitrogen near the hydrogen ionization front.

Atomic calculations show that charge exchange with hydrogen, at a rate of $10^{-9}\,cm^{+3}\,s^{-1}$ or larger, is common for many ions. The typical recombination coefficients for metals, at $T_e = 10^4\,K$, are about $10^{-12}\,cm^{+3}\,s^{-1}$, and charge exchange with hydrogen is therefore an important process for all regions where $N(H^0)/N(H^+) \geq 10^{-3}$. For example, the charge exchange of H^0 with Fe^{++} is an important factor in increasing the Fe^+ fraction in AGN clouds.

Charge exchange can proceed via excited states, leading to some line emission. There is not yet any clear evidence that this is important in AGN clouds.

4.3.6 H^- and molecules. In large column density clouds, a trace amount of hydrogen can be in the form of H^- and H_2. This can be an important opacity source at infrared frequencies that increases the amount of infrared heating. The expected amount of H^- and H_2 must therefore be considered.

An important creation mechanism for H^- is radiative attachment. The reverse process, photodetachment, is a destruction mechanism for H^- and a heating source for the gas. The two are described, schematically, by

$$H^0 + e \rightleftharpoons H^- + h\nu \,. \tag{24}$$

Out of the collisional processes, associative detachment

$$H^- + H^0 \rightleftharpoons H_2 + e \,, \tag{25}$$

is a significant destruction source for H^-, while the reverse process, is the collisional dissociation of H_2. Another destruction mechanism for H^- is collisions with free electrons

$$H^- + e \rightleftharpoons H^0 + e + e \,. \tag{26}$$

Several other processes, such as charge exchange of H^- with positively charged ions (*charge neutralization*), can be important too. Out of the possible heating-cooling processes applicable to H^- and H_2, those associated with bound-free and free-free transitions in H^- are probably the most important.

The H^- and H_2 related processes are of marginal importance in AGN clouds, unless the column density is extremely large ($\sim 10^{25}\ cm^{-2}$).

4.4 Radiative Transfer

The photoionization calculations described above are relatively straight forward, provided the internally produced radiation can freely escape the cloud. This is the case in most galactic nebulae, and possibly also in AGN NLR clouds, but not in the BLR, where the optical depth to line and continuum radiation is significant. In this case the calculations must be modified, and the following pages describe some commonly used ways to do so.

4.4.1 Continuum transfer. The internally produced recombination (bound-free) radiation can have a large effect on the degree of ionization by interacting with the gas far from its point of creation. Sophisticated, iterative methods have been developed to account for the propagation of this radiation in the cloud. They depend on the gas distribution and geometry and cannot be applied, in a simple way, to all configurations.

Approximate methods have been developed too, to shorten the calculations and reduce the number of iterations. The "on-the-spot" approximation is based on the assumption that the diffuse radiation is absorbed very close to its point of creation. This is normally a good assumption for the ground level recombination of helium and hydrogen, in nebulae that are very optically thick to the

Lyman ionizing radiation. In this case the mean free path of the ground level recombination radiation is very short, and the assumption of local absorption works very well. The approximation is easy to apply since all that is required is to omit recombination to the ground level from the total recombination coefficient in the ionization (3) and thermal equilibrium (10) equations. The method cannot be used for treating recombination to excited levels, where the mean free path of the bound-free radiation is of the order of the cloud size or larger. It also fails near the boundaries of the cloud, where the optical depth to the surface is short even for the ground level recombination radiation.

There are ways to improve this simplified treatment. In the "modified on-the-spot" approximation, a correction factor is applied to each of the recombination coefficients, depending on the optical depth to the surface at the relevant frequency. The application is particularly simple in a slab geometry, where the radiation can be divided into inward going and outward going beams, and only optical depths to two surfaces need to be computed. For example, the recombination coefficient for hydrogen in equation (3), at a point inside the cloud where the Lyman continuum optical depth to the inner (illuminated) side is τ_{in} and to the outer side is τ_{out}, can be written in the following way:

$$\alpha(H^+) = 0.5\alpha_1[\exp(-a_1\tau_{\text{in}}) + \exp(-a_1\tau_{\text{out}})] + \alpha_B , \qquad (27)$$

where α_1 is the recombination coefficient to $n = 1$ and α_B is the sum of all recombination coefficients to levels with $n > 1$. The factor a_1, which is of the order of 2, takes into account the oblique escape of the ground level recombination photons and the frequency dependence of the optical depth. A modification of α_2, α_3 etc. can be included in a similar way. Here the frequency dependence of the optical depth must be calculated with great care, which means that a_1 is a strong function of the location in the cloud.

In a second approximation, named "outward only", the locally produced diffuse radiation is added to the incident flux and carried into the cloud in one, or more directions. Its obvious limitations is near the illuminated surface, where no diffuse radiation is allowed to escape. The process puts much of the heat deep in the cloud, causing an unrealistic temperature structure.

The free-free optical depth is never very large and the above approximate transfer methods are not adequate in this case. In many cases the optical depth is so small that no correction term is required. In other cases the free-free optical depth, $\tau(ff)$, must be calculated at all frequencies and included in the free-free heating integral (11). The free-free cooling rate is then modified using the $\exp(-h\nu_{\text{cut}}/kT_e)$ factor mentioned in 4.2.2. This local treatment is only a first order approximation to the rather complex full treatment of the free-free radiation transfer.

4.4.2 Line transfer. Standard radiative transfer techniques require a numerical solution of the radiation field everywhere in the gas. Each individual line profile is divided into several frequency bins, and the redistribution in frequency, following an absorption-emission process, is taken into account at all points. This is successfully applied in stellar atmosphere calculations, where conditions are close to LTE. Under such conditions, the local temperature and the level populations are not very sensitive to the emitted line flux and good solutions are obtained even when a small number of transitions are considered.

This is not the case in gaseous nebulae, where conditions are far from LTE, and a complete solution of the statistical equilibrium equations is required in order to calculate the temperature. Realistic photoionization calculations for AGN clouds involved the computation of several hundred emission lines, the large majority of which are optically thick. Neglecting some lines in the energy balance calculations, for the sake of treating the transfer of others in a more complete way, may result in a poor estimate of the kinetic temperature and wrong line ratios. Combining the two types of treatments, by solving *the full* radiative transfer in *all lines*, is beyond the capability of the most sophisticated computer codes available. We are thus faced with the choice of treating the radiative transfer in detail, at the expense of the atomic physics, or vice versa.

The alternative, so far preferred in most advanced calculations, is to treat the atomic physics in the most accurate way and use a simplified method for the line transfer. The method is known as *the escape probability method* and is demonstrated here for the simple case of a two level atom.

Consider a two level atom with an energy separation between the levels of E_{12} and a normalized line profile Φ_ν, which is assumed to be identical for both absorption and emission. Let I_ν be the radiation intensity and J the intensity averaged over angles and frequencies

$$J = \frac{1}{4\pi} \int \int I_\nu \Phi_\nu d\nu d\Omega . \tag{28}$$

Consider only radiative processes; spontaneous emission, with a rate of $n_2 A_{21}$, absorption, with a rate of $n_1 B_{12} J$ and induced emission, with a rate of $n_2 B_{21} J$. The rate equation for the level population is:

$$\frac{dn_2}{dt} = -\frac{dn_1}{dt} = -n_2 A_{21} - J(n_2 B_{21} - n_1 B_{12}) . \tag{29}$$

For isotropic line emission the emission coefficient is:

$$\epsilon_\nu = \frac{1}{4\pi} n_2 A_{21} h\nu_{12} \Phi_\nu \tag{30}$$

and the absorption coefficient is:

$$\kappa_\nu = \frac{1}{4\pi} (n_1 B_{12} - n_2 B_{21}) h\nu_{12} \Phi_\nu , \tag{31}$$

where stimulated emission is counted as negative absorption. The line source function, S_ν, is therefore

$$S_\nu = \frac{A_{21} n_2}{B_{12} n_1 - B_{21} n_2} = \frac{2h\nu^3}{c^2} \frac{1}{(n_1 g_2 / n_2 g_1) - 1} , \tag{32}$$

where we have made use of the fact that $A_{21} = (2h\nu^3/c^2) B_{21}$ and $g_1 B_{12} = g_2 B_{21}$.

Let β_{21} be the probability of a line photon to escape the cloud and $(1-\beta_{21})$ the probability to be trapped. In the escape probability method we assume that

$$J = S(1 - \beta_{21}) , \qquad (33)$$

which, by using the definition of S (32) and substituting into the rate equation (29), simplifies to

$$\frac{dn_2}{dt} = -\beta_{21} A_{21} n_2 . \qquad (34)$$

In the same way the emergent line flux, per unit volume, is:

$$\beta_{21} A_{21} n_2 h \nu_{12} . \qquad (35)$$

This is the essence of the escape probability method. It shows that the equations are similar to the optically thin case except that an *effective Einstein coefficient*, $\beta_{21} A_{21}$, replaces A_{21}. The method ensures local energy conservation and the local temperature is well determined. The scheme is easily generalized to a many level atom, by replacing A_{ji}, for each transition, by $\beta_{ji} A_{ji}$.

The escape probability approximation gives a correct solution where all scatterings are local and there is little diffusion in space (i.e. the photon is scattered many times close to its point of creation and then escapes the cloud without any further intaraction). It is also formally correct for the uniform case, where the temperature and degree of ionization are the same throughout the clouds. In such cases β is a "mean escape probability" which is a function of the total cloud optical depth.

Most realistic nebulae are not uniform throughout. Moreover, the line scattering process cannot be entirely local and some diffusion in space must occure. Thus the escape probability describing the trapping of the radiation, in the statistical equilibrium equation (34), is not necessarily the same function needed for calculating the emergent flux (35). Despite this, the advantage of this technique, especially the ability to treat hundreds of optically thick transitions simultaneously, is so great that it is currently being used in many photoionization calculations. The emphasis so far has been on getting reliable estimates of β for different line profiles and cloud geometries.

The scattering of resonance line photons is a well studied problem and various excellent calculations are available to estimate it under a variety of conditions. The number of scattering depends on the geometry, the optical depth and the line profile (or more accurately, the "redistribution function"). A general result of such calculations is that the scattering of line photons is mostly local (i.e. little diffusion in space) if the re-emitted photon is in the core of the line, within 3 Doppler widths of the line center. Such photons escape the cloud by diffusion into the line wings where scattering is coherent and a small number of scattering carry the photon a large distance in space. It means that the "local scattering" assumption, used in the escape probability approximation, is quite adequate for all resonance lines whose optical depth does not exceed about 10^4 (the optical depth corresponding to 3 Doppler widths).

The main result of the numerical transfer calculations mentioned above is expressed as the number of scatterings before escape, $Q(\tau)$. This is related to β via

$$\beta(\tau) = \frac{1}{1 + Q(\tau)} . \tag{36}$$

It is usually found that the number of scatterings is roughly linear with the *line center* optical depth, calculated from

$$d\tau_{12} = \frac{0.015 f_{12} \lambda_{12}}{v_{Doppler}} (n_1 - g_1 n_2/g_2) ds, \tag{37}$$

where f_{12} is the oscillator strength and $v_{Doppler}$ is the line Doppler width. [3] Thus

$$\beta = \frac{1}{1 + k(\tau)\tau} , \tag{38}$$

where $k(\tau)$ is a weak function of τ and is of the order of 2-5.

The total path length traveled by the photon before escape is also proportional to τ, with a different dependence factor, $k'(\tau)$. Numerical calculations show that $k'(\tau) \simeq k(\tau)$, i.e. the time it takes optically thick line photons to escape the cloud is several times longer than the time it takes the optically thin photons. This is important for dynamical reasons, since the trapped line radiation increases the internal pressure in the cloud (see chapter 9). The implication is that the radiation pressure in optically thick lines is enhanced by a factor of $\sim$5, almost regardless of the optical depths.

The method most commonly applied in modeling the broad line clouds is the "local escape probability", whereby the escape probability at each point is a function of the optical depth at that location. Thus, in a slab model, at a point in the cloud where the line center optical depth to one surface is τ and to the other surface is $(\tau_{tot} - \tau)$, the local escape probability is:

$$\beta(\tau_{tot}, \tau) = \frac{\beta(\tau) + \beta(\tau_{tot} - \tau)}{2} . \tag{39}$$

Obviously, τ_{tot} is not known a-priori and two or more iterations are required for a complete convergence of the calculations.

The following expressions for β are similar to what is used in most current calculations:

a: For $L\alpha$, $HeI\lambda10830$ and most resonance lines a good approximation is

$$\beta = \frac{1}{1 + 2\tau} , \tag{40}$$

b: For all other hydrogen lines, and non-resonance transitions

$$\beta = \frac{1}{1 + 2\tau} + \frac{0.25 a^{1/2}}{\tau^{1/2}} , \tag{41}$$

---

[3]   A component of microturbulence has been suggested to increase the line width and to reduce the optical depth. These are not considered here. We also do not consider velocity gradient (e.g. expansion) inside the clouds, that require a different escape probability function

where a is the damping constant for the line. The notable difference from the resonance line case is the dependence on $\tau^{-1/2}$ at large (~ 5000) optical depths. This different functional form is a question of some debate and is of great significance for lines like $H\alpha$.

c: Stark broadening of the upper levels of some transitions, changes the escape probability at high densities. Some calculations of the modified β are available for hydrogen. They must be inccorporated in the calculations for $N_e \gg 10^{10}\ cm^{-3}$.

4.4.3 Line and continuum fluorescence. Wavelength coincidences between emission lines ("line fluorescence") can be an important source of radiative excitation. The best known examples are the $HeII - OIII$ Bowen Fluorescence, at a wavelength of about $304\mathring{A}$, and the $OI - L\beta$ fluorescence at $1025\mathring{A}$. The first involves the excitation of OIII lines by the absorption of HeII $L\alpha$ photons. It is important in both the NLR and the BLR clouds, as indicated by the observed OIII Bowen lines, at wavelengths around $3000\mathring{A}$. The second is illustrated in Fig. 6 and results in extra excitation of the OI $^3D^0$ level by the hydrogen $L\beta$ line. It is very important in the BLR, where the scattering of the $H\alpha$ photons increases the $L\beta$ radiation intensity in the part of the cloud where the $H\alpha$ optical depth is large. Observable lines that are enhanced by this process are $OI\lambda1302$ and $OI\lambda8446$ (see diagram).

Wavelength coincidences among FeII lines can be important too. There are *several hundred* such coincidences and some may be more important than others. Another interesting possibility is a wavelength coincidence between $L\alpha$, which has a broad profile due to its large optical depth, and several FeII lines.

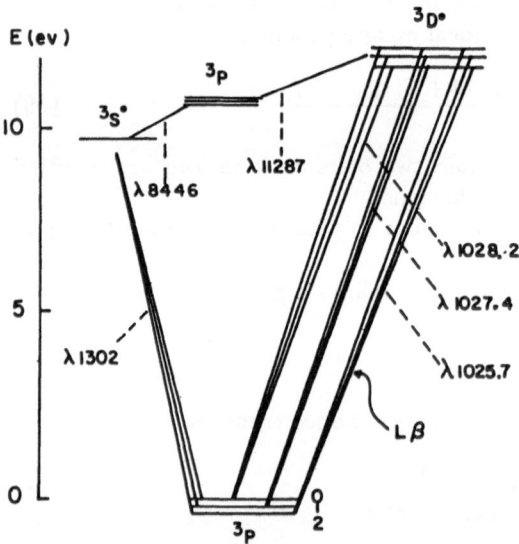

Fig. 6. The energy level diagram of OI showing the possible fluorescence with $L\beta$. Observable lines that are mostly affected are at $\lambda1302$ and $\lambda8446$. The process is most important in regions of large $H\alpha$ optical depth

Other possibilities that have been mentioned involved $MgII\lambda2798$, $NV\lambda1240$ and more.

Accurate treatment of line fluorescence requires a complete transfer calculation. There is also a local, less accurate solution, based on the escape probability method, that is simple to use and easy to incorporate into the statistical equilibrium equations. It involves the assumption of rectangular line profiles, (or more accurately a constant source function across the line profile), and gives quite good results. Its main disadvantage is the local treatment and the poor approximation at the line wings, where the source function is not constant.

Line fluorescence in AGNs has been a source of some confusion. Such processes are efficient in removing line photons from one transition, and pumping them into another, at frequencies where the radiation field is most intense, i.e. close enough to the line center for the source function to be constant. Further out into the line wings the source function is smaller and the pumping efficiency reduced. A separation of only a few Doppler widths between lines, can result in almost a zero fluorescence efficiency. This is the case even in very large optical depth lines, such as $L\alpha$, where the line profile is many Doppler widths wide.

Line photons can be destroyed by continuum absorption processes. This is sometimes called "continuum fluorescence" and is particularly important for optically thick lines, where the effective absorption optical depth is increased by the increased path length of the photons (see the $k'(\tau)$ factor mentioned earlier). Important examples are the ionization of hydrogen $n = 1$ by resonance lines with $\lambda < 912\text{\AA}$, the ionization of hydrogen from the $n = 2$ and $n = 3$ levels by $L\alpha$, $MgII\lambda2798$, $H\alpha$ and $FeII$ lines, and the ionization of neutral helium from the 2^1S and 2^3P levels. Absorption by dust grains, that are mixed in with the gas, and by H^-, are other examples.

The escape probability method provides a simple local treatment for this situation. Consider again the two level atom, a line absorption cross sections of κ_l and a continuum absorption cross section, at the line frequency, κ_c. Define

$$X_l = \frac{\kappa_l}{\kappa_l + \kappa_c} \quad ; \quad X_c = \frac{\kappa_c}{\kappa_l + \kappa_c}, \tag{42}$$

and

$$\beta_{lc} = \beta(\tau_l + \tau_c). \tag{43}$$

The escape probability formalism suggests that in the presence of a continuum opacity source, an effective escape probability

$$\beta_{\text{eff}} = X_c + X_l\beta_{lc}, \tag{44}$$

is to replace β_{21} in the statistical equilibrium equation (34). In this case, the emergent line flux, per unit volume, is

$$n_2 A_{21}\beta_{lc}h\nu_{21}, \tag{45}$$

and the number of continuum absorptions (e.g. photoionizations) is

$$n_2 A_{21} X_c(1 - \beta_{lc}). \tag{46}$$

This treatment is local and does not take into account the absorption of line photons away from their point of creation. A possible way to improve it, in cases of large continuum optical depth, is to multiply Eqn. (45) by $\exp(-\tau_c)$, where τ_c is a typical continuum optical depth, e.g. toward the inner surface of the cloud. The extra amount of continuum absorption should then be added to the expression in (46). This is not the only way to treat the continuum absorption process, and other, rather different methods, have also been suggested.

Absorption of external continuum radiation by spectral lines can be computed with the same formalism. Consider a point in the cloud where the optical depth to the illuminated surface, in a certain line, is $\tau_{\rm in}$. The probability of a photon emitted towards the continuum source to escape is $\beta(\tau_{\rm in})$, which is also the probability of the external radiation to reach that point in the cloud. The local J (33) is thus increased by an amount corresponding to the unattenuated external flux multiplied by $\beta(\tau_{\rm in})$. In the case of a central point source with luminosity L_ν, the increase in J is

$$\beta(\tau_{\rm in})L_\nu/16\pi^2 r^2 \,, \tag{47}$$

and the rate equation, omitting collisional and ionization processes, takes the form

$$\frac{dn_2}{dt} = -\beta_{21}A_{21}n_2 - \beta(\tau_{\rm in})A_{21}\frac{c^2}{2h\nu^3}\frac{L_\nu}{16\pi^2 r^2}(n_2 - g_2 n_1/g_1)] \,, \tag{48}$$

where β_{21} is the two-directional escape probability of equation (33). The process is important in cases of large ionization parameter. It can become significant in the partly neutral zone, where continuum absorption in spectral lines is immediately followed by collisional ionization from an excited state.

4.5 The Emergent Spectrum

4.5.1 Input parameters. A complete model for an isolated cloud requires that the following parameters be defined:

a: The frequency dependence of the incident continuum, L_ν.
b: The gas density (or pressure) and column density, $N_{\rm col}$.
c: The gas chemical composition.
d: The geometrical shape of the cloud.
e: The ionization parameter, U.

The incident continuum is known over a large range of frequencies. Unfortunately, the most important part, the Lyman continuum up to the soft X-ray energies, is not observed except in some high redshift quasars. This is a major drawback and indirect methods (chapter 10) must be used to estimate L_ν in that gap. Another uncertainty is the high energy tail ($h\nu > 30$ keV) which is not yet observed in many AGNs. This has important implications in models of large $N_{\rm col}$. Finally, free-free absorption at low frequencies may be a significant heating source for the BLR, but it is not at all clear that the observed far infrared radiation originates in a small, point-like continuum source (see R. Blandford's contribution). If this radiation comes from an extended disk, or any other structure large in comparison with the BLR size, then its effect on the broad line gas, via free-free heating, is largely reduced.

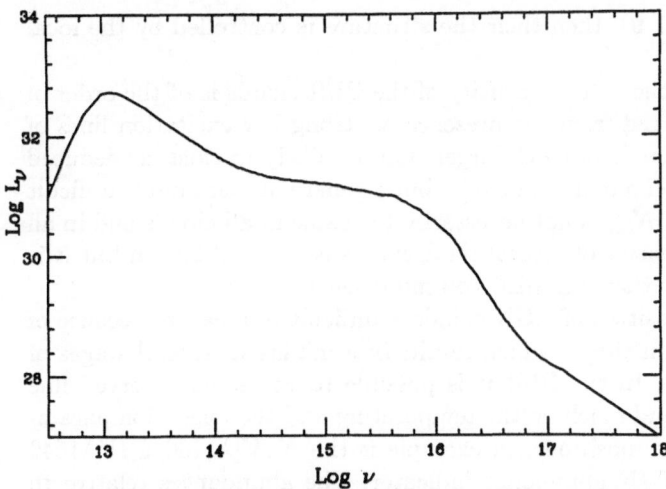

Fig. 7. A characteristic AGN continuum

Fig. 7 shows a typical continuum which is used in photoionization calculations and is consistent with our knowledge of many radio quiet AGNs. The
characteristic features are the steep infrared $(1 - 3\mu m)$ slope (a spectral index
of about 1.5), the flattening at optical and ultraviolet wavelengths, sometimes
referred to as "the big blue bump", the cutoff at about 3-10 Ryd (implied
by the models but not directly observed) and the flat X-ray continuum. The
drop at $\lambda > 30\mu m$ is artificially introduced, because of the above mentioned
uncertainty about the origin of this component.

The gas density can be measured, or at least estimated, from the observed spectrum. Standard methods of nebular analysis, involving the relative strength of several forbidden lines, suggest that $N_e \leq 10^8 cm^{-3}$ in all the
NLR clouds, with a typical density of about $10^4 cm^{-3}$. The density of the
BLR clouds is more difficult to estimate. The absence of strong broad forbidden lines of $[OIII]\lambda5007$ indicates that collisional de-excitation takes place,
and therefore $N_e \geq 10^8 cm^{-3}$. On the other hand, the strong intercombination lines of $CIII]\lambda1909$, $NIII]\lambda1750$, $NIV]\lambda1486$ and $OIII]\lambda1663$ suggest
$N_e \leq 10^{12} cm^{-3}$. Detailed calculations confirm these limits.

It is customary to fix either the density or the pressure at the illuminated
face of the cloud, and proceed by assuming constant density, constant pressure,
or some other assumption. This depends, of course, on conditions outside the
cloud. For example, in the "two phase model" (chapter 9) the BLR clouds are
embedded in a hot $(\sim 10^8 K)$ intercloud gas that provides the external confining
pressure. Constant pressure models are appropriate in this case. On the other
hand, the cloud internal pressure, near a variable continuum source, varies in
time following the changes in T_e, N_e and the line radiation pressure. Since
the sound crossing time (chapter 9) of the BLR clouds is comparable to the
variability time of the central radiation source, a stable pressure equilibrium
may never be achieved. Constant density models may be more appropriate in
this case. Finally, if AGN clouds are stellar atmospheres or the outer parts of

accretion disks (chapter 9), then their the structure is controlled by the local gravity.

The lower limit on the column density of the BLR clouds is of the order of 10^{22} cm^{-2}. This is derived from the presence of strong low excitation lines of MgII and FeII. Some clouds of much larger N_{col} are likely to exist, as deduced from the frequently observed lines of $CaII$, but a general upper limit is difficult to establish. Obviously, N_{col} is not necessarily the same in all clouds and in all objects. The column density of typical NLR clouds is not well known but it is estimated to be smaller than the BLR column density.

The chemical composition of AGN clouds is difficult to measure because of the nonstellar continuum shape, which results in a mixing of several stages of ionizations in one zone. In the BLR it is possible to use some observed line ratios, that do not depend much on the temperature and the ionization parameter, to estimate the composition. An example is the $NIV]\lambda1486/CIV\lambda1549$ ratio which is a good C/N abundance indicator. The abundances relative to hydrogen are ill determined because of the uncertainty in the calculated intensity of the hydrogen lines. There is a similar difficulty in the NLR, due to the difficulty in determining the electron density and temperature from optical forbidden lines. The hope is to use ultraviolet HST measurement of dielectronic recombination lines, and to combine them with collisionally excited lines, to determine the temperature. Thus the line pair ($CII\lambda1335$, $CIII]\lambda1909$), can be used to measure T_e in the C^{++} zone. This temperature, combined with the measured intensity of $NIII]\lambda1750$ that comes from the same region, will enable us to directly measure the C/N abundance ratio.

The comparison of observations of many line ratios with the best model calculations suggest that the following *cosmic abundances* (Table 2) are within a factor 2 of the abundances in many AGNs.

Table 2. Cosmic abundances used in AGN models

Element	N/N(H)
helium	0.1
carbon	3.7×10^{-4}
nitrogen	1.2×10^{-4}
oxygen	6.8×10^{-4}
neon	1.0×10^{-4}
magnesium	3.3×10^{-5}
aluminium	2.5×10^{-6}
silicon	3.2×10^{-5}
sulphur	1.6×10^{-5}
argon	3.8×10^{-6}
calcium	2.0×10^{-6}
iron	2.6×10^{-5}

The shape of the clouds is important because it determines the escape of line and continuum photons. The escape probability function in a sphere is different from that in a slab and there are "skin effects" to be considered. An *infinite slab* model has been adopted in many cases and will be used here. Later (chapter 5) we consider spherical BLR clouds but retain the slab escape probability function. This is well within the general limitation and uncertainty of using such transfer methods.

Finally, the value of the ionization parameter has to be determined (alternatively, the cloud distance and L_ν at some frequency). This is often used as the variable parameter in the calculations.

4.5.2 Examples. Fig. 8 shows calculated line fluxes, as a function of the ionization parameter, for an isolated BLR cloud. The input continuum is the one shown in Fig. 7, the density is constant at $10^{10}\ cm^{-3}$, the column density is $10^{23}\ cm^{-2}$ and the composition is as in Table 2. The line intensities are given relative to $H\beta$. Calculations for an isolated NLR cloud, with the same continuum source and abundances and $N = 10^4\ cm^{-3}$, are shown in Fig. 9. In this case the calculations stop at $\tau_{912} = 10^{3.5}$.

As evident from the diagrams, some line ratios are good ionization parameter indicators. In the BLR model, this is $OVI\lambda1035/CIV\lambda1549$; in the NLR model $[OII]\lambda3727/[OIII]\lambda5007$. The $CIII]\lambda1909/CIV\lambda1549$ ratio, that was thought to be a good ionization parameter indicator for the BLR, is not so, because of the big changes in the ionization structure at large U that limit the extent of the C^{+++} zone. The low excitation lines of MgII and FeII (not shown here) depend on U in a very complicated way and are not good ionization parameter indicators. They are further discussed in chapter 6.

The observed intensities of the broad high excitation lines (Table 1) suggest that the typical ionization parameter for the BLR is at least 0.1, if all lines are to be produced in the same population of clouds. There are, however, other

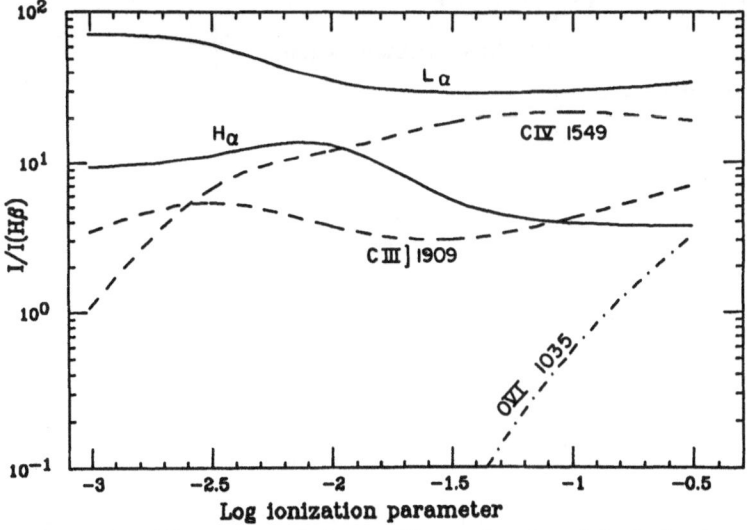

Fig. 8. Broad line ratios, relative to $H\beta$, as a function of the ionization parameter

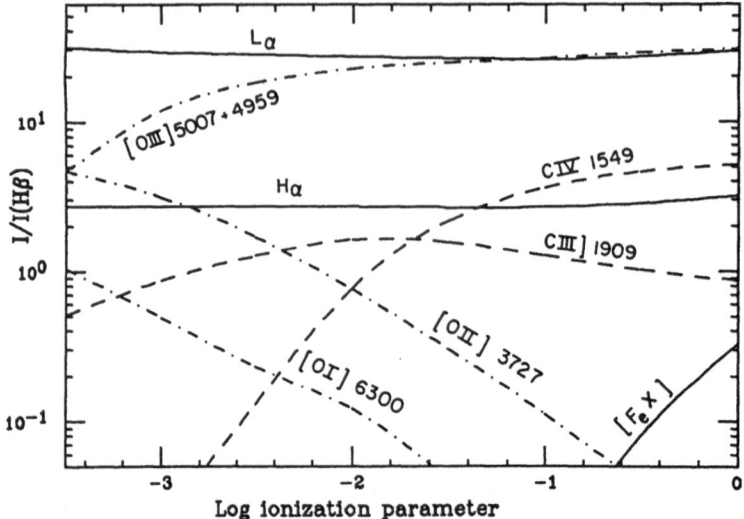

Fig. 9. Narrow line ratios, relative to $H\beta$, as a function of the ionization parameter

ways to produce strong, high excitation lines, such as optically thin material. A typical ionization parameter for the NLR is about 0.01, but there is a large diversity in this value, as discussed in chapter 11. The observed strength of $[FeX]\lambda6734$ and other narrow $[FeX]$ and $[FeXI]$ lines presents a problem, since such lines are calculated to be too weak in clouds with the ionization parameter required to give the observed $[OII]\lambda3727/[OIII]\lambda5007$ ratio. The origin of these lines may be some transition zone between the BLR and the NLR, or perhaps the interstellar matter of the host galaxy.

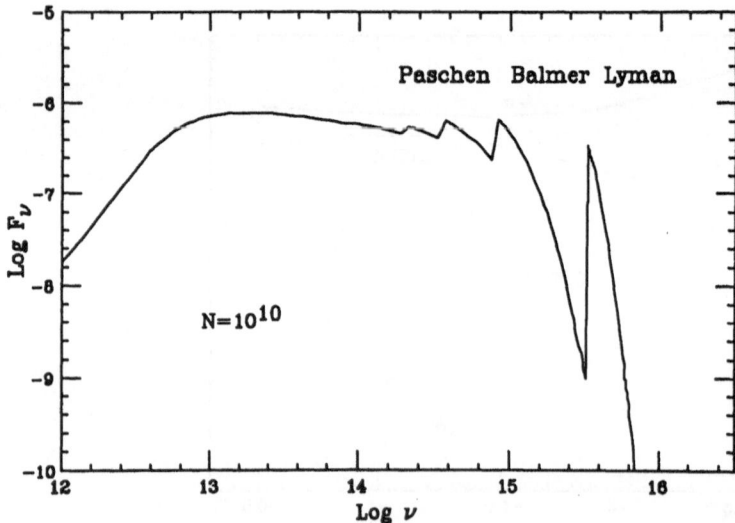

Fig. 10. Diffuse continua emitted by a broad line cloud with $U = 0.3$ and all other parameters as specified in the text

Regarding the broad hydrogen lines, the calculated change in $L\alpha/H\alpha/H\beta$ is mainly due to the increase of the Balmer optical depth with U. This is related to a well known problem of AGN study to be discussed in chapter 6. Finally, the diffuse bound-free and free-free continua, emitted by one of the broad line clouds considered here, are shown in Fig. 10.

The next step involves the combination of many one-cloud models in order to compare them with observed AGN spectra. This requires some knowledge of the gas distribution and is discussed in the following chapter.

4.6 Bibliography

There are several excellent text books on physical processes in low density nebulae. The most recent one is by Osterbrock (1989), where high density processes in AGNs are also described. A complete description of atomic processes is given in Mihalas (1978). Comprehensive discussions of radiative transfer methods are given in a book edited by Kalkofen (1984, see Canfield et el. there for application to AGNs) and in a soon to be published book on Masers by Elitzur. Many of the theoretical aspects dealt with in this chapter have been discussed in several review papers, where a more complete lists of references can also be found. The situation prior to 1979 is summarized by Davidson and Netzer (1979). Later, exhaustive reviews are by Kwan and Krolik (1981), Ferland and Shields (1985) and Osterbrock and Mathews (1986). Some detailed papers, on more specific topics, are:

X-ray related processes: Halpern and Steiner (1983), Kallman and McCray (1982), Shull and Van Steenberg (1985, heating and ionization by secondary electrons) and Ferland and Rees (1988).

Chemical composition: Shields (1976, broad lines, C/N and O/C), Uomoto (1984), Gaskell et al. (1981, metal deficiency in dusty BLRs), Wills, Netzer and Wills (1985, iron abundance), Binnet (1985, narrow lines in LINERs).

H^- **and molecules:** The most recent and complete paper is by Ferland and Persson (1989), where references to earlier papers are given.

Very high densities: See Ferland and Rees (1988) and Rees, Netzer and Ferland (1989).

Continuum transfer: See Netzer and Ferland (1984) and refrences therein for a modified on-the-spot approximation. See Collin-Souffrin and Dumont (1989) for treatment of hydrogen $n > 1$ levels. Puetter and Hubbard (1985) describe the formal solution of the free-free radiation transfer.

The escape probability method: The basic formalism is given by Elitzur (1984). The local escape probability is discussed in Kwan and Krolik (1981), and in many later papers. Useful approximations for β and more references can be found in Rees, Netzer and Ferland (1989). For Stark broadening see Puetter (1981). Continuum destruction is discussed in Netzer et al. (1985). Their treatment is however different from the one suggested by Hummer (1968) and at this stage it is not clear which is more applicable to AGN clouds. Puetter and collaborators (see references in Hubbard and Puetter 1985) discussed various aspects of the escape probability method, as well as some more complete transfer methods, as applied to AGN clouds. Most of their calculations are for a pure hydrogen case. A similar approach, including more elements, is described in Collin-Souffrin and Dumont (1986), where earlier references to such work can be found. Avrett and Loeser (1988) described an attempt to perform a complete transfer calculation for some lines, combined with a simplified treatment for many others. Numerical calculation of the scattering of resonance line photons, and references to earlier papers on the subject, can be found in Hummer and Kunasz (1980).

Line and continuum fluorescence: For the escape probability approximation of this process see Elitzur and Netzer (1984, lines), Netzer, Elitzur and Ferland (1985, line and continuum) and Lockett and Elitzur (1989, a critical discussion of the escape probability approximation). A more complete transfer approach to this problem is decribed by Weymann and Williams (1969) and Eastman and MacAlpine (1985). Line fluorescence in FeII is described in Netzer and Wills (1983), Wills, Netzer and Wills (1985) and Elitzur and Netzer (1985). For fluorescence with $L\alpha$ see Penston (1987).

Complete photoionization calculations: There are quite a few of these, starting from the works of Davidson (1972) and MacAlpine (1972). Some recent ones are: Kwan (1984, BLR, dependences on density and ionization parameter), Mushotzky and Ferland (1984, BLR, dependence on ionization parameter), Netzer, Elitzur and Ferland (1985, BLR and NLR), Ferland and Osterbrock (1986, NLR), Netzer (1987, BLR, continuum shape, angle dependence and large ionization parameter), Krolik and Kallman (1988, continuum shape), Rees, Netzer and Ferland (1988, BLR, including very high densities), Ferland and Persson (1989, BLR, large ionization parameters and extreme column densities), Collin-Souffrin and Dumont (1990) and Dumont and Collin-Souffrin (1990, very low ionization parameter and high densities) and Korista and Ferland (1989, interstellar high ionization lines).

5 Photoionization Models for a System of Clouds

Having calculated the emergent spectrum of isolated clouds, we now consider the gas distribution and the different ways to combine the emission from such clouds into a composite spectrum.

5.1 Spherical BLR Models

Consider a spherically symmetric system of clouds, extending from r_{in} to r_{out}. The mass of the individual clouds is conserved, but it is not necessarily the same for all clouds. The following physical parameters are all functions of r: the cloud number density $n_c(r)$, the cloud particle density N, the cloud emissivity $j_c(r)$, and the cloud velocity $v(r)$.

Most cases of interest involve optically thick clouds, where energy conservation requires that the total emergent line and diffuse continuum luminosity equal the energy absorbed by the gas. In this case, the geometrical cross section of the cloud as seen from the center, $A_c(r)$, is another useful parameter.

Designate $\epsilon_l(r)$ as the flux emitted by the cloud, in a certain emission line l, per unit projected surface area ($erg\ s^{-1}\ cm^{-2}$), we have the following relation for the cloud emission:

$$j_c(r) = A_c(r)\epsilon_l(r)\,. \tag{49}$$

We now make the following simplified assumptions about the radial dependence of the different parameters:

$$\epsilon_l(r) \propto r^{-m}\,, \tag{50}$$

$$n_c(r) \propto r^{-p}\,, \tag{51}$$

$$A_c(r) \propto r^{-q}\,, \tag{52}$$

$$N(r) \propto r^{-s}\,, \tag{53}$$

and

$$v(r) \propto r^{-t}\,. \tag{54}$$

Using these definitions, the radial dependence of the ionization parameter is:

$$U(r) \propto r^{s-2}\,. \tag{55}$$

5.1.1 Covering factor. For optically thick clouds, it is useful to introduce the *covering factor*, $C(r)$, which is the fraction of sky covered by photoionized gas clouds, as seen from the center. For a single cloud

$$C(r) = \frac{A_c(r)}{4\pi r^2}\,, \tag{56}$$

and for a thin shell of thickness dr

$$dC(r) = A_c(r)n_c(r)dr \propto r^{-(p+q)}dr\,. \tag{57}$$

The integrated covering factor can be estimated from the Lyman continuum observations (chapter 2) or from comparing the line and continuum luminosities (chapter 6). For the more luminous AGNs it is about 0.1 for the BLR and less than that for the NLR. It is thus justified to neglect obscuration and proceed on the assumption that the flux reaching the clouds depends only on r.

5.1.2 Integrated line fluxes. Given the calculated $\epsilon_l(r)$ for individual clouds (chapter 4), we can now integrate over r to obtain cumulative line fluxes. In the most general case

$$E_l(r) \propto 4\pi \int_{r_{in}}^{r_{out}} n_c(r)A_c(r)\epsilon_l(r)r^2 dr$$

$$\propto \begin{cases} \frac{1}{3-(p+q+m)}[(\frac{r_{out}}{r_{in}})^{3-(p+q+m)} - 1] & (p+q+m \neq 3) \\ ln(\frac{r_{out}}{r_{in}}) & (p+q+m = 3) \end{cases} \tag{58}$$

There are interesting cases where the number of radius dependent parameters is considerably reduced. An important example is a system of spherical clouds, of radius $R_c(r)$, in pressure equilibrium with a confining medium of pressure P. The kinetic temperature of a photoionized gas is only a weak function of U, and we can safely assume that the radial dependences of P and N are identical ($\propto r^{-s}$). Since $R_c^3 N = const.$, the cloud cross-section is

$$A_c(r) \propto R_c^2 \propto r^{2/3s} \tag{59}$$

(i.e. $s = -3/2q$) and the column density is

$$N_{\text{col}}(r) \propto R_c N \propto r^{-2/3s} . \tag{60}$$

Assume further that the clouds are moving *in or out* with their virial velocity. Mass conservation ($n_c v r^2 = const.$) requires that $p = 2 - t$ and substituting $t = 1/2$ we get the following radial dependence of the covering factor:

$$dC(r) \propto r^{2/3s - 3/2} dr . \tag{61}$$

General considerations (chapter 9) suggest that $1 \leq s \leq 5/2$ in many cases of interest. In particular, $s = 9/4$ corresponds to a case where the covering factor of a thin shell is proportional to the shell thickness.

The *total* flux, in all forms of emission, from a shell of thickness dr is proportional to $dC(r)^4$, and the cumulative flux is proportional to

$$\frac{1}{2/3s - 1/2}[(\frac{r_{\text{out}}}{r_{\text{in}}})^{2/3s - 1/2} - 1] \quad (s \neq 3/4) . \tag{62}$$

This is, in fact, a good approximation for many individual lines (i.e. $m = 2$ is a good approximation for many lines).

We conclude that in those cases with $s > 3/4$, the cumulative covering factor, and therefore the integrated line emission, increases rapidly outward, and the innermost parts of the atmosphere do not contribute much to the line emission. For $s \leq 3/4$ much of the line emission takes place close to the center and there is a natural boundary to the cloud distribution. Note, however, some potential difficulties. For example, the column density of ionized material depends on U and r, and clouds that are optically thick near r_{out} may become transparent closer in, or vice versa.

5.1.3 Specific models. Fig.11 shows cumulative BLR line fluxes, as a function of r_{out}, for a constant ionization parameter model, $s = 2$. They were calculated as explained in chapter 4, and added according to the prescription explained above. In this particular case, $L(\text{ionizing continuum}) = 10^{46} \ erg \ s^{-1}$, $U = 0.3$ for all clouds and the normalization is such that $N_{\text{col}} = 2 \times 10^{23} \ cm^{-2}$ where $N = 10^{10} \ cm^{-3}$. The integration starts at $r_{\text{in}} = 10^{16.5} \ cm$, where the density is $10^{12} \ cm^{-3}$, and carried out to large radii. Vertical cuts in the diagram correspond to a certain r_{out}, and give the integrated line fluxes up to that radius.

The narrow line fluxes can be integrated in a similar way, and Fig. 12 shows a model with $s = 1$, normalized such that $U = 3.6 \times 10^{-2}$ where $N = 10^5 \ cm^{-3}$

---

[4]   The radial dependence for individual lines is different, unless $m = 2$

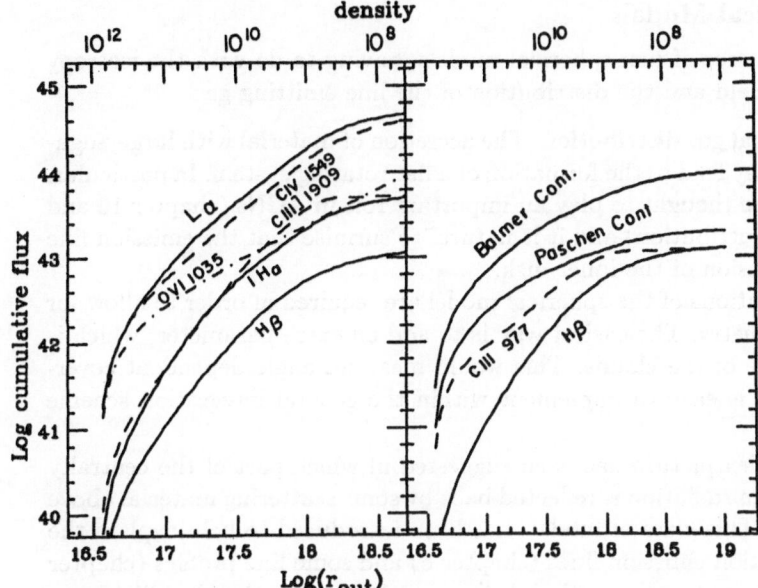

Fig. 11. Cumulative broad line fluxes, as a function of r_{out}, for a model with $U = 0.3$ and $s = 2$. The model is calculated for the continuum shown in Fig. 7, assuming $L(ionizing\ luminosity) = 10^{46}\ erg\ s^{-1}$. The value of r_{out} for other luminosities is obtained by noting that $r_{out} \propto L^{1/2}$

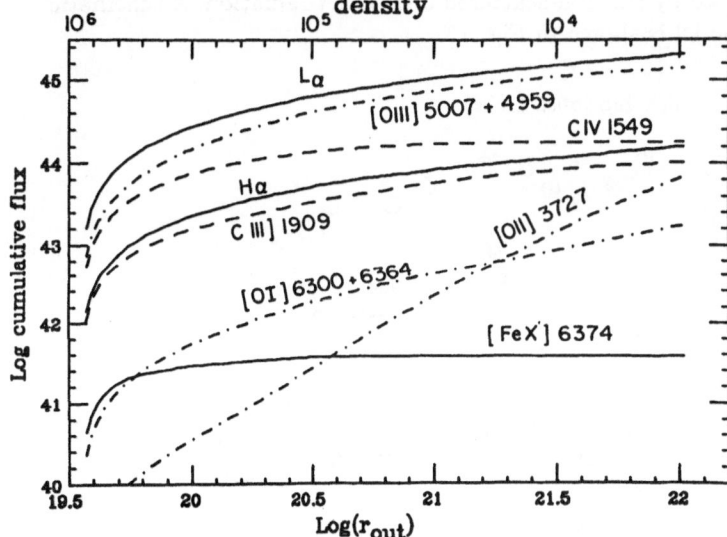

Fig. 12. Cumulative narrow line fluxes, as a function of r_{out}, for a model with $s = 1$ and parameters as explained in the text

and $N_{col} = 10^{22}\ cm^{-2}$. Here again $L(ionizing\ luminosity) = 10^{46}\ erg\ s^{-1}$, the inner radius is at $10^{19.5}\ cm$ and the density there is $10^6\ cm^{-3}$. Note that here, and in the BLR model of Fig. 11, the line intensity shown is normalized in such a way that the covering factor is unity at the outermost radius shown.

5.2 Non-spherical Models

There are two classes of non-spherical models, having to do with the isotropy of the radiation field and the distribution of the line emitting gas.

5.2.1 Non-spherical gas distribution. The accretion of material with large angular momentum may lead to the formation of a flat rotating system. In particular, accretion disks are thought to play an important role in AGNs (chapter 10 and R. Blandford's contribution) and it is natural to surmise that the emission line region is an extension of the inner disk.

Some modifications of the spherical model are required in order to allow for the different geometry. The easiest way is to add an extra parameter, which is the latitude angle of the clouds. This would mean an angle dependent covering factor, which is easy to implement within the general integration scheme outlined above.

A more complex picture has been suggested in which part of the centrally emitted continuum radiation is reflected back by some scattering material above the disk surface. This class of models has been introduced to help explain the strong low ionization emission lines (chapter 6) and some line profiles (chapter 9). It assumes a very large accretion disk, up to 10^6 gravitational radii ($R_G = GM/c^2$). The lines are emitted from the outer parts of the disk, where they are excited by the back-scattered hard ionizing radiation. The high ionization lines, and most of the $L\alpha$ flux, come from a more "normal", spherical system of clouds, that are ionized by the non-scattered ultraviolet radiation. A schematic illustration of the model is shown in Fig. 13.

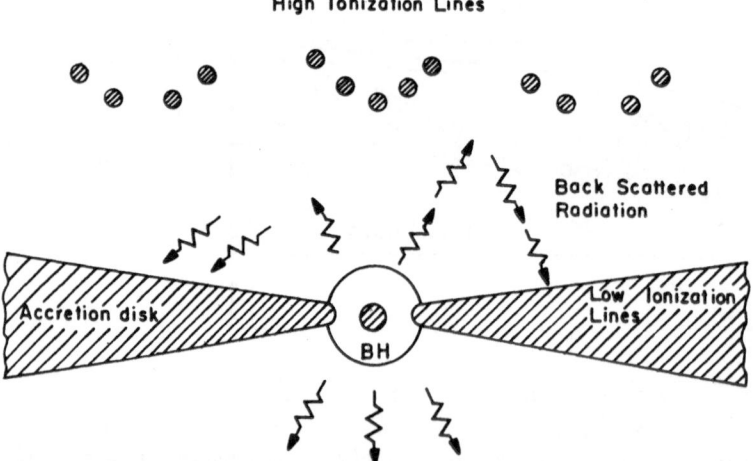

Fig. 13. A schematic two-component model for the BLR. The high ionization lines are emitted in a spherical system of clouds, and are excited by the direct ultraviolet radiation of the central source. The low ionization lines come mainly from the outer regions of the central disk, where most of the line excitation is due to back-scattered, hard ionizing photons. (After Collin-Souffrin, Perry and Dyson(1987), Collin-Souffrin (1987) and Dumont and Collin-Souffrin (1990))

There are several important implications to the model, to do with the line profiles (chapter 9) and covering factor. A possible difficulty is the very large disk radii postulated to explain the line strength. As explained in chapter 10, and in R. Blandford's contribution, the structure of geometrically thin AGN disks is not well understood. Current models suggest that such disks are radiation pressure dominated, and their self gravity radius is of the order of $10^3 - 10^4 R_G$. Much of the postulated low ionization lines originate outside this radius, where the disk, if it exists, is probably fragmented. This is, perhaps, not a strong objection considering the large uncertainties in the disk model. Moreover, a fragmented disk is as likely location for such lines as a uniform, continuous disk.

5.2.2 Anisotropic radiation field. A spherical or a flat cloud system can be illuminated by radiation from an anisotropic central source. This can be caused by relativistic beaming or by the combined effects of inclination and limb darkening in an accretion disk. The anisotropic disk radiation can be supplemented by an isotropic X-ray source, or perhaps some other components. Several examples of this continuum are shown in chapter 10. The ionization and excitation of the gas depends, in this case, on both r and the latitude angle of the clouds. This must be included in a modified integration scheme for calculating the cumulative line emission. Currently, there are too few observational constraints to test these ideas.

5.3 Bibliography

Relatively little has been done on combining isolated cloud models in the way described here. Rees, Netzer and Ferland (1989) suggested the general scheme and calculated many models for different values of s. Some more examples are given in Netzer (1989). See also Viegas-Aldrovandi and Gruenvald (1984) for references on multi-cloud NLR models. The idea of low ionization disk emission lines is described in several papers by Collin-Souffrin and collaborators, see for example Dumont and Collin-Souffrin (1990) and references therein. Nonspherical models related to thin accretion disks are discussed in Netzer (1987).

6 Comparison with the Observations

As noted earlier, there are some 20 broad, isolated emission lines, and a similar number of narrow lines, that can be measured and compared with the model calculations. Each observed line ratio conveys different information about the line emission region, and there are more than enough constrains on the models. At this stage we are only interested in integrated line intensities. Later we discuss the information obtained from line variability (chapter 8) and line profiles (chapter 9).

6.1 The Broad Line Region

Fig. 14 shows a comparison of the composite quasar spectrum of Fig. 2 and the theoretical model of Fig. 11. The calculated line intensities, up to $r_{out}=2 \times 10^{18}$ cm, are put on top of an artificial continuum, similar in shape to the observed one, and normalized to give the observed $L\alpha$ equivalent width.

Given the uncertainty in the model parameters, the difficult radiative transfer and the unknown gas distribution, it is quite remarkable that photoionization models for the BLR, like the ones describe here, give a good overall fit to the observed spectrum. Many of the observed line ratios are reproduced by the models and can be used to deduce important physical properties. There are, however, some line ratios that badly disagree with the model predictions, suggesting that some important ingredients are still missing. Below is a brief account of the present status of the theory, and the outstanding problems in this area.

6.1.1 Resonance metal lines. The strong ultraviolet lines of $CIV\lambda1549$, $NV\lambda1240$ and $OVI\lambda1035$ are good indicators of the gas temperature and ionization parameter. As seen from Table 1 and Fig. 14, the calculated strength of these lines, relative to $L\alpha$, are in fairly good agreement with the observations. The agreement is not as good in models of smaller U, although, a harder ionizing continuum can compensate, somewhat, for that. It has been proposed that optically thin BLR clouds, combined with much smaller ionization parameters, can explain the strong high excitation lines. Such clouds are expected to have very strong $NV\lambda1240$, $OVI\lambda1035$ and $HeII$ lines. Their contribution to the total broad line flux is large only if their covering factor is much greater than the covering factor of the optically thick clouds.

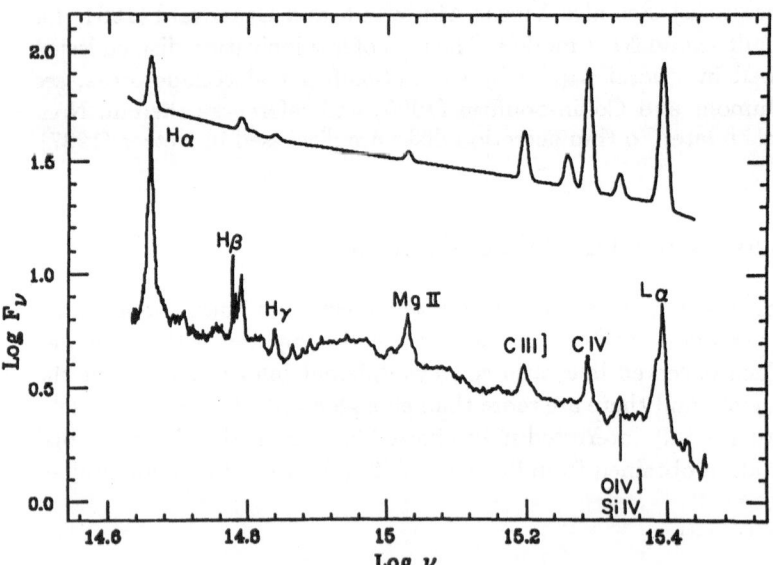

Fig. 14. A comparison of a composite quasar spectrum (bottom) with model calculations

The $CIII]\lambda1909/CIV\lambda1549$ line ratio has been used, in the past, to deduce the ionization parameter. The models presented here clearly show this line ratio to be insensitive to the exact value of U. In fact, the line ratio is in good agreement with the observations over most of the range calculated, and disagreement appears only where the contribution of lower density material starts to dominate the spectrum.

6.1.2 Intercombination lines. Semi-forbidden lines such as $CIII]\lambda1909$, $NIII]\lambda1750$, $OIII]\lambda1663$ and $NIV]\lambda1486$, are weak in the inner part of the BLR, where the density is above their critical density ($\simeq 10^{9.5}\ cm^{-3}$ for $CIII]\lambda1909$). Further out, where the density is lower, such lines can be important coolants, and the energy distribution among the different cooling agents is changed. The presence of strong intercombination lines is a sign that the contribution of $N_e > 10^{10}\ cm^{-3}$ material to the emitted spectrum is not significant.

6.1.3 Broad forbidden lines. With the $N \propto r^{-s}$, $s > 0$ density law considered here, some forbidden lines, such as $[OIII]\lambda4363$, are predicted to be strong in large r clouds, where the density drops to their critical density ($\simeq 10^8\ cm^{-3}$). Strong, broad forbidden lines are never observed, although there are hints to the presence of weak, broad $[OIII]\lambda5007$ in some objects. Thus, there seems to be a natural limit to the extent of the BLR. This may be due to the radial dependence of the covering factor at large r, to clouds becoming optically thin, or to some other reasons.

6.1.4 The hydrogen spectrum. The calculated intensity of $L\alpha$ in the $s = 2$ model (Fig. 11) increases with r much like the predicted $r^{5/6}$ dependence of (62). This reflects, mostly, the increase in the covering factor, and suggests that a fixed proportion of all ionizing photons are converted to $L\alpha$. It resembles the so called "recombination Case B flux", occurring in lower density nebulae, where each absorbed Lyman continuum photon results in the emission of a $L\alpha$ photon. In AGN BLR clouds, the situation is more complicated, due to the high density and large optical depth. However, in many models the calculated $L\alpha$ flux is within a factor 2 of the simple, "Case B" value.

Despite the simple atomic configuration, the good atomic data and the big improvements in the treatment of line transfer, the hydrogen line spectrum of AGNs is not yet well understood. This is demonstrated in Fig. 14 where it is evident that the calculated $H\alpha$ and $H\beta$ lines are much weaker, relative to $L\alpha$, than in the observations. This has come to be known as the "$L\alpha/H\beta$" problem. It is not yet clear whether it reflects wrong physical assumptions, the inaccuracy of the calculations or, perhaps, some reddening.

Regarding wrong physical assumptions, there are two proposed explanations. The first invokes a very strong, hard X-ray continuum, extending to MeV energies, and the second, extreme column densities ($\simeq 10^{25}\ cm^{-2}$). The two are not without difficulties. A strong X-ray-γ-ray continuum is observed only in very few AGNs (there are only very few such observations) while the "$L\alpha/H\beta$" problem seems to be common to most objects. There are also problems in violating the γ-ray background if all AGNs have such a hard continuum.

Large column densities are appealing for some reasons (see below) but the large Compton depth makes the line transfer calculations questionable, and there are difficulties associated with the physical size of the clouds in low luminosity objects (chapter 8).

The most likely cause for inaccurate calculations is the simplified escape probability treatment. Typical BLR clouds are expected to have huge $L\alpha$ and $H\alpha$ optical depths ($\tau(L_\alpha) \sim 10^8, \tau(H_\alpha) \sim 10^4$), and the local nature of this transfer method may not be adequate for such extreme conditions. Among the present dustless models, that use as an input the typical observed continuum, some get close to explaining the observed $L\alpha/H\alpha$ ratio and some manage to reproduce the observed Balmer decrement, but none is successful in explaining both.

Line reddening is another possible explanation which is not without its difficulties. It is discussed in chapter 7.

6.1.5 The helium spectrum. The optical depth in many HeI lines must be large because of the high population of the HeI 2^3S metastable level. The $HeI\lambda5876$, $HeI\lambda10830$ and other HeI line intensities are likely to be affected by that, and accurate transfer calculations are required.

To date most accurate calculations consider an up to 100 level HeI atom, with optical depth in all lines. Such a large number of levels is needed since three-body recombination is important in populating the high energy HeI levels at the BLR densities. The calculated line intensities are quite reliable, but not reliable enough to use the model helium/hydrogen line ratios to determine the helium abundance.

The calculations of the HeII spectrum are much simpler. The optical depth in all lines, except for the Lyman series and, perhaps, $HeII\lambda1640$, is small, and the three-body recombination process is not as important as in hydrogen. A notable problem is the $HeII\lambda304$ $L\alpha$ line, which is a major ionization source for hydrogen and a major fluorescence excitation source for $OIII$. The approximate methods (chapter 4) that are used leave much to be desired and the calculated line intensity is rather uncertain. The observation of this line is a major challenge of space astronomy and a real comparison with the calculations is still to come. Another complication is the wavelength coincidence between the hydrogen $L\alpha$ and the HeII $H\beta$ lines (separation of 0.498Å). This is a potential pumping source for the HeII $n = 4$ level but it is thought to be unimportant because of the small optical depth in the HeII $H\beta$ line, and the relatively large wavelength difference. The result of the small optical depth, and the good atomic data, is that the $HeII\lambda\lambda1640/4686$ line ratio is easy to calculate. This line ratio is an important reddening indicator and its use in determining the reddening in AGN clouds is explained in chapter 7.

6.1.6 FeII and MgII lines. The low excitation lines of FeII and MgII are produced in the partly neutral region of the BLR clouds. Such regions are thought to be heated and ionized by X-ray photons. They are characterized by $T_e \simeq 10^4 K$ and $N_{H^0}/N_{H^+} \simeq 10$.

While $MgII\lambda2798$ is a relatively simple line to calculate, this is not the case for the FeII lines, because of the extremely complicated energy level configuration of Fe^+. There are several thousand FeII transitions to be considered, many with a large optical depth. The atomic data for this ion is poorly known and reliable cross sections are only starting to become available. An additional complication is the large number of wavelength coincidences of different FeII lines; more than 300 (!) with separation less than $10\ km\ s^{-1}$. This is a major population process for the levels that must be taken into account. Other potentially important processes are the absorption of incident continuum radiation in FeII lines and the continuus opacity due the hydrogen $n = 2$ level.

The large number of FeII lines form several distinct emission bands at 2200-2600Å, 3000-3400Å, 4500-4600Å and 5250-5350Å. The strongest ultraviolet FeII lines originate from some odd parity levels with energies of $\sim$5 eV above the ground. Other ultraviolet lines, out of energy levels as high as 9 eV, are also observed. Such lines are not consistent with collisional excitation at the deduced electron temperature of $\sim 10^4 K$ and fluorescence or some other unknown processes must be responsible for that. All this is not unique to AGNs. The same FeII lines are known to be strong in the spectrum of symbiotic stars; galactic objects with no hard X-ray continuum.

Fig.15 demonstrates the complicated nature of the FeII spectrum. It shows a calculated FeII spectrum, for AGN clouds, with more than 3000 FeII lines. There is no way to isolate most of these lines, because of their large number and the broad line profiles. The convolution of the theoretical spectrum with a typical observed line profile (bottom part of the diagram) form broad, shallow

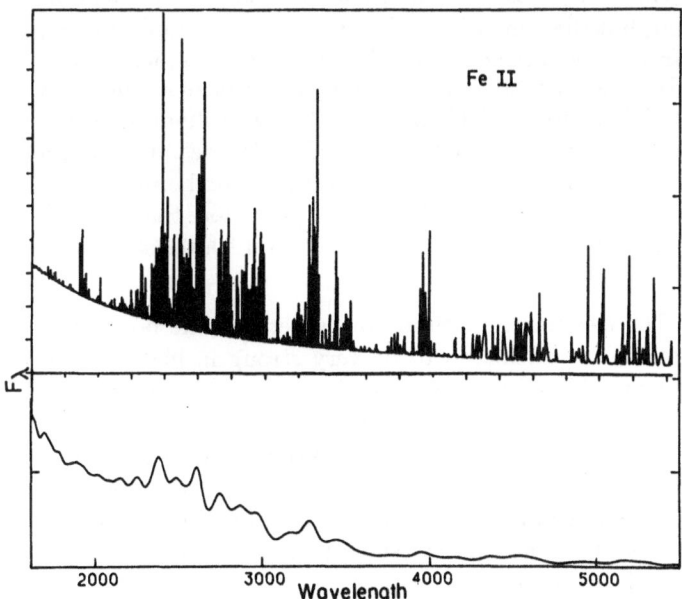

Fig. 15. Top: A theoretical FeII spectrum on top of a power-law continuum. Bottom: the same spectrum but lines are 4000 $km\ s^{-1}$ gaussians

emission features that demonstrate the difficulties in measuring the continuum luminosity, and the intensity of lines such as $MgII\lambda2798$ and $H\beta$, in spectral regions rich in FeII lines. The conglomerations of the strong FeII lines, the Balmer continuum, and other spectral features, creates a noticeable energy excess between the wavelengths of 2000 and 4000Å. This feature is sometimes referred to as the "small bump" and was confused, in the past, with the underlying nonstellar continuum of AGNs.

The FeII spectrum is one of the unsolved problems of AGN study. The total observed strength of these lines can equal the $L\alpha$ intensity, while the calculated strength is only about 1/3 or 1/2 of that. The ratio of the optical FeII lines to the hydrogen Balmer lines presents a similar, or even bigger problem, and there is also a difficulty in explaining the observed ratio of optical FeII lines to ultraviolet FeII lines. Suggested explanations, within the general framework of photoionization, include very high densities, large iron abundances and emission from the outer regions of central accretion disks. There was also a suggestion of a different model, based on the idea that the lines are formed in a thick, warm medium which is mechanically heated. Such models have the extra degree of freedom of not being directly associated with the central radiation source.

6.1.7 CaII lines. These are the lowest ionization lines observed in the spectrum of AGNs. The strongest feature is the infrared triplet at 8498, 8542 and 8662Å. The lines are observed in about 1/3 of all objects, while other CaII lines, such as the H, K and the forbidden lines near 7300Å, are weak or absent. Theoretical modelling shows that the internal CaII line ratios, and their strength relative to $H\beta$, requires very large column densities, $N_{col}\sim 10^{25}\ cm^{-2}$. Such models are appealing for some theoretical reasons (energy budget, to be explained below, the $L\alpha/H\beta$ problem) but there are difficulties as well. For example, objects with very weak CaII infrared lines are not very different, spectroscopically, from objects with strong CaII lines. In particular, strong CaII emitters are not very different in their $L\alpha/H\beta$ from weak CaII emitters. If large column densities are essential to explain both the observed $L\alpha/H\beta$ and CaII lines in AGNs, it is not very clear why the CaII lines are not more common. Furthermore, very large column density clouds, with typical BLR densities, are more than $10^{15}\ cm$ thick, a dimension which is of the order of the cloud-central source separation in low luminosity AGNs (chapter 8).

6.1.8 Diffuse continua. The free-free continuum and several of the bound-free continua (Paschen, Balmer) are calculated to be very strong in high density, small r clouds, where many bound-bound transitions are collisionally suppressed (Fig. 11). The observations of the Balmer and Paschen continua in low redshift broad-line AGNs suggest that the contribution of $N_e \gg 10^{10}\ cm^{-3}$ clouds to the line spectrum is not significant. The Lyman continuum emission is more difficult to measure because of possible confusion with Lyman absorption, the numerous absorption systems at this wavelength range, and the limited wavelength coverage of such observations.

6.1.9 Very high excitation metal lines. These lines ($CIII\lambda977$, $OIII\lambda835$ etc.) are calculated to be strong at small r high density clouds, where the temperature is high due to the collisional suppression of other cooling agents, such as $CIV\lambda1549$. The lines are weak or unobserved in most AGN spectra which is another argument against having a large contribution from very high density clouds to the broad-line spectrum.

6.1.10 Small dense BLR. Some gas clouds may survive in the innermost part of the BLR, where the radiation field is most intense. These must have very high densities ($\sim 10^{12} \ cm^{-3}$) since low density material will not achieve thermal equilibrium at this enviroment (chapter 9). The clouds may be associated with the inflow of gas from the BLR, or perhaps produced near the central black hole. At such high densities the gas must be close to thermal equilibrium, most cooling is via bound-free and free-free emission and no line emission is likely to be important. The clouds are therefore reprocessing the central continuum radiation, absorbing it at some frequencies and re-emitting in others. The resulting spectrum can resemble, in some ways, the spectrum of a thin accretion disk, showing a "blue bump" in optical and ultraviolet energies and a strong edge at the Lyman limit. Currently there are too few observational constraints, and too many theoretical uncertainties, to put this idea into a serious observational test.

6.2 The Narrow Line Region

The observations of narrow lines in high luminosity AGNs are not nearly as good as those of the broad lines. In particular, it is difficult to separate the broad and the narrow ultraviolet lines and there are no reliable measurements of the narrow $CIV\lambda1549$ and $CIII]\lambda1909$ lines in quasars and Seyfert 1 galaxies. The situation is likely to be improved with the HST observations but so far the only narrow ultraviolet lines that have been measured are in Seyfert 2 galaxies.

There are very good observations of optical narrow lines that can be compared with the model predictions. The overall agreement is very good and narrow line models, like the one shown in Fig. 12, reproduce the relative strength of $[OIII]\lambda5007$, $[OII]\lambda3727$, $[OI]\lambda6300$ and $H\beta$ quite well. This is not the case for $[FeX]\lambda6734$, and the line is observed to be much stronger than predicted. A similar, although somewhat smaller discrepency, occurs for the lines of $[NeV]$ and $[FeVII]$.

Line profile observations (chapter 9) indicate a large density gradient in the NLR. The validity of the model in Fig. 12, where the density gradient is quite small, is thus questionable and the very high ionization lines may come from a much denser part of the NLR. Another, very different suggestion is that the high ionization lines come from the interstellar medium of the host galaxy.

Lower excitation spectra, such as in LINERs, cannot be explained by the relatively high ionization parameter model of Fig. 12. Such spectra are discussed in chapter 11.

The intensity of the narrow Balmer lines are easy to calculate. The $H\alpha$ optical depth is not likely to be large, and the $H\alpha/H\beta$ ratio is closed to the

Case B value. A comparison with the theoretical $L\alpha$ intensity is somewhat less reliable. First, the line is likely to be collisionally excited by a density dependent amount. In addition, the typical NLR density is close to the critical density of the 2-photon transition ($\sim 1.5 \times 10^4\ cm^{-3}$) and the relative population of the hydrogen 2s and 2p levels may be different in different clouds. Because of this the recombination $L\alpha/H\beta$ ratio can vary from about 23 (low density limit, the 2s and 2p levels are not coupled) to 34 (high density limit, the 2s to 2p population ratio is 1:3). Combined with the collisional enhancement of $L\alpha$, the overall expected range in the $L\alpha/H\beta$ ratio is about 25-100. Accurate modelling is required for comparing this line ratio with the observations.

6.3 The Energy Budget of the BLR

While the individual line ratios are the best indicators for the physical conditions in the BLR clouds, a simple energy conservation argument provides another strong constraint on the models.

Consider optically thick clouds absorbing *only* ionizing radiation, and no reddening by dust. The total energy emitted by the clouds is simply the energy absorbed by them, which is the product of the ionizing radiation and the covering factor. Photoionization calculations for the BLR indicate that the emitted $L\alpha$ flux is not very different from the "Case B flux" mentioned earlier, whereby each absorbed ionizing photon results in one $L\alpha$ photon. In this case the number of $L\alpha$ photons is the product of the number of ionizing photons and the covering factor. Combining the two we have a simple observational way to measure *the mean energy* of an ionizing photon, $\bar{\nu}$ (in Ryd.):

$$\bar{\nu} = \frac{total\ energy\ emitted\ by\ the\ clouds}{number\ of\ L_\alpha\ photons} \ . \tag{63}$$

This is a most important information about the shape of the Lyman continuum that cannot be obtained by direct observations of the Lyman continuum.

The integrated flux emitted by the BLR clouds is not easy to measure since some of it is in broad spectral features, such as the "small 2000-4000Å bump", the Paschen continuum and several infrared lines. It is estimated to be 5-9 times the $L\alpha$ intensity which means, according to the above relation, $\bar{\nu} \simeq 4 - 7\ Ryd$.

To illustrate this further assume that the ionizing continuum is a simple power-law in energy, $L_\nu = C\nu^{-\gamma}$, extending up to a cut-off frequency ν_{cut}, where ν is in Rydberg and C is a constant. The mean energy of an ionizing photon is

$$\bar{\nu} = \frac{\int_1^{\nu_{\text{cut}}} C\nu^{-\gamma}d\nu}{\int_1^{\nu_{\text{cut}}} C\nu^{-\gamma-1}d\nu} = (\frac{\gamma}{1-\gamma})\frac{1 - \nu_{\text{cut}}^{1-\gamma}}{\nu_{\text{cut}}^{-\gamma} - 1} \ , \tag{64}$$

where we have assumed $\gamma \neq 1, 0$. This expression should be compared with (63) to obtain the value of γ. For example, for $\nu_{\text{cut}} = 10\ Ryd$, which is consistent with the observational constraints mentioned in chapter 4, we get $\gamma \simeq 0$. This is in conflict with the observations that show a typical observed slope, at a rest wavelength of 1000Å, of about $\gamma = 0.6$ and an even steeper slope at shorter wavelengths. The discrepancy has been named "the energy budget problem".

There are several suggested solutions to this problem. First, only the soft ($\nu \leq 10 \, Ryd$) Lyman continuum photons have been considered here while high energy photons are observed in almost all AGNs. Such photons hardly interact with the gas unless the column density is much greater than $10^{23} \, cm^{-2}$. Very thick clouds have been suggested, in which a large fraction of the high energy continuum is absorbed by the gas. Thick clouds can also absorb the infrared continuum, which helps too. Second, the models may be wrong, in particular the assumption about the number of $L\alpha$ photons and its relation to the ionizing flux. Also, the observed lines may come from two distinct parts of the BLR (the surface of the central disk?). Third, the above argument makes use of the intrinsic properties of AGNs, but the observed fluxes may be different from the intrinsic fluxes. For example, reddening by dust can change the observed line ratio and the inferred mean photon energy. None of these explanations is entirely satisfactory and it is likely that the real solution involves some combination of all.

A somewhat related problem is the ratio of the high and low excitation lines. Generally speaking, much of the "soft" ionizing flux ($\nu \leq 20 \, Ryd.$) is converted to recombination and high ionization lines, while the harder photons, that can penetrate much deeper, are converted to low excitation lines. It can thus be argued that the flux ratio of high to low ionization lines is a measure of the flux at different wavelengths. One can use it to formulate a "second order energy budget problem" related to the fact that the observed low excitation lines of MgII and FeII are too strong relative to $H\beta$. This cannot be solved by reddening but the argumet is based, to a large extent, on the observed FeII lines, that are not well understood.

There are other methods for estimating the shape of the ionizing continuum. In particular, the equivalent width (EW) of the recombination lines can be used for that. For example, the "Case B" $L\alpha$ equivalent width, for a system of optically thick clouds with a covering factor $C(r)$, around the above power-law continuum, is:

$$EW(L_\alpha) = \frac{1215}{\gamma}(3/4)^\gamma [1 - \nu_{\text{cut}}^{-\gamma}]C(r) \, \mathring{A} \quad (\gamma \neq 0) \ . \tag{65}$$

A similar ratio can be calculated for $EW(HeII\lambda1640)$, where the integration in this case is for $\nu \geq 4 \, Ryd$. The observed EW of these two lines can be compared with this theoretical prediction in order to estimate the continuum slope around 1-4 Ryd. Alternatively, the EW of the HeII lines, combined with an assumption on the covering factor, can be used to estimate $L_\nu(1640\mathring{A})/L_\nu(228\mathring{A})$ etc. These arguments cannot be simply used in the disk-like geometry mentioned in chapter 5.

6.4 Bibliography

Many of the theoretical papers mentioned in chapters .4 and 5 include also a detailed comparison with the observations. Specific references addressing topics in this chapter are:

General BLR models: Rees, Ferland and Netzer (1989, very high density models), Ferland and Persson (1989, very large U and N_{col} models, the most complete discussion of the CaII lines), Krolik and Kallman (1988, line ratio dependences on continuum shape), Collin-Souffrin, Hameury and Joly (1988, thick Compton heated BLR clouds).

FeII lines: Netzer and Wills (1983, FeII line fluorescence, general FeII line calculations), Wills, Netzer and Wills (1985, more detailed FeII models and a comparison with the observations), Penston (1987, $L\alpha$ fluorescence with FeII), Joly (1987, detailed FeII calculations for a medium which is mechanically heated), Collin-Souffrin et al. (1988, FeII lines from a Compton heated media), Dumont and Collin-Souffrin (1990, FeII lines from scattered radiation on accretion disks). Many detailed articles on FeII lines can be found in the proceedings of the 94th IAU colloquium held in 1986 (Viotti, Vittone and Friedjung editors, Reidel publishing company).

Small dense BLRs: Ferland and Rees (1988).

Energy budget and continuum slope: Netzer (1985, first presentation of the energy budget problem), Collin-Souffrin (1986, further ellaboration and separation into low and high excitation lines), Ferland and Persson (1989, application to a large N_{col} clouds), MacAlpine et al. (1985, determination of the Lyman continuum slope from HeII lines).

NLR models: Ferland and Osterbrock (1986).

7 Dust and Reddening

Dust is associated with cool astrophysical gas in almost all known nebulae and it is hard to imagine that AGNs are exceptional in this respect. There are three ways to discover the dust, via its thermal emission, through its effect on the observed spectrum (extinction) and by light polarization. This review is concerned with AGN emission lines and the information they reveal about the physical conditions in the nucleus. Therefore we concentrate on the extinction and mention the two other dust properties only to the extent that they are likely to be correlated with the dust causing the extinction.

7.1 Thermal Emission and Polarization

Consider an optical-ultraviolet continuum source with integrated luminosity L_{46} ($10^{46} erg\, s^{-1}$), surrounded by a cloud of dust particles at a distance r_{pc} parsecs. The particles absorb the optical and ultraviolet continuum radiation and re-emit it at infrared energies. Assume spherical particles and an absorption-emission infrared coefficient which is proportional to λ^{-1}. The equilibrium dust grain temperature, T_{dust}, is given to a good approximation by

$$T_{\text{dust}} \simeq 1700\, L_{46}^{0.2} r_{pc}^{-0.4}\, K\,. \tag{66}$$

The maximum temperature at which dust can survive is about 1700 K, thus the evaporation distance, r_{ev} is roughly

$$r_{\text{ev}} \simeq L_{46}^{1/2}\, pc\,. \tag{67}$$

This distance is just outside the estimated BLR size (chapter 8). Thus, dust particles cannot survive in the BLR unless they are shielded from the central radiation source. They can survive almost anywhere outside the BLR, in particular in the NLR.

There are several theoretical suggestions, as well as some observational evidence, that hot dust, just outside the BLR, is responsible for at least some of the observed infrared radiation of luminous AGNs. One possibility is that the dust is in a flat disk-like configuration, which is the extension of the inner accretion disk. Spherical dust distribution has been considered too. There is at least one bright Seyfert galaxy, F-9, where the dimension of the dust cloud has been measured directly, since in this case the large changes in the optical-ultraviolet radiation of the central source induced a similar, but delayed variation in the infrared dust emission. The time lag between the optical and the infrared continuum variation, when converted to physical dimension through the speed of light, indicates that the nearest dust grains are at a distance which is comparable to the evaporation distance of equation (67).

7.2 Internal and External Dust

The dust in external galaxies may not have exactly the same extinction properties as in our interstellar medium. In particular, some of the broad extinction features, such as the one centered at 2200Å, may be weak or absent in other galaxies. The amount of extinction will be estimated assuming a galactic dust-to-gas ratio and a simple, λ^{-1} extinction law. Thus, for a dust on the line of sight to the source, the extinction in magnitude, for a gas column density N_{col}, is

$$m_\lambda \simeq 0.5\,\Big[\frac{N_{\text{col}}}{10^{21}\, cm^{-2}}\Big]\Big[\frac{5000\text{Å}}{\lambda}\Big]\,. \tag{68}$$

Dust can also be mixed in with the gas, absorbing both the external incident radiation and the internally produced line photons. The first and large effect on the emergent spectrum is the extinction of the ionizing radiation in a wavelength dependent way. This can be incorporated into the photoionization calculations provided the extinction properties of the dust at $\lambda \leq 912$Å are known.

Internal dust can also destroy line photons with an efficiency that depends on the wavelength and the optical depth of the line in question. For forbidden lines, intercombination lines, and all other lines of neglible optical depth, the absorption probability is simply $[1 - \exp(-\tau_{\text{dust}})]$ and depends only on the line frequency. This is not the case for resonance lines and other lines of considerable optical depth, where the lengthening of the path before escape is considerable (about a factor of 5, see section 4.4.2) due to the large number of scatterings. Such line photons are easily destroyed by dust and the result is a considerable weakening of the large optical depth lines compared with all other lines. AGN observations do not show any large reduction in the strength of $L\alpha$, $CIV\lambda1549$, and other optically thick lines, compared with the calculated intensity of the intercombination lines like $CIII]\lambda1909$. Therefore, the amount of internal dust, at least in the BLR clouds, cannot be large.

It is easy to incorporate these effects into the calculations using the formalism described in chapter 4 (equations 42-46). The main complication is the unknown dust distribution, which may not be uniform. In particular, the neutral gas zone is a more likely location for the dust particle to survive the intense radiation of the central source. Finally, internal dust can also change the hydrogen line spectrum in a low density gas, by providing a de-excitation mechanism for some high energy levels, decreasing, in this way, the effective optical depth of the Lyman lines.

7.3 Dust and Reddening in the NLR

The reddening of the narrow emission lines can easily be measured from the observed intensity of the Balmer hydrogen lines. The only modification that is required is a small correction to the predicted $H\alpha/H\beta$, since the intrinsic narrow-line ratio in Seyfert 1 and Seyfert 2 galaxies is about 3.1, rather than the 2.8 calculated from simple recombination theory (possibly a small collisional contribution to $H\alpha$). Other potential reddening indicators for the NLR are the HeII lines at 4686, 3204 and 1640Å, the [SII] lines at 10320, 6716+6731 and 4069+4076Å, and perhaps some [OII] lines. The HeII lines are the most promising in this respect, since the theoretical ratios are well known. However, so far only the hydrogen lines have been used, because of the low sensitivity of the pre HST ultraviolet experiments. A further observational limitation in Seyfert 1s and quasars is the uncertainty associated with the separation of the broad and narrow line profiles.

Observations of many narrow line AGNs clearly show a significant amount of line reddening. The amount range from very small to extremely large. In particular, some of the NLXGs show indications for a very large amount of line and continuum extinction. Some papers suggest that the amount of extinction is correlated with the inclination of the host galaxy, being larger in edge-on galaxies. There are other reasons to believe that the obscuring gas has a flattened distribution, thus the amount of extinction can vary a lot from object to object.

The narrow emission line profiles provide another indication for dust in NLRs. Many lines have a noticable blue asymmetry which is interpreted as a combination of radial motion and dust obscuration. Line profiles are further discussed in chapter 9.

7.4 Dust and Reddening in the BLR

As discussed in the previous chapter, the agreement between models and observations of the BLR is very good for some lines and poor for others. For luminous AGNs, the comparison between the observations and the models (Fig. 12) suggests a smooth trend, in a sense that calculated ultraviolet line intensities, relative to $L\alpha$, are in better agreement with the observations compared with the calculated optical line intensities. The disagreement may thus be a continuus function of wavelenth, which is typical of reddening by dust. A small amount of line-of-sight extinction ($m_V \simeq 0.6\ mag.$) can considerably reduce this discrepency. Unfortunately, the situation is far from being simple. The uncertainty in current BLR models is more than enough to account for the above discrepency and there are several other explanations for this discrepency.

As for the low luminosity AGNs, there are strong indications that the faint, broad Balmer line wings observed in many of those, are heavily reddened. In view of the observational and theoretical limitations, the idea of reddening in BLRs must be investigated by looking for emission line ratios that are good reddening indicators.

Most broad emission lines are optically thick and their calculated intensity somewhat uncertain. In particular, the $L\alpha/H\alpha/H\beta$ ratio cannot be used as a reddening indicator because of the complicated line transfer and other uncertainties already discussed. The situation regarding the hydrogen Paschen lines is somewhat better, and the uncertainty involved in the calculated ratio of lines originating from a common upper level, such as $P\beta$ and $H\gamma$, is perhaps not as large. Out of all other emission lines, only two line pairs seem to be adequate reddening indicators. These are the HeII and the OI line pairs discussed below.

7.4.1 Reddening from HeII lines. As discussed in the previous chapters, the HeII spectrum of broad line AGNs is relatively simple and reliable calculations are already available. The complicated HeII $L\alpha$ transfer does not influence the HeII Balmer and Paschen lines, and even the small optical depth in the HeII $H\alpha$ line, at 1640Å, does not change the line ratios by too much. Thus, the theoretical Paschen and Balmer HeII lines can be compared with the observations to check for reddening. In particular, the theoretical calculations predict that

$$8 \leq \frac{HeII\, \lambda1640}{HeII\, \lambda4686} \leq 11\ . \tag{69}$$

Currently, there are not enough reliable measurements of this line ratio to make any general statement about reddening in BLRs. There are some indications, in a few objects, that a small amount of reddening is indeed present. The observations are difficult because of the weakness of the HeII lines and the blending with nearby spectral features. They are also hampered by the large time variation in the intensity of the HeII lines in low luminosity AGNs.

The $HeII\, \lambda10123$ line has now been measured in several sources and can be used, given adequate infrared spectral resolution, as another reddening indicator for low redshift AGNs.

7.4.2 Reddening from OI lines. The calculated OI line spectrum is somewhat model dependent because of the fluorescence with $L\beta$. However, most of the emission in the OI lines at 8446 and 1302Å (Fig. 6) is due to this process and there is only a little extra contribution to the $OI\lambda1302$ line due to collisional excitation. As a result, the line ratio is easy to calculate and it is a useful reddening indicator. Recent photoionization calculations suggest the following range for the theoretical line ratio:

$$4.5 \leq \frac{OI\,\lambda1302}{OI\,\lambda8446} \leq 6.5\,. \tag{70}$$

Here again the measurements are very difficult to perform and there are only a hand ful of those. The little information available so far suggest some broad line reddening, by an amount which is highly uncertain.

7.5 Continuum Reddening

Continuum reddening has been investigated too. This has been clearly seen in some narrow line AGNs but the situation regarding broad line objects is not very clear.

The diversity in the shape of the optical-ultraviolet continuum of Seyfert 1 galaxies and quasars is not very large. This would argue against dust extinction since the effect on the continuum slope is likely to be large, even for a very small amount of dust. Arguments to the contrary have also been raised and there is no conclusive evidence yet. It is however clear that a small proportion of AGNs show a very steep ultraviolet continuum, with a shape that is entirely consistent with dust extinction. It is also clear that the Lyman continuum of some quasars, including low redshift ones (i.e. no significant intergalactic absorption) is too steep to account for the observed emission line strength in these objects.

The dust, if it exists, may have a very different effect on the line and continuum spectra. The dust location is thus of primary importance and further progress in this area must await until this is better understood.

7.6 Bibliography

Comprehensive reviews on dust and reddening in AGNs can be found in MacAlpine (1985) and in Osterbrock's book (1989).

Thermal emission and polarization: For model involving dust emission see Barvainis (1987) and Sanders et al. (1989) and references therein. The delayed infrared continuum response in F-9 is described in Clavel, Wamsteker and Glass (1989). There are several infrared polarization studies of quasars, but most of the polarization in such cases is due, probably, to a non-thermal source. A convincing case of dust polarization is described by Rudy and Schmidt (1988). For references on the polarization of Seyfert galaxies see chapter 11.

Internal dust: See Ferland and Netzer (1979), Martin and Ferland (1980) and Cota and Ferland (1988).

Narrow line reddening: Many references are given in MacAlpine (1985). See also De Zotti and Gaskell (1985).

Broad line reddening: See MacAlpine (1985) for references until 1985. Also Ward et al. (1987, use of Paschen and HeII lines), Wamsteker et al. (1990, HeII lines in NGC 5548) and Goodrich (1990, BLR reddening in Seyfert type 1.8-1.9).

Continuum reddening: See references in Lawrence and Elvis (1982) and MacAlpine (1985). Also Sanders et al. (1989).

8 Line and Continuum Variability: Mapping the BLR

The energy output of most AGNs is changing with time, in some cases by a large factor. Such changes in ionizing luminosity drive similar changes in emission line intensities, and the two are expected to be well correlated. This has been observed in many broad line objects and is a clear indication that photoionization is the main source of excitation for the BLR gas. Variable narrow lines are not usually observed, which is explained by the much larger extent of the NLR. The correlated line and continuum variability, and the mapping of the gas distribution, are the subject of this chapter.

8.1 Line and Continuum Light Curves

Consider the case where the continuum luminosity is changing with time. We assume that the recombination time (chapter 4) is much shorter than the light travel time across the line emitting region, and the dynamical time (chapter 9) is much longer than the typical variability time. [5] Thus the gas distribution and the covering factor are not changing with time. We neglect changes in continuum shape and drop the frequency dependence of L_ν, thus $L(t)$ is the *continuum light curve*.

The observed flux in a certain emission line depends on the gas distribution and is also a function of time. Consider the line l whose integrated luminosity, for a constant continuum output, is E_l. The time dependence of E_l is a result of both the varying ionizing continuum and the light travel time across the BLR. A distant observer sees different parts of the BLR responding to a certain continuum variation at different times. If the continuum varies much faster than the light crossing time, the line will be roughly constant, and no information on the geometry will be obtained. For a slower varying continuum, in a symmetric system of clouds, there will be a time lag between the continuum change and the line response which is roughly r/c, where c is the speed of light and r is a "typical" BLR size, to be defined later.

In what follows we assume a similar response for *all* emission lines and drop the l. For optically thick clouds, the last statement is equivalent to the

---

[5]   The dynamical time for Seyfert 1 galaxies is only a few years (chapter 9) and this assumption may not always hold.

assumption that the line flux is linearly dependent on the continuum flux, or, in the notation of chapter 5, $\epsilon_l(r) \propto r^{-2}$. We define $E(t)$ to be the *line light curve* and note that because of the time dependence this is not entirely consistent with the definition of E_l in chapter 5 (equation 58).

The relation between $L(t)$ and $E(t)$, and the information derived from it, will be demonstrated using the specific case of NGC 4151. This source is a well known Seyfert 1 galaxy, whose line and continuum variability are well documented. It was monitored extensively, from the ground, in 1988, and clear variations of the nonstellar continuum and the hydrogen Balmer lines were seen. This is illustrated in Fig. 16.

The $H\alpha$ and optical continuum light curves of NGC 4151, during 1988, are shown in Fig. 17. There are 55 observations, obtained over a period of 216 days. The mean sampling interval is about 4 days, but there are some large gaps, as well as periods of more frequent sampling.

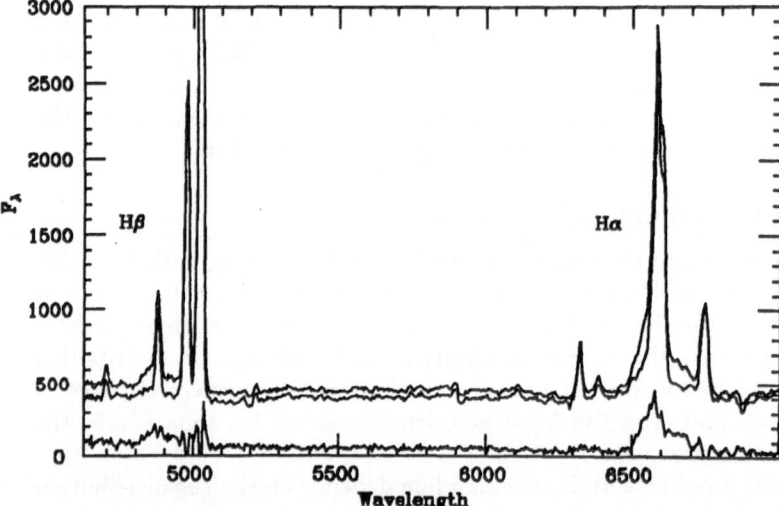

Fig. 16. Two spectra of NGC 4151, obtained at the Wise observatory on Jan. 20 and July 17, 1988. The lowest curve is a difference spectrum, showing the variation in broad emission lines and continuum fluxes (from Maoz et al. 1991)

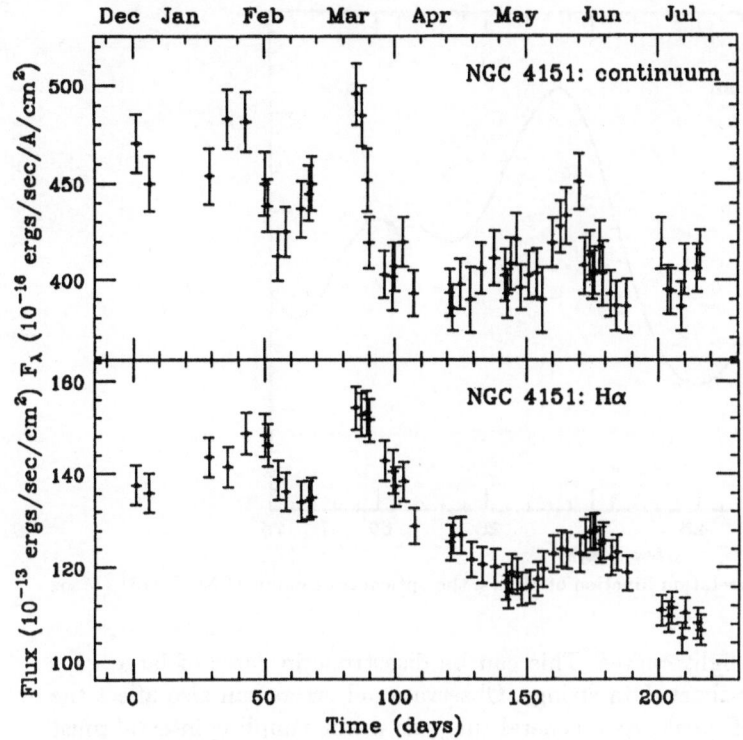

Fig. 17. The optical continuum (4730Å) and the $H\alpha$ light curves of NGC 4151 in 1988 (Maoz et al. 1991)

8.2 Cross Correlation

A main aim of studying the line and continuum light curves is to measure the "BLR size". This quantity may be defined in several different ways. For example, an emissivity-weighted average radius, r_{av}, is defined by

$$r_{av} = \frac{\int_{r_{in}}^{r_{out}} r\,dE(r)}{\int_{r_{in}}^{r_{out}} dE(r)} \ . \tag{71}$$

Other measures of the "BLR size" are mentioned below.

Cross correlating the line and continuum light curves is a simple way of estimating the dimension of the line emitting region. This is a well known technique, designed to give the most likely time-lag between the line and continuum pulses. The Cross Correlation Function (CCF) is a set of correlation coefficiets, giving a measure of the correlation between the line and continuum light curves for certain chosen time lags. Fig. 18 shows the CCF for the two light curves of NGC 4151. A clear broad, significant peak (a correlation coefficient of 0.85 for 55 points) is seen at a lag of 9 days. This time lag, multiplied by c, can be defined as "the cross-correlation size" of the BLR.

There are several problems in applying the cross correlation method to real data sets. Some interpolation is required in order to use this procedure for

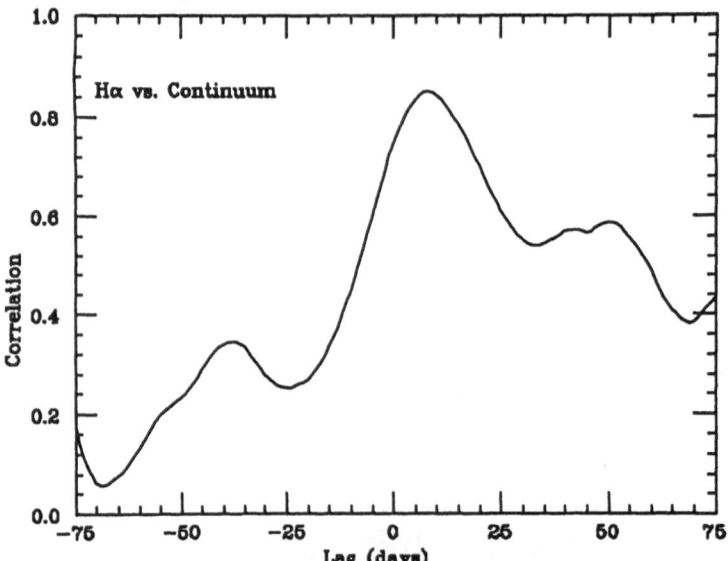

Fig. 18. The cross-correlation function of $H\alpha$ vs. the optical continuum of NGC 4151 (Maoz et al. 1991)

unevenly sampled light curves. This can be disastrous in cases of large gaps in the data and/or short data strings. Observational errors can also affect the position of the CCF peak. As a general rule, the mean sampling interval must be considerably shorter than the typical continuum variability time scale. This reduces the interpolation errors and the resulting uncertainty in the position of the peak. Finally, the cross-correlation peak position depends on the nature of the continuum light curve and the response (linear or non-linear) of the lines. Thus the above mentioned "cross-correlation size", obtained from the CCF of one emission line, is not very well defined.

A serious limitation of the cross correlation analysis is the lack of a solid criterion for estimating its uncertainty. Some formulae have been given for the case of even sampling, but their application to a real, unevenly sampled data set is questionable. The probability of obtaining a certain "cross correlation size" from a given light curve is a function of the observational errors, the interpolation method and the number of observations. The probability distribution is not necessarily normal, and must be carefully calculated given an assumed BLR geometry and observational errors. A variant of the method, "the Discrete Correlation Function (DCF)", avoids the interpolation all together by binning the data according to the mean time between observations. Here, again, the meaning of the derived time-lag uncertainty is somewhat ambiguous.

A main disadvantage of the cross correlation method is the fact that the entire data set, sometimes the result of one or two observing seasons, is used to obtain *one number* (the time lag). Weather, telescope scheduling and other limitations result in a random sampling of the intrinsic light curve which, if sampling is not frequent enough, means an uncertain result. Many more experiments, of similar duration, must in principle be performed in order to asses the accuracy of such results. This is not practical and numerical simulations, mimicking the real observing conditions, are used instead. One method developed for this purpose utilizes Monte-Carlo simulations to find the Cross Correlation Peak Distribution (CCPD) which gives the probability of finding the peak of the CCF at a certain lag, given an assumed continuum light curve and a certain geometry. Obviously, the CCPD is narrower, and the results more meaningful, for cases of frequent sampling and high signal/noise data.

Finally, the time lag obtained from the peak of the CCF is not necessarily a good indicator of the gas distribution, even for well sampled light curves. It tends to weight the inner parts of the line emission region more than the outer parts, in a way which depends on the continuum light curve. For thick geometries, the cross-correlation size is considerably smaller than the emissivity-weighted radius, r_{av} of (71).

8.3 The Transfer Function

A more ambitious task is to try and map out the entire BLR, using information obtained from the light curves.

Given the assumption of linear response of the line to the continuum pulse, we can formulate the relation between $L(t)$ and $E(t)$ using a "transfer function", $\Psi(t)$, (called also a "response function"):

$$E(t) = \int_{-\infty}^{\infty} L(\tau)\Psi(t-\tau)d\tau , \qquad (72)$$

i.e. $E(t)$ is the convolution of $L(t)$ with $\Psi(t)$.

As can be seen from this equation, $\Psi(t)$, in appropriate units, equals the $E(t)$ that would result from $L(t)$ which is a δ-function at $t = 0$ (a continuum "flash"). For gas which is distributed in a thin shell of radius r, the transfer function is a "boxcar" shaped pulse, lasting from $t = 0$ until $t = 2r/c$, with a constant value of $c/2r$. The rise at $t = 0$ is due to the fact that the gas along the line of sight appears to respond immediately to the continuum pulse, and the information about the continuum and line variation arrived to the observer simultaneously. The constant value of $\Psi(t)$ results from the time delay of a ring at a polar angle θ, $r(1 - \cos\theta)/c$, and the emissivity of the ring which is proportional to its surface area. In a similar fashion the transfer function of a circular ring, inclined at an angle i to the line of sight, is non-zero between times $r(1 - \sin i)/c$ to $r(1 + \sin i)/c$, with its center at time r/c. This is illustrated in Fig. 19.

The transfer function of a thick shell is obtained by integrating the thin shell $\Psi(t)$ over all radii and weighting the contribution at each radius according to the emissivity. The case of a shell of inner radius r_{in} and outer radius r_{out} is shown in Fig. 20. As seen from this diagram, $\Psi(t)$ is a constant in time between $t = 0$ and $t = 2r_{in}/c$, and declines to zero between $t = 2r_{in}/c$ and $t = 2r_{out}/c$. The shape of the declining part depends on the gas distribution and emissivity. We can find this shape in the simple, optically thick case, using the notation of chapter 5 and the radial dependence of the covering factor from equation (57),

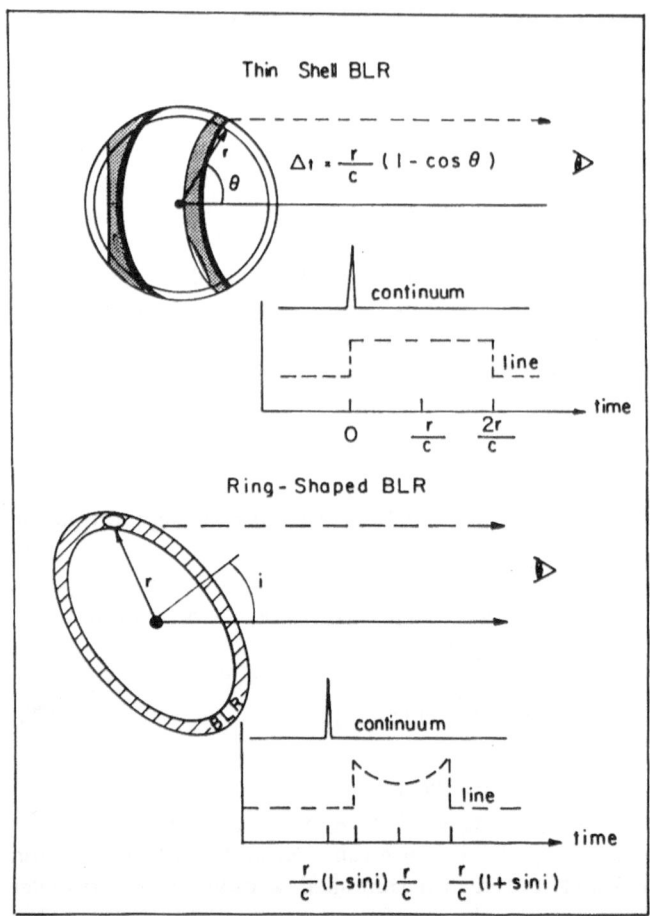

Fig. 19. The response of a thin shell and an inclined ring emission line regions to a δ-function continuum light curve. (Shadded area in the top half shows the narrow ring at a polar angle θ.) For such a continuum variation, the shape of transfer function Ψ(t), is identical to the shape of the line light-curve

$$\Psi(t) \propto \int \frac{dE(r)}{r} \propto \int \frac{dC(r)}{r} \propto r^{-(p+q)} \quad (p+q \neq 0) . \tag{73}$$

This is illustrated in Fig. 20 for the cases of $(p+q)$ =-2, 0 and +2. In a similar fashion, the transfer function of a thick disk is obtained from integrating over rings.

We see that valuable information about the gas distribution can be obtained from $\Psi(t)$ and it is desirable to find this function and investigate its shape. In principle, this is not a difficult task since $\Psi(t)$ can be recovered from the data by applying the convolution theorem,

$$\tilde{\Psi}(\omega) = \frac{\tilde{E}(\omega)}{\tilde{L}(\omega)} , \tag{74}$$

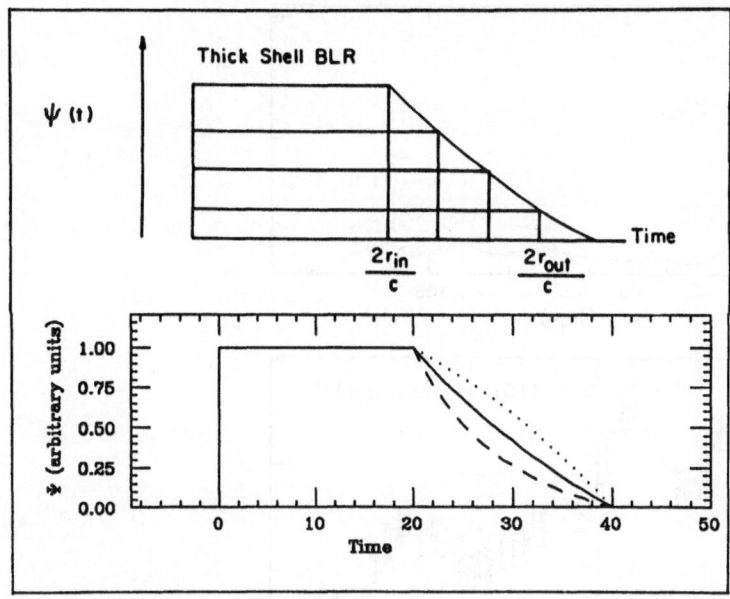

Fig. 20. Bottom: Transfer functions of a thick spherical shell of inner radius 10 time units and outer radius 20 time units, for cases of $(p+q) = -2$ (dotted line), 0 (solid line) and +2 (dashed line). The top half demonstrate the contributions of different thin shells to $\Psi(t)$, for the case $(p+q) = 0$.

where ˜ designates the Fourier transform. Performing this operation, using the observed $L(t)$ and $E(t)$ and transforming back to the time domain, recovers $\Psi(t)$. In practice, this is not a trivial task. $E(t)$ and $L(t)$ are often unevenly sampled in time, have large gaps and span a relatively short period. Under such conditions, Fourier methods become problematic, and a meaningful transfer function may become hard to obtain. Frequently sampled data, with a sampling interval shorter than the typical variability time scale, can be quite useful, provided the light curve is long enough and the measurement error small compared with the variability amplitude. There are improved statistical methods of recovering $\Psi(t)$ from the data (e.g. the maximum entropy method), using additional constraints on its expected shape at different times.

The transfer function for NGC 4151, obtained from applying the maximum entropy method to the data presented here, is shown in Fig. 21. The diagram also shows the $H\alpha$ light curve which is obtained by convolving this function with the continuum light curve. The fit of the $H\alpha$ light curve is quite satisfactory, suggesting that this $\Psi(t)$ is not a bad approximation to the real transfer function. The empirical $\Psi(t)$ rises sharply at $t = 0$ and drops to zero, in a gradual way, over ~ 30 days. This is consistent with a thick shell geometry with a very small inner radius and an outer radius of about 15 light-days. It is also consistent with an edge-on disk (and other geometries) of similar dimensions.

A word of caution is in order. There are cases where several different transfer functions can fit the data equally well. This depend on the numerical method used and the quality of the light curves. For example, the features in the NGC

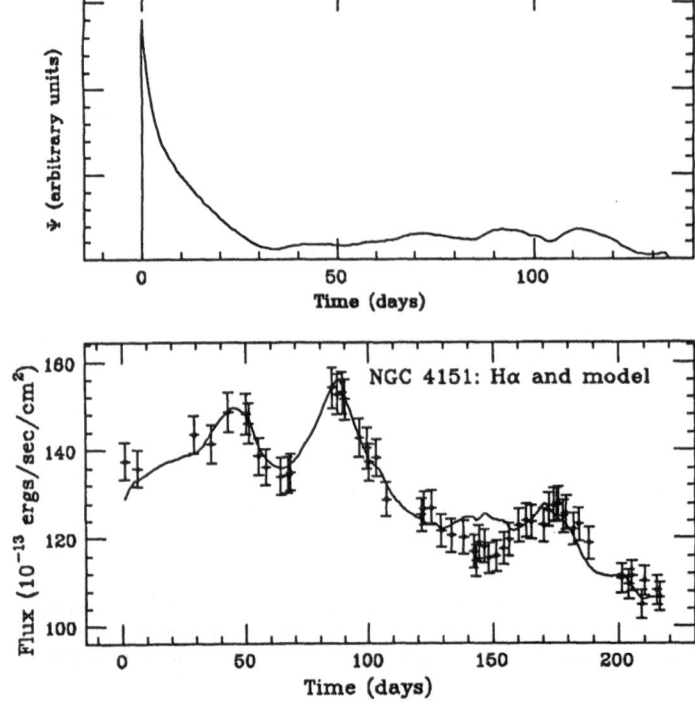

Fig. 21. Top: Transfer function for NGC 4151, obtained from the line and continuum light curves in Fig. 17, using a maximum entropy deconvolution. Bottom: A model emission line light curve obtained from the transfer funtion, on top of the $H\alpha$ light curve

4151 transfer function at $t \sim 50 - 100$ days can be interpreted as due to some line emitting material far away from the nucleus. However, they can also be due to the numerical method used, given the freedom to put the emitting material anywhere around the central source. Physical constraints, such as an imposed upper limit on the BLR extension, should be used in such cases.

The experimental limitations are so severe that today the BLR transfer function is only known in one or two cases. The main problem in ground-based observations is the large amount of telescope time needed for proper sampling of the light curve, and the requirement of flux calibrated data. Space-born instrument are more suitable for the task but the aperture size of most of these is small and it has been extremely difficult to obtain enough observing time to perform the experiment. The first, complete ultraviolet data set was obtained in 1989 by a large group of IUE observers, who monitored the Seyfert 1 galaxy NGC 5548 for a continuous period of eigth months. Much of our understanding of AGN variability is based on this data set.

The main theoretical limitation of the method is the assumption of linearity. While the *total* line and diffuse continua emission of an optically thick gas is indeed proportional to the continuum flux, this is not case for *individual* lines. This is well illustrated in Fig. 8 that shows the response of different emission lines to variations in the ionization parameter. The use of $\Psi(t)$ obtained for a certain line to deduce the gas distribution must therefore be done with great care. In addition, the geometry deduced from the empirical transfer function is not unique, and more information is required to choose among all possible alternatives. Finally, all present day studies make the specific assumption that the observed optical, or ultraviolet continuum variations are proportional to the ionizing continuum variations. This is not necessarily the case and alternatives must also be investigated. Line profiles can provide additional constrains on the gas distribution, as discussed in the next chapter.

8.4 Light Curve Modeling

An alternative empirical way to find the gas distribution, is by fitting models to the observed emission line light curve. This is done by convolving the observed continuum light curve with several assumed transfer functions. The most likely geometry is the one resulting in the best fit to the observed line light curve. The result of this methods is not unique and several different geometries can give satisfactory fits. It can be very useful in eliminating some geometries, and reducing the parameter space. The main limitations are again the uncertain interpolation procedure and the non-linear response of the lines.

8.5 The BLR Size

Having discussed the observations and several potential ways of analysis, we are now in a position to estimate the BLR size. In the time of writing there are only a hand-full of Seyfert 1 galaxies where the cross correlation size has been realiably measured. The range in luminosity is about 500 and the range in cross correlation size about 25. This incomplete data set, combined with a guess that the average BLR size is about double the cross-correlation size, suggest the following approximate relation for low luminosity AGNs:

$$r_{av} \simeq 0.1 \, L_{46}^{1/2} \, pc. \tag{75}$$

While the $L^{1/2}$ dependence is only a guess at this stage, it is in line with the asumption that the average density and ionization parameter in different BLRs is not a strong function of the continuum luminosity. Further discussion of this relation is given in chapter 10.

8.6 Anisotropic Emission and Beaming

Some bright quasars are known to vary, by a large amplitude, over short periods of time. These are named OVVs (Optically Violent Variables). They are all radio sources that show some of the characteristics of the BL Lac objects, such as a non-thermal continuum and high degree of polarization. It is believed that relativistic beaming is important in this case, and the objects are seen close to the direction of the beam.

Recent observations show rapid emission line variability in some of these objects, on time scales shorter than estimated from the $L - r$ relations of other AGNs. Since beaming is important, the ionizing flux is highly anisotropic, and line emission from clouds along the beam is much more intense than in other directions. The situation is very similar to the one illustrated in the top half of Fig. 19, where the beam illuminates only a small part of the shell-shaped BLR, and θ is the angle between the beam and the observers' direction. The time lag, $r(1 - \cos\theta)/c$, is short since θ is small.

There are several difficulties with this picture. First, for a double sided beam, there is line emission from the opposite side of the source. Unless it is obscured, this should arrive to the observer at a much later time. No such event has yet been observed. Second, the observed line profiles are usually smooth and symmetric. This is difficult to reconcile with a one-sided, jet-like system, whose center of gravity coincides with the central radiation source. Perhaps other mechanisms, such as radiation pressure (see next chapter), control the shape of the line profile. Nevertheless, beaming is a potentially important process that ought to be studied further since it may play a role in many objects whose ionizing beam is pointed away from us.

8.7 Bibliography

Observations: See chapter 2.

Theory: The first theoretical paper on the subject is by Bahcall, Kozlovsky and Salpeter (1972). More recent and detailed calculations are described in Blandford and McKee (1982). The application of the cross-correlation method to line variability studies is discussed in Gaskell and Sparkes (1986) and Robbinson and Perez (1990). The uncertainties in the cross-correlation method are discussed by Gaskell and Peterson (1987), Maoz and Netzer (1989, the CCPD method) and Edelson and Krolik (1989, the Discrete Correlation function). Deconvolution and light-curve modeling are discussed in Maoz et al. (1991).

9 Gas Motion and Dynamics

The discussion in the previous chapters gave support to the accepted model of AGN emission line regions. Such regions are thought to contain numerous optically thick clouds, that are exposed to the intense ionizing flux of a central source. The origin, stability and motion of these clouds, and the resulting line profiles, are the subjects of this chapter.

9.1 Confined Cloud Models

9.1.1 Important time scales: Several time scales must be considered in relation to the formation, stability and motion of the clouds.

The recombination time: This is the time required to reach ionization equilibrium. It is given by

$$t_{\rm rec} = 1/(\alpha N_e) \simeq 10^{-5} N_{10}^{-1} \; years, \qquad (76)$$

where N_{10} is the density in units of $10^{10} \; cm^{-3}$.

The dynamical time: This is the crossing time (or orbital time) of the line emitting region. For the BLR

$$t_{\text{dyn}} = \frac{r_{\text{av}}}{<v>} \simeq 30 L_{46}^{1/2} \ years, \qquad (77)$$

where the previous relation between the average dimension and the luminosity (75) has been used.

The sound crossing time: This is the time required to establish pressure equilibrium within a single cloud. The total cloud thickness is not well determined but the thickness of the ionized gas is estimated to be $10^{12-13} \ cm$ for the BLR clouds. This gives

$$t_{\text{sc}} = R_c/c_s \simeq 2 \times 10^{-12} T_e^{-1/2} R_c \ years \qquad (78)$$

where R_c is the cloud radius in cm and $c_s = \sqrt{2kT_e/m_p}$ is the sound speed.

9.1.2 Confinement. The mass of individual BLR clouds, as deduced from the models, is well below their Jeans mass, i.e. self gravity is negligible. In the absence of confinement, such clouds will disintegrate on a time scale of t_{sc}. Since this time is much shorter than the dynamical time, the clouds must be confined or else be continuously produced throughout the BLR, which requires extremely large mass flux through the line emission region.

One way to confine the clouds is by a Hot Intercloud Medium (HIM), much like in the interstellar medium. Such a "two-phase model" has been the subject of extensive investigations. Magnetic confinement is another possibility that needs to be considered. Below is a short description of the hot intercloud medium model.

9.1.3 The two-phase model. In this model the cool ($T_e \simeq 10^4 K$) gas is in pressure equilibrium with a lower density, much hotter gas, of temperature T_{HIM}, filling the volume between the clouds. Heating and cooling of the hot gas is mainly by Compton and inverse Compton scattering, and T_{HIM} is given by the Compton temperature

$$T_C = \frac{h\bar{\nu}}{4k}, \qquad (79)$$

where $h\bar{\nu}$ is the mean photon energy, weighted by the cross section. In this limit the temperature is independent of the gas density and is determined only by the energy distribution of the incident continuum. As a result, T_{HIM} is the same throughout the line emitting region.

The special feature of the two-phase model is that a single parameter controls the ionization and thermal properties of both components. The parameter, called Ξ, is defined by

$$\Xi = const. \times \frac{U}{T_e} \propto \frac{incident \ flux}{NT_e} \propto \frac{radiation \ pressure}{gas \ pressure}. \qquad (80)$$

The value of this parameter is the same for the hot and cool gas at a certain location. [6]

The Ξ vs. T_e dependence is the key to the understanding of the two-phase model. At small Ξ the temperature is kept around $10^4\,K$ due to bound-free and collisional excitation cooling. At this value of Ξ there is only one solution to the kinetic temperature. At much larger Ξ, the ionization increases, the cooling is less efficient, and no equilibrium can be achieved until Compton cooling become dominant, at about $10^7 - 10^8\,K$. In some cases there is a small range in Ξ where the thermal equilibrium conditions may allow *both* Comptonized gas and low temperature gas to coexist and there is a two temperature solution for the same pressure. This is demonstrated in Fig. 22 that shows the calculated T_e

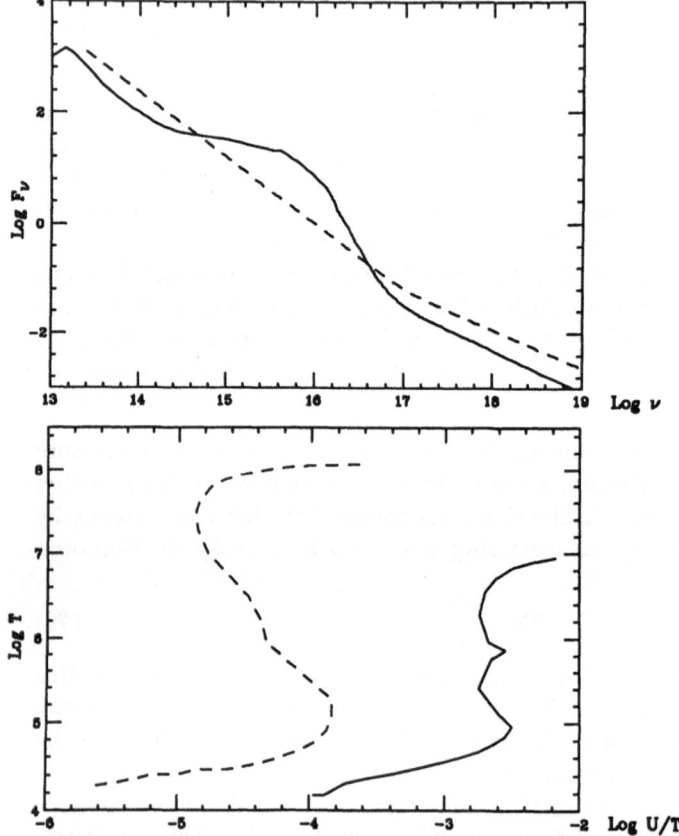

Fig. 22. Calculated T_e vs. U/T_e (bottom) for the two incident continua shown in the upper panel. The solid line is the "standard" continuum (Fig. 7) adopted in this work. The dashed line is a two-component power-law continuum, with spectral indices of 1.2 in the infrared-ultraviolet and 0.7 in the X-ray. A two-phase solution (S shaped part of the curve) can only be obtained for the power-law continuum (courtesy of T. Kallman)

---

[6] $\quad\Xi$ is referred, in some papers, as "the ionization parameter". This should not be confused with U, as defined here, which has no temperature dependence.

vs. U/T_e curve for two different continua, one with a two phase solution and one without.

A two-phase equilibrium can only exist at high enough T_{HIM}, where the T vs. U/T curve is S shaped. In this case there are two stable branches, at low ($\sim 10^4 \, K$) and high ($\sim 10^8 \, K$) temperatures, allowing thermal instabilities to grow and develop into a two-component atmosphere. Only one of the continua shown in the diagram (in dashed line) gives this kind of curve. Older photoionization models assumed this continuum to be typical of AGNs. It was consistent with $T_{\mathrm{HIM}} \simeq 10^8 \, K$ and a two-phase model for the gas. Moreover, estimates of the ionization parameter, obtained from emission line ratios, agreed roughly with the U/T required for the two-phase instability. Since this is the only range in which hot and cold gas can coexist, it suggests a specific value of U for the BLR clouds. This was considered to be the explanation for the fact that the ionization parameter in faint Seyferts and bright quasars, a factor of a thousand in luminosity apart, is roughly the same.

Unfortunately, more realistic continuua, like the one used throughout this work, changed this result. This continuum has a large ultraviolet excess which results in more Compton cooling. The highest temperature is around $10^7 \, K$, too low to allow a two-phase solution (solid line in Fig. 22). [7] Thus the idea of a stable two-phase BLR, in the form presented here, is questionable. Today it is also known that the allowed range in U is in fact quite large (chapters 4-6) and the motivation for a specific value of U is not as strong.

A hot inter-cloud medium with $T_{\mathrm{HIM}} \sim 10^7 \, K$ presents some other problems. First, the optical depth to electron scattering may be large. We can estimate this from the pressure equilibrium condition and the pressure-radius dependence of chapter 5, $P \propto r^{-s}$, with $s \sim 2$:

$$\tau_{\mathrm{es}}(HIM) = \int_{r_{\mathrm{in}}}^{r_{\mathrm{out}}} \sigma_{\mathrm{es}} N_{HIM} dr \simeq \frac{7 \times 10^{-10} r_{\mathrm{in}}}{(s-1)T_{HIM}} \simeq \frac{7 \times 10^7 L_{46}^{1/2}}{T_{HIM}} . \qquad (81)$$

In making this estimate we have assumed that $r_{\mathrm{out}} \gg r_{\mathrm{in}}$, the BLR density at r_{in} is $10^{11} \, cm^{-3}$, the temperature of the cold clouds is $10^4 \, K$, and the luminosity-size relation is as in equation (75). This shows that luminous objects, with low T_{HIM}, are opaque to electron scattering. This would have observable consequences. Large amplitude continuum variations will be smeared out and broad electron scattering wings will appear in all emission lines. Moreover, the K-shell opacity in iron and other elements can exceed the electron scattering opacity, and will show up as strong absorption X-ray edges. Such effects are not observed.

There are serious dynamical implications to a cool stationary HIM. Typical velocities in the BLR are several thousands $km s^{-1}$, much larger than the sound speed in a cool HIM. A super-sonic motion through this medium must result in drag forces, Rayleigh-Taylor instability and breakup of the clouds. This leads to fragmentation into optically thin filaments, in contradiction with the observed

---

[7]   This temperature may be angle dependent, if the ultraviolet source is a thin accretion disk continuum (chapter 10).

strong lines of MgII and FeII. There are additional difficulties and more reasons to abandon the idea of a stable, two-phase model.

The HIM, if it exists, need not be stable. There are additional heating mechanisms, (superthermal particles, radio frequency heating) and the pressure equilibrium may not be exact since cloud evaporation will tend to raise the cloud pressure over the HIM pressure. Line radiation pressure is known to be important in the cold clouds and can exceed the cold gas pressure in cases of a large ionization parameter. A simple pressure equilibrium may not be achieved if this is the dominant pressure inside the clouds. A relativistic intercloud medium can help to solve some of the difficulties, especially the cloud instability problem, since in this case the drag force is very small.

Finally, we should comment on the probable scaling laws for the pressure and density for a stable, or unstable hot confining medium. Using the notation of chapter 5, the value of s is likely to be in the range of 3/2 to 5/2, i.e. the assumption of $P \propto r^{-2}$ is quite adequate for such cases.

9.1.4 Magnetic confinement. This is a new idea that has received little attention so far. The field strength required to confine the BLR clouds is about 1 G and its possible origin may be in a relativistic wind, an accretion flow or an accretion driven wind. Magnetically-confined clouds would have a filamentary structure and would tend to elongate along the field lines. The pressure laws expected are $P \propto r^{-5/2}$ in an accretion flow and $P \propto r^{-2}$ in a relativistic wind.

9.2 Other Models

Below is a short account of models not requiring external confinement of the clouds.

9.2.1 Winds. Several wind models have been proposed for the origin and motion of the clouds. The main problem in most of these is the source of the cloud gas. The possibilities of mass loss from giants and from the surface of accretion disks have been considered. In one such model, a hot supersonic wind shocks against some obstacles (supernovae remnants or OB stellar winds) to form clouds in the cooled postshock gas.

9.2.2 Orbiting clouds. Massive clouds, in parabolic orbits, have been proposed for the line emitting gas. The clouds are confined by a central relativistic wind and lose mass as they fall in. Tidal forces break the clouds into smaller fragments, at about 1 pc. Such clouds are assumed to produce observed line emission at all distances, from 1 kpc to 0.1pc. The line core is produced at large distances, where the velocities are small, and the line wings are produced by fragments, at the standard BLR distance.

9.2.3 Disks. There have been several models relating the line emission to the central, hypothetical accretion disk (chapters 5). Such a configuration provides a large amount of cool gas that needs no external confinement, and can produce line emission provided a suitable energy mechanism exist. Standard thin accretion disk models suggest that the temperature and the flux emitted by the disk, at the typical BLR distance, are too low to be observed. Thus an additional heating source for the gas is required. As explained in chapter 5, this can be achieved if part of the central continuum radiation is back-scattered onto the disk surface at large radii (Fig. 13).

9.2.4 Stars. Stellar atmospheres have been proposed for the BLR clouds. Given the size and density of the broad line clouds, these must be giant size atmospheres. Stellar atmospheres require no confinement, which is the main virtue of this model. The difficulty is the large number of giants ($\sim 10^7$) needed to explain the observed covering factor in BLRs. Given a normal stellar population, the total cluster mass is enormous and can dominate the dynamics near the center. It has been suggested that radiation pressure driven winds, or accelerated stellar evolution, keep the required number of stars small.

If BLR clouds are indeed stellar atmospheres, the density and pressure profiles may be very different from the ones assumed so far. The implications to the line intensity, variability and profiles are still to be investigated.

9.3 Observed Line Profiles

The large diversity in line profiles has already been demonstrated in chapter 2 and several more examples are given in Fig. 23. The broad line profiles of many objects are smooth and symmetric. Their shape is well fitted, in many cases, by a logarithmic function ($E_\lambda \propto -Log \mid \lambda - \lambda_0 \mid$, see Fig. 23). Profile differences can be large or small. There are cases of almost identical profiles for lines of very different excitation. A likely explanation here is that the emission line spectrum of all clouds is the same. There are opposite examples too, and a clear trend is still to be found. It has been claimed that the high excitation lines are systematically broader than the low excitation lines, but there are definitely some exceptions to that. One common trend is observed between the $H\alpha$ and $H\beta$ line profiles, where the latter is relatively stronger away from the line center (see diagram). Another trend is for the $HeI\lambda5876$ line to be somewhat broader than $H\beta$.

Broad, asymmetric lines are seen in many objects, but no clear tendency for the blue or the red wing to be stronger. Some line profiles are bumpy, and their shape varies in time. The best known examples are some steep spectrum radio sources whose line profiles are extremely broad and disturbed. It has been suggested that the lines are emitted in an edge-on, disk-like system, with axis parallel to the direction of the radio jet.

Finally, there is a clear tendency for the high excitation lines in luminous AGNs to be systematically blue-shifted relative to the low excitation lines. The effect can be large, up to 3000 $km\ s^{-1}$, and seems to increase with luminosity. The few detailed studies of this phenomenon suggest that the systemic redshift is given by the low excitation lines. An example is shown in Fig. 23.

As for the narrow lines, the profiles are smooth with a clear tendency for the blue wing of some lines to be stronger. The blue asymmetry is most noticeable in lines of higher excitation and/or critical de-excitation density. This part of the line originates in high density, fast moving NLR clouds. The phenomenon is observed in most Seyfert 1 and Seyfert 2 galaxies.

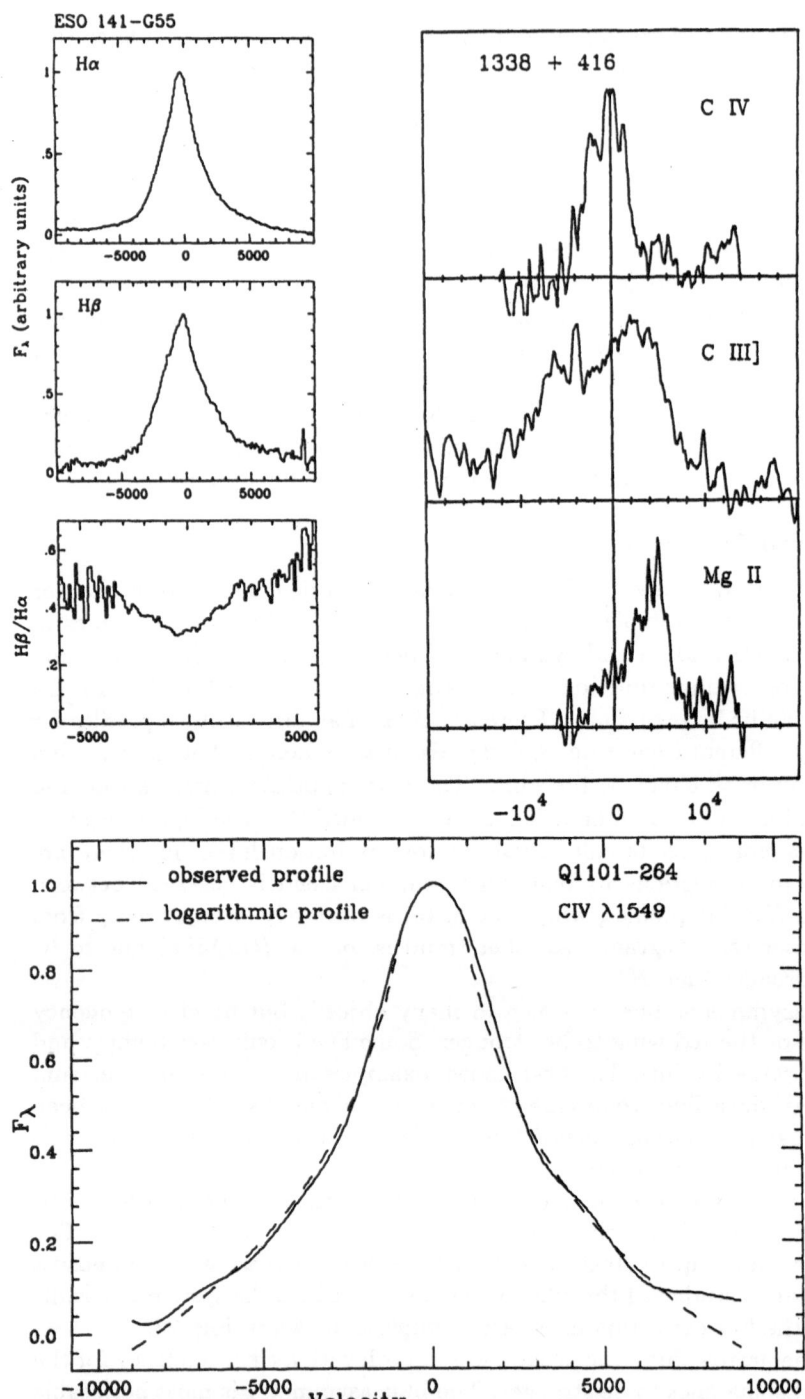

Fig. 23. Observed broad line profiles. Bottom: A symmetric $CIV\lambda1549$ line (solid curve) fitted with a logarithmic profile (dashed line) (courtesy of B. Wilkes). Top left: $H\alpha$, $H\beta$ and their ratio, showing the larger width of $H\beta$ (courtesy of G. Stirpe). Top right: Emission line redshift differences (Corbin 1990)

9.4 Theoretical Line Profiles

9.4.1 Radial forces. We consider the general case of small clouds, moving through a confining medium. The radial motion of a cloud is determined by the gravitational acceleration, g_G, the radiative acceleration, g_{rad}, due to the absorption of the central continuum radiation, and the drag force, f_d. The equation of motion is:

$$\frac{dv}{dt} = -g_G + g_{\text{rad}} + \frac{f_d}{M_c} . \tag{82}$$

The radiation acceleration can be treated in a very general way by noting that the clouds are in ionization and thermal equilibrium. Energy conservation implies that the absorption of a photon of frequency ν is associated with cloud emission (*all* lines and diffuse continua) of $h\nu$. The momentum transfer to the cloud, per unit time, is $h\nu/c$. Thus, the radiative accelerating force is proportional to the total cloud emission. In the notation of chapter 5, the cloud emission is $j_c(r)$, the mass of the cloud is M_c, and the radiative acceleration is proportional to $j_c(r)/M_c$.

We shall now be more specific about radiative acceleration. Consider first a fully ionized gas and assume, for simplicity, the pure hydrogen case. The mean energy of an ionizing photon is $h\bar{\nu}$ and there are αN_e^2 ionizations and recombinations per unit volume and time, each associated with a momentum transfer of $h\bar{\nu}/c$ to the cloud. We neglect the absorption of non-ionizing continuum photons and line photons. The radiative acceleration is

$$g_{\text{rad}}(r) = \frac{h\bar{\nu}}{c} \frac{\alpha N_e^2}{m_p N_e} = const. \times N_e , \tag{83}$$

where m_p is the proton mass and we have neglected the temperature dependence of the recombination coefficient α. Thus, the radiation acceleration of a fully ionized gas is proportional to the gas density and practically independent of the column density. (This is identical to the previous result since $j_c(r) \propto N_e^2 V_c$ and $M_c \propto N_e V_c$, where V_c is the volume of the cloud.)

The other extreme situation is the case where all the ionizing flux is absorbed by an optically thick cloud. Here the amount of flux absorbed is proportional to the cloud cross section, $A_c(r)$, and

$$g_{\text{rad}}(r) = \frac{A_c(r)}{cM_c} \int_{\nu_0}^{\infty} \frac{L_\nu}{4\pi r^2} d\nu . \tag{84}$$

The properties of realistic clouds must be between these two cases. The ultraviolet radiation is absorbed in the fully ionized part and the harder radiation in the neutral zone. In particular, the cloud must be transparent at some frequency and the amount of momentum absorbed depends on the column density. It can be shown that the acceleration associated with the absorption of X-ray photons is proportional to $1/R_c^2(1 + N_{\text{col}}/N_0)$, where $N_0 \simeq 2 \times 10^{20}$ cm^{-2}.

Finally, the drag force exerted on a radially-moving cloud, depends on the cross-sectional area, the intercloud medium density and the relative velocity between the cloud and the intercloud medium.

125

Given a typical AGN continuum, we find, for optically thin gas,

$$\frac{g_{\text{rad}}}{g_G} \simeq 20N_{10}\frac{r_{18}^2}{M_9} , \qquad (85)$$

where $r_{18} = r/10^{18}\,cm$ and $M_9 = M_{\text{BH}}/10^9\,M_\odot$. This ratio is larger than unity even for very large M_{BH}, and in the absence of strong drag forces, the radial motion of optically thin clouds is governed by outward radiation acceleration. As for optically thick clouds, g_{rad}/g_G depends on the density and column density of the clouds and must be calculated for the given situation.

9.4.2 Pancake clouds. If clouds move radially through an intercloud medium, their shape and optical depth will be modified. Detailed calculations show that fully ionized, low mass clouds, approach a nearly spherical shape as they move out under the influence of radiation pressure acceleration. More massive clouds adopt a "pancake" shape, having much larger dimensions perpendicular to the direction of motion. This has important consequences to the emission line spectrum since relative line strength may depend on the cloud location. Moreover, not all clouds are accelerated outward and those that form closer in, where the ambient density is larger, may fall in under the influence of gravity. The net result is that the simple approximation adopted for $A_c(r)$ in chapter 5 may not be valid in such cases. There are additional implications to the line profiles, as discussed below.

9.4.3 General line profiles. Consider a sperical system of isotropically emitting clouds, with a number density $n_c(r)$ and emission per cloud $j_c(r)$. We further assume that all of the cloud emission is represented by a *single emission line*. The wavelength dependence luminosity ("profile") in a line of rest wavelength λ_0 is

$$E_\lambda = 2\pi \int_{r_{\text{in}}}^{r_{\text{out}}} \int_{-1}^{1} n_c j_c r^2 \delta[\lambda - \lambda_0(1 + v\mu/c)]d\mu dr , \qquad (86)$$

where $\mu = \cos\theta$ in a spherical coordinate system with its z axis parallel to the line of sight. Integration over the δ-function gives $c/v\lambda_0$, provided $0 <| \lambda - \lambda_0 | c/\lambda_0 < 1$.

We now make the simplifying assumptions of pure radial motion and a constant mass flow (mass conservation),

$$\dot{M} = 4\pi r^2 n_c M_c v(r) = \text{const.} \qquad (87)$$

All clouds are assumed to form near r_{in} and experience a radial acceleration of $g(r) = vdv/dr$. Substituting into the line profile equation we get,

$$E_\lambda = \frac{\dot{M}c}{2\lambda_0} \int_{v_1}^{v(r_{\text{out}})} \frac{j_c}{g(r)M_c} \frac{dv}{v} , \qquad (88)$$

where v_1 is the largest of $v(r_{\text{in}})$ and $| \lambda - \lambda_0 | c/\lambda_0$. The line profile is logarithmic, $E \propto Log(| \lambda - \lambda_0 |)$, in those cases where $j_c/g(r)M_c$ is constant. Assuming mass conservation (87) this is equivalent to the following condition:

$$r^2 n_c j_c \propto \frac{dv}{dr} . \qquad (89)$$

9.4.4 Line profiles for radiatively driven clouds. The general argument about the relation between the amount of radiation absorbed and the rate of momentum transfer (beginning of 9.4.1) suggests that in this case the radiative acceleration is proportional to j_c/M_c. Thus if the radiation pressure force is the dominant force, the line profile has a logarithmic shape. Below we demonsrate this for the two extreme cases considered earlier.

In the optically thin case, $j_c \propto V_c N_e^2$, $M_c \propto N_e V_c$ and $g_{rad} \propto N_e$. Thus $j_c/g(r)M_c = const.$ and the line profile is logarithmic. In this case, the radial velocity at a distance where the density is N is approximately

$$v(r) \sim (Nr)^{1/2} \propto N^{1/2} L^{1/4} , \qquad (90)$$

where we have used the previously obtained radius-luminosity relation (75). With the estimate of g_{rad}/g_G (85), and the known density in the BLR, it can be shown that velocities of more than $10,000 \, km \, s^{-1}$ can be obtained.

As for the optically thick case, the radiative acceleration is given by Eqn. (84) and the cloud emission by

$$j_c = A_c(r) \int_{\nu_0}^{\infty} \frac{L_\nu}{4\pi r^2} d\nu . \qquad (91)$$

Thus, $j_c/g(r)M_c = const.$ and we recover the logarithmic line profile. Note again the assumption that all emission comes out in *one* emission line (or is divided among all lines in the same way at all distances) which is crucial for obtaining this particular line profile.

9.4.5 General logarithmic profiles. The conditions for a logarithmic line profile can be investigated in a more general way, using the radial dependences of the cloud parameters. Adopting Eqn. (89) as the basic requirement for a logarithmic shape, and the previous parametric descriptions, $n_c(r) \propto r^{-p}$, $A_c(r) \propto r^{-q}$ and $v(r) \propto r^{-t}$, we get the following requirement for a logarithmic line profile:

$$p + q + m - t = 3 . \qquad (92)$$

The line emissivity parameter, m, is approximately 2 for many emission lines, so an almost as general requirement is

$$p + q - t = 1 . \qquad (93)$$

As an example, consider infalling optically thick clouds, no radiation pressure and no drag force. Here $t = 1/2$ and mass conservation requires that $2 - p = t$ or $p = 3/2$. Logarithmic line profiles will result if $q = 0$ (also $s = 0$), i.e. constant radius clouds. In the two-phase scenario, this means a constant confining pressure throughout the emission line region.

The $s = 0$ situation is illustrated below in more detail for a few velocity laws and complete photoionization models. The models are similar to the ones described in chapters 4 and 5, except that in this case the cumulative line fluxes are calculated with the assumption of a constant confining pressure and a constant gas density, $s = 0$. These integrated fluxes, as a function of r, are shown in Fig. 24.

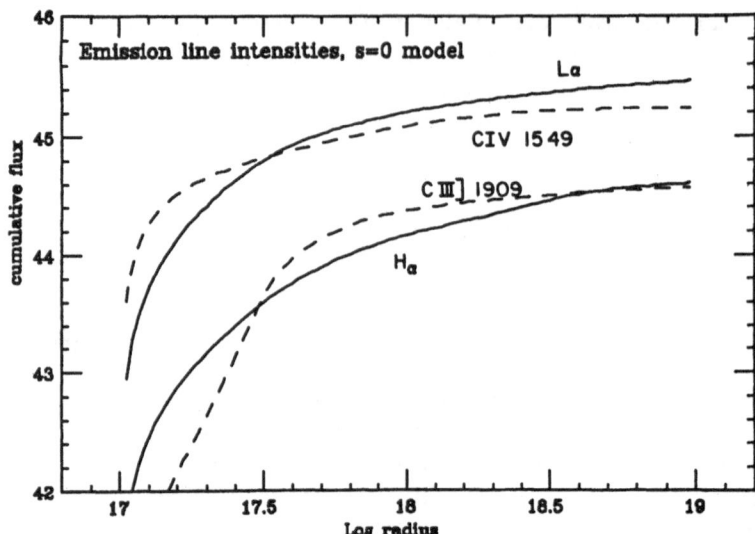

Fig. 24. Cumulative line fluxes, as a function of the outer radius, for a photoionization model with $s = 0$ (constant confining pressure) and $N = 10^{10} \, cm^{-3}$. The normalization of the model is such that $U = 0.1$ at $N_{col} = 10^{23.5} \, cm^{-2}$

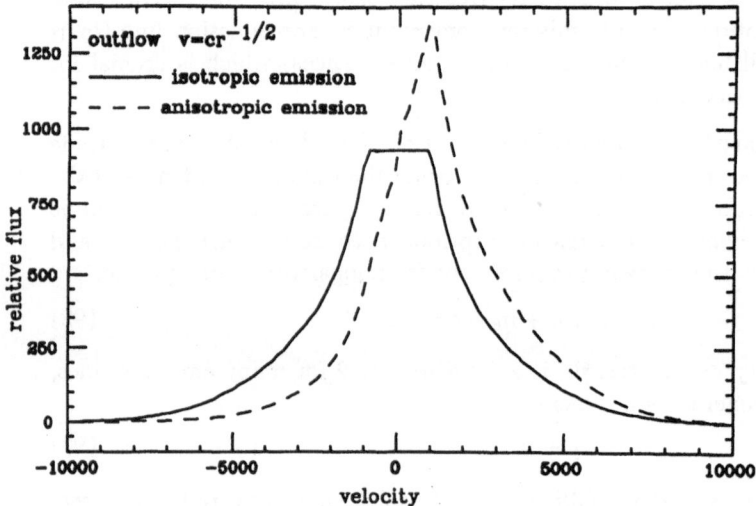

Fig. 25. $L\alpha$ profiles obtained by integrating the line emission in the photoionization model of Fig. 24 with the velocity law indicated. Solid line: isotropic $L\alpha$ line emission. Dashed line: anisotropic line emission (90% of the photons escape from the illuminated surface).

Line profiles for this cloud system have been calculated under several different assumptions. The first case is a decelerating outflow, with isotropic line emission and $v \propto r^{-1/2}$. This velocity law satisfies the condition for a logarithmic profile (93) as is evident also from the $L\alpha$ profile shown in Fig. 25. The profile is very similar to the $CIV \lambda 1549$ profile shown in Fig. 23, except for the flat top, which is the result of the finite r_{out} used (i.e. no low velocity material). A flat-topped profile is a signature of radial flow motions with a weak v/r dependence and/or a small r_{out}/r_{in}. In the particular case shown here, the requirement for lines with smooth wings between 200 and 10,000 $km\,s^{-1}$, implies $r_{out}/r_{in} > 2500$. This is an extreme and unrealistic ratio for the BLR.

A major uncertainty is the dependence of the cloud cross-section on r. It depends on the (hypothetical) inter-cloud medium and the velocity law and may be very different from the simple r^{-q} parametric dependence assumed here. In pancake clouds, which is about the only case that was studied in detail, A_c has a complicated radial dependence. Such clouds may be optically thin around their edge and their column density changes continuously throughout the motion. Needless to say, the resulting line profiles must be calculated with great care, taking the real emissivity into account.

9.4.6 Orbital and chaotic motion. Chaotic (random orientation) cloud motion can produce line profiles that are in good agreement with some observations. Orbital motion can give good agreement too, at least for some emissivity laws. For example, it has been suggested that the BLR clouds are moving in parabolic orbits, with some net positive angular momentum. Reasonable assumptions about the distribution of clouds in angular momentum require that the cloud density be given by $n_c(r) \propto r^{-1/2}$. Making the additional assumption of constant confining pressure (no change of the cloud cross section with r) we obtain $j_c(r) \propto r^{-2}$ which, upon substituting into the line profile equation (86), gives:

$$E_\lambda \propto \int_r n_c j_c r^2 v^{-1} dr \;\propto\; \int_v v^{-3} dv \;\propto (\lambda - \lambda_0)^{-2}\,. \tag{94}$$

Such profiles seem to fit nicely the far wings of many lines. Thus there is more than one dynamical model that can explain the observed profile.

As for the accretion disk model, in this case the motion is in a plane and a characteristic profile, made up of two humps and a central dip, results. The relative intensity of the profile components depends on the emissivity as a function of r and the size of the disk. Small disks would tend to give a large central dip, while in very large ones the dip is filled by emission from the outer, slowly rotating gas.

A few AGNs show disk-type line profiles and it has been suggested that this is a common phenomenon, except that the central dip in most other cases is filled in by emission from low velocity material. The distance of the low velocity material can be calculated, given the central mass. In some cases it is way beyond the outer boundary of the BLR, and may be inside the NLR. Such outer parts are well beyond the self gravity radius of a thin, radiation pressure supported accretion disk (chapters 5 and 10).

Much of the effort in fitting AGN line profiles by thin disk models has focused on the Balmer hydrogen lines in BLRGs. The specific disk models used are the ones discussed in chapter 5, where the low excitation lines are the result of back-scattering of the X-ray radiation onto the surface of the disk, at large radii. Such models predict little or no high ionization line radiation from the disk and it remains to be seen whether the observed $L\alpha$ and $CIV\,\lambda1549$ lines are indeed different from the $H\alpha$ and $H\beta$ line profiles in those objects.

9.4.7 Line asymmetry and wavelength shifts. All examples so far considered assume isotropic line emission. This is not necessarily the case for lines whose optical depth structure, within the cloud, is nonuniform. The most notable example is $L\alpha$, whose optical depth is very different in the ionized and neutral parts. Most of the emitted $L\alpha$ photons escape through the illuminated cloud surface, and the radiation emitted in the outer direction is almost totally absorbed. Photoionization calculations predict that in a plane-parallel geometry, more than 95% of the broad $L\alpha$ emission is emitted from the illuminated surface of the clouds. The effect is smaller, but not negligible, in other broad lines such as $H\alpha$, $H\beta$, and $MgII\lambda2798$.

Another example of anisotropic line emission is obscuration by dust. The intercloud medium may be dusty, causing line emission from the farther hemisphere to be fainter. Alternatively, the dust may be embedded in the clouds, mainly in the neutral part. The line emission from the back side of the clouds (the side away from the ionizing source), is weaker in this case. This is a likely situation in NLR clouds.

Anisotropic line emission has a direct consequence on the observed line profile of radially moving clouds. It introduces a profile asymmety whose magnitude depends on the degree of anisotropy and the velocity pattern. For example, in outflow motion the $L\alpha$ profile would show a strong red asymmetry (red wing stronger than blue wing). A similar asymmetry is obtained for outflowing dusty clouds, whose dust particles reside in the back of the clouds. An outflow motion through a dusty intercloud medium gives a blue asymmetry. Fig. 25 shows the broad $L\alpha$ profile resulting from the outflowing $s = 0$ atmosphere, when the $L\alpha$ anisotropy is taken into account. The strong red asymmetry is clearly visible in this case.

Asymmetric broad line profiles are indeed observed in some cases, but in many AGNs the line profiles, including $L\alpha$, are quite symmetric. There are several possible explanations for this. The first and most obvious one is that there is little, if any, radial motion of BLR clouds. There are other evidences (section 9.5) to support this claim. Alternatively, some of the $L\alpha$ emission may originate in outflowing (or infalling) optically thin clouds, whose emission pattern is much more isotropic. The obvious difficulty is the presence of strong, low excitation lines, that require large neutral hydrogen column densities. Pancake shaped clouds have a variable column density, and may be thin along their rim. This helps to reduce the profile asymmetry in a radially moving system.

An alternative explanation for symmetric line profiles in a radially moving system is the presence of some scattering material in the vicinity of the clouds. The scatteres may be the hot electrons in the HIM or dust particles in the NLR. The main cause of asymmetry is the weak flux from the hemisphere nearer to the observer and there are several ways by which this can change in the presence of scatterers. First, the scattered ionizing radiation can hit the back, neutral side of the clouds, producing more ionization and line emission. Second, if the intercloud medium is optically thick, the radiation observed from the farther hemisphere is reduced. Third, the inward emitted line photons in the near hemisphere can be scattered back into the line of sight. Fig. 26 shows

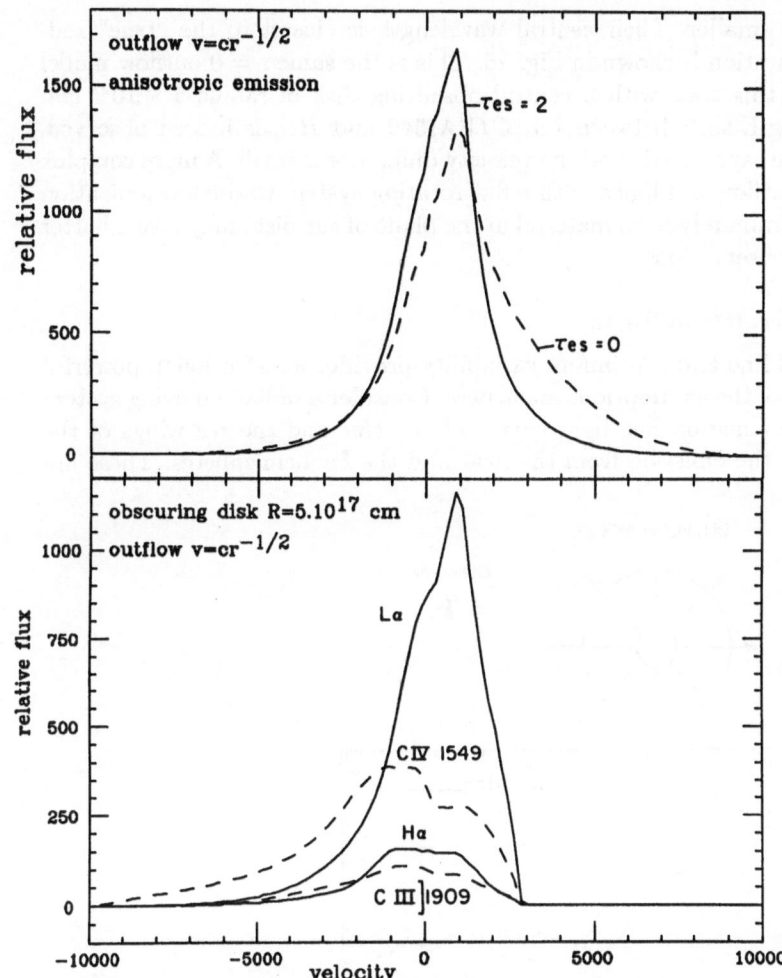

Fig. 26. Top: Unnormalized $L\alpha$ profiles for the $s = 0$ outflowing atmosphere, showing the effect of a Compton thick medium. The $L\alpha$ emission from the clouds is highly anisotropic but the line asymmetry is reduced, compared to the Compton thin case, due to scattering related effects. Bottom: The same model, with $\tau_{es} = 0$, combined with a central obscuring disk of radius 5×10^{17} cm. Note the wavelength shift of the line centers.

two $L\alpha$ profiles, resulting from the same cloud system as before, but moving in this case through a Compton thick HIM. The line profiles are calculated for the cases of $\tau_{es} = 0$ and $\tau_{es} = 2$, and the latter is indeed more symmetric.

The wavelength shift of some broad lines in high luminosity AGNs is definitely of great significance. In the time of writing there is no satisfactory explanation for this. One idea is that the shift is caused by a combination of obscuration and outflow motion. Consider for example a deccelerating outflow around a large accretion disk, with U decreasing outward. The high excitation lines, like $CIV\lambda1549$, are formed near the disk and much of their red-wing (emission from the other side of the disk) is not observed. Lower excitation lines, such as $MgII\lambda2798$, are produced further away from the disk, where the obscuration is smaller. Their central wavelength is closest to the "true" red-shift. An illustration is shown in Fig. 26. This is the same $s = 0$ outflow model as before but this time with a central obscuring disk of radius 5×10^{17} cm. Some wavelength shift between $L\alpha$, $CIV\lambda1549$ and $H\alpha$, is indeed observed, but all lines are asymmetric and the velocity difference is small. A more complex situation, of outflow combined with a flat rotating system whose low ionization lines are predominately from material in the plane of the disk, may give a better match to the observations.

9.5 Cross-correlation Tests

The correlated line and continuum variability provides an additional, powerful test on several of the assumptions made here. Consider a radially moving system of clouds. The emission line light curves of the *blue* and the *red* wings of the lines represent the emission from the near and the far hemispheres. These are

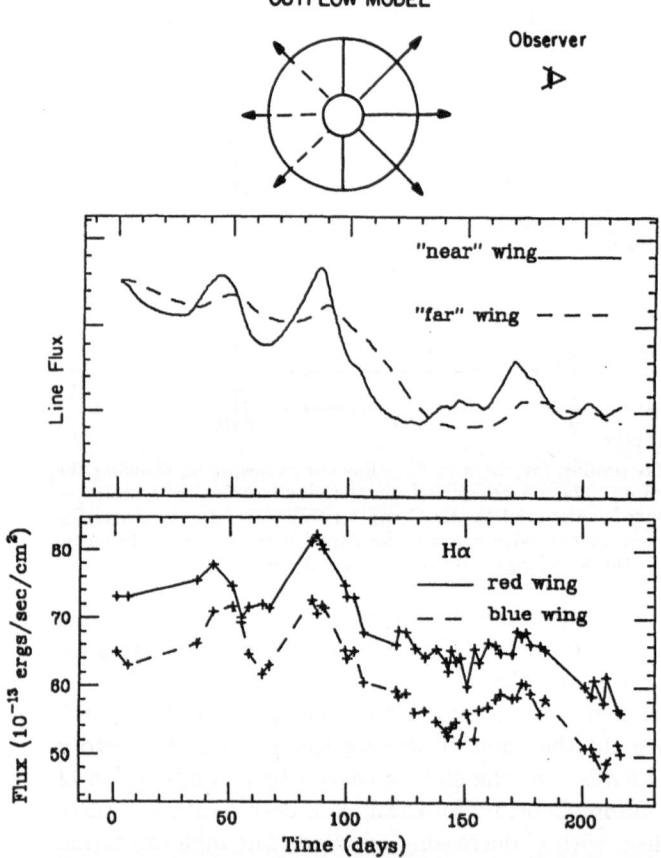

Fig. 27. Top: A schematic representation of a spherical outfowing system of clouds. Middle: Theoretical blue and red wing light curves obtained by assuming constant emission per unit covering factor, and the continuum light curve of NGC 4151 (Fig. 17). Bottom: Observed light curves for the blue and red wings of $H\alpha$. (Maoz et al. 1991)

very different, because of the different distance to the observer. The response of the near hemisphere is almost immediate, with sharp features, while the response of the far hemisphere is an average over a longer time, with smooth, shallower features.

The situation is illustrated in Fig. 27. for the case of NGC 4151 discussed in chapter 8. The top part of the diagram shows theoretical blue and red wing light curves, given the variable continuum of this source (Fig. 17). This should be compared with the measured light curves of the the blue and red halfs of the $H\alpha$ line, shown in the lower panel. The difference between the model and the observations is very large indeed: The measured red and blue wing light curves are identical, within the errors (and their cross-correlation gives a zero time lag) and do not resemble the theoretical curves. This suggests that a radial gas motion can be ruled out at a high confidence level, in this source.

Variable line profiles are important since they provide a way for mapping out the velocity field in the BLR. In principle, the transfer function should be constructed for each velocity, and a full 6-dimensional phase-space analysis should be performed. Obtaining profile information at each phase of activity is even more difficult than following the light curve behavior, and at time of writing no such analysis has been carried out.

9.6 Bibliography

The two-phase model: The basic paper is by Krolik, McKee and Tarter (1981). Other detailed papers on this topic are by Lepp et al. (1985), and Kallman and Mushotzky (1985). Krolik (1988) has discussed transient cool clouds embedded in hot gas. The difficulties with the model, especially in relation to the hot medium temperature, are discussed by Fabian et al. (1986) and Mathews and Ferland (1987). The last reference addresses the dynamical implications in great detail. For cloud formation, in relation to thermal instability, see Mathews and Doane (1990).

Other models: Magnetic confinement is discussed in Rees (1987). Cloud formation in winds is discuused by Smith and Raine (1985), Perry and Dyson (1985, winds and obstacles) and Shlosman et al (1985, radiation pressure driven wind from the surface of a disk). Collin-Souffrin, Dyson, Dowell and Perry (1988) suggested a two component model for the BLR, including high ionization broad line clouds, and the outer parts of a central accretion disk. More on accretion disks as a source of line emission can be found in a series of papers by Collin-Souffrin, Dumont and collaborators (See Dumont andCollin-Souffrin 1990 and references therein). The orbiting cloud model has been suggested by Kwan and Carroll (1982), Carroll and Kwan (1985) and Carroll (1985). The implications to the NLR are discussed in Carroll and Kwan (1983). Critical discussions of these and other models are given in review articles by Mathews and Capriotti (1985) and Osterbrock and Mathews (1986). BLR models involving stars are discussed by Voit and Shull (1988), Penston (1988), Kazanas (1989) and Scoville and Norman (1989).

Line profiles: For line profile observations see chapter 2. The basic work on acceleration by radiation pressure is by Mathew (1974, 1982a) and Blumenthal and Mathew (1975, 1979). These authors have also investigated the shape of the clouds, in particular "pancake" clouds. Different radial dependences are discussed by Capriotti et al (1980, 1981), Carroll and Kwan (1983, 1985), Bradley and Puetter (1986) and Penston et al. (1990). Shields has investigated disk line profiles in a number of papers; see Shields (1989) for references. More work on disk-type profiles is in Mathews (1982b). Application of disk profiles to the case of Arp 102B are discussed in Chen et al. (1989). For reviews see Mathews and Capriotti (1985), Osterbrock and Mathews (1986) and Shields (1989).

Line asymmetry and wavelength shifts: The first theoretical discussion of line asymmetry is by Ferland, Netzer and Shields (1979). Further work is by Capriotti et al. (1981) and Puetter and Hubbard (1987). Kallman and Krolik (1986) investigated line profiles in a presence of a Compton thick medium. For the observations, see chapter 2.

Cross-correlation tests of radial motion: See Maoz et al. (1991) and references therein.

10 Line and Continuum Correlations

So far we have only considered line intensities and line profiles in individual objects. More insight on AGNs can be gained by comparing different objects, and by analyzing the statistical properties of large samples. It enables us to search for luminosity dependences, to compare bright and faint objects in a variety of ways, and to examine the consequences of the photoionization theory.

In this chapter we discuss several observed correlations between the broad emission lines and the nonstellar continuum of AGNs. We use them to obtain further estimates on the ionization parameter, the covering factor, the velocities and the masses of AGNs. Thin accretion disks are also discussed. A parallel discussion, based on observations of the narrow lines, is given in chapter 11. Throughout the chapter we use the symbol L to designate the integrated continuum luminosity over some specified frequency range, and L_λ the monochromatic luminosity per Å.

10.1 Line Intensity vs. Continuum Luminosity

The intensity of most broad emission lines is strongly correlated with the continuum luminosity. An example is shown in Fig. 28, where the $L\alpha$ intensity of 328 AGNs is compared with their continuum luminosity. An almost perfect correlation, with a best slope of 0.88, is found. This gives further support to the idea that photoionization is, indeed, the main sourse of excitation for the lines.

The slope of the correlation in Fig. 28 is close to, but not exactly 1.0. This can be interpreted in several ways. One possibility is that the integrated covering factor decreases with increasing luminosity. If this is correct, and there are no other factors involved, then the decrease in covering factor amounts to

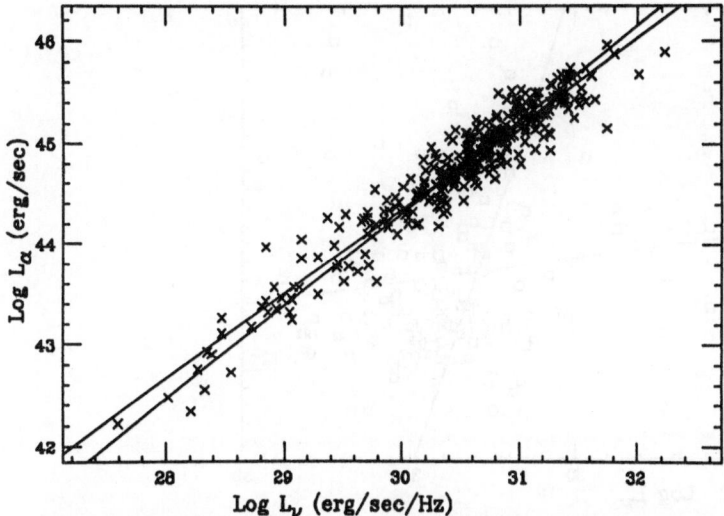

Fig. 28. $L\alpha$ luminosity vs. continuum luminosity for 328 AGNs. (Here and in the following examples $H_0 = 75\ km\ s^{-1}\ Mpc^{-1}$ and $q_0 = 0.5$)

a factor of about 3.5 over the range of $10^{4.5}$ in continuum luminosity. Another possibility is a change in the shape of the ionizing continuum, such that the number of ionizing photons (mainly in the 1-4 Ryd. range), compared with L_{1215}, is smaller in brighter objects; i.e. a softer ionizing continuum in more luminous objects. The change must be small, as the relative intensity of the high excitation lines does not seem to be a function of luminosity. There are other possibilities, such as changes in the amount of the optically thin material and/or the optical depth of the clouds, with luminosity.

10.2 Line - Line Correlations

Emission line ratios are the best ionization parameter indicators, and the correlation of line ratios with the continuum luminosity can be used to check the U vs. L dependence.

Fig. 29 shows the $CIV\lambda1549/L\alpha$ line ratio, as a function of the continuum luminosity, for a sample of 165 AGNs. The scatter in this ratio is large, but the tendency is for the line ratio to decrease with increasing continuum luminosity. This indicates, perhaps, that the ionization parameter in bright quasars is smaller than in Seyfert 1s (Fig. 8). Although the correlation is statistically significant, it is not clear how representative it is of the AGN population. The objects under study were randomly selected from the literature, and there are several potential selection effects to be considered. Moreover, despite the tendency of decreasing $CIV\lambda1549/L\alpha$ with increasing luminosity, some bright quasars are definitely exceptional in this respect.

Line ratio diagrams have been constructed for other emission lines, some in carefully selected samples, where selection effects are not likely to dominate. Such well selected samples are rare, and the number of objects in them rather small. They cover a limited range in luminosity, and general tendencies are

135

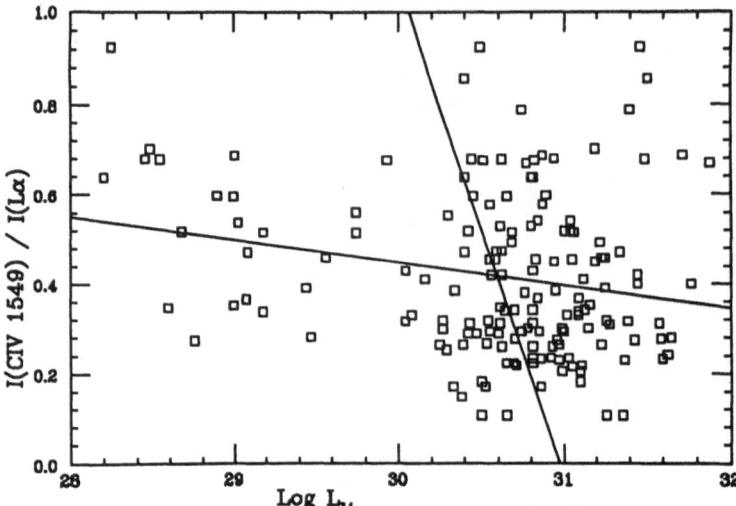

Fig. 29. $CIV\lambda1549/L\alpha$ vs. continuum luminosity at 1549Å for 165 AGNs

hard to discover. Of the more interesting findings we note the increase of the $CIII]\lambda1909/CIV\lambda1549$ line ratio with L, suggesting, perhaps, a change in U and/or density with luminosity.

10.3 Line Width vs. Continuum Luminosity: M/L for AGNs

10.3.1 Observed correlations. A superficial study of AGN spectra reveals the large diversity in line widths. The FWHM (Full Width Half Maximum) of a certain line, in objects of similar continuum luminosity, can differ by a factor of two or more. For example, the typical FWHM in radio-quiet Seyfert 1s is about 4000 $km\ s^{-1}$, while in radio-loud Seyfert 1s (the BLRGs), with similar continuum luminosity, it can exceed 10,000 $km\ s^{-1}$. Extremely broad emission lines seem to be typical of steep spectrum radio quasars. Many of these objects show also a bumpy, asymmetrical line profile.

The $CIV\lambda1549$ FWHM versus L_{1549} correlation is shown in Fig. 30. There is a weak tendency for brighter objects to have somewhat broader lines which, for the data in the diagram, is expressed by

$$FWHM(CIV\lambda1549) \propto L_{1549}^{0.3} \ . \tag{95}$$

The correlation is noisy and only marginally significant. More important, the above sample, which was collected from published line lists, does not include the BLRGs. Having a few such objects in the sample would increase the average FWHM at the low luminosity end, causing the above correlation to weaken, or disappear.

This and several published line width vs. L correlations, are all affected by the lack of well selected samples. It will take more observational effort, and better defined samples, to verify whether or not the line-widths are indeed correlated with the continuum luminosity. Since no firm conclusion can be

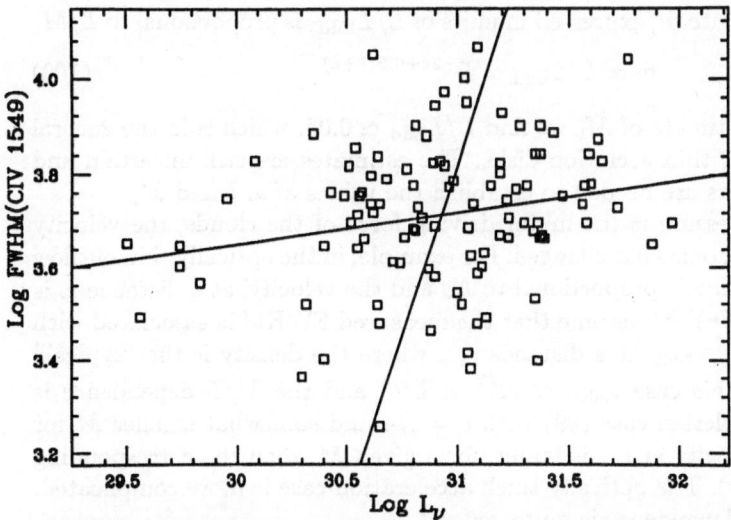

Fig. 30. The FWHM of $CIV\lambda1549$ vs. continuum luminosity at 1549Å for 104 AGNs

drawn at this stage, we proceed by examining the consequences of two likely possibilities: FWHM independent of L and FWHM$\propto L^{1/4}$.

10.3.2 M/L for AGNs. Assume that the emissivity weighted radius, r_{av} in (71), is the typical dimension of the BLR. Assume further that the Keplerian velocity at r_{av}, measured in units of 3000 $km\ s^{-1}$, is

$$v_{3000} = \sqrt{3}/2\ FWHM\ . \tag{96}$$

For a bound Keplerian motion around a central mass M,

$$M \simeq 7 \times 10^{-10} v_{3000}^2 r_{av}\ \ M_\odot\ , \tag{97}$$

and from the definition of the ionization parameter (5)

$$r_{av} \simeq 3.5 \times 10^{17} \Big[\frac{L_{46}(ion)}{U N_{10}\bar{\nu}}\Big]^{1/2}\ cm, \tag{98}$$

where $L_{46}(ion)$ is the ionizing luminosity in $10^{46}\ erg\ s^{-1}$, $N_{10}=N/10^{10}\ cm^{-3}$ and $\bar{\nu}$ the mean energy of an ionizing photon, in Rydberg.

Assume now $U \propto L^a$, $v_{3000} \propto L^b$ and $(N_{10}\bar{\nu})$ which is independent of $L(ion)$. Thus

$$M = M_1[L_{46}(ion)]^{2b+0.5(1-a)}\ . \tag{99}$$

The previous estimate of r_{av} (75) obtained from line reverberation studies, combined with typically observed FWHM, enable us to estimate that for Seyfert 1 galaxies $M_1 \sim 10^9 M_\odot$.

The value of a is not well known, but it is likely to be in the range 0 to -1/2. For the case of $a = 0$ (U independent of L) there are two interesting possibilities: $b = 0$ (line width independent of luminosity), which results in $M \propto L^{1/2}$, and $b = 1/4$, that gives $M \propto L$.

The accretion rate $\dot{m}$, expressed in units of L/L_{Edd}, is proportional to L/M,

$$\dot{m} \propto L/L_{\mathrm{Edd}} \propto L^{-2b+0.5(1+a)} . \tag{100}$$

With the above estimate of M_1 we find $L/L_{\mathrm{Edd}} \simeq 0.05$, which is in the general accepted range for thin accretion disks. The estimates are still uncertain and more measurements are needed to establish the values of a, b and M_1.

If radiation pressure is the major driving force of the clouds, the velocity field and L/M are somewhat changed. For example, in the optically thin outflow case, the acceleration is proportional to N_e and the velocity, at a distance r, is proportional to $(Nr)^{1/2}$. Assume that the measured FWHM is associated with a terminal velocity v_{3000} at a distance r_{av}, where the density is the "typical" BLR density. In this case $v_{3000} \propto r_{\mathrm{av}}^{1/2} \propto L^{1/4}$ and the M/L dependence is similar to the Keplerian case (99) with $b = 1/4$ and somewhat smaller M for a given L (the velocity at r_{av} is larger, for a given M, than the corresponding Keplerian velocity). The optically thick acceleration case is more complicated, but the $v \propto r^{1/2}$ dependence is quite general.

10.3.3 Line width and radio properties. The radio emission of AGNs is observed to come from either a compact or an extended radio source. The compact source coincides with the optical continuum source. Its radiation is thought to be beamed, due to relativistic motion, and its apparent luminosity depends on the observer's viewing angle. The extended radio emission originates in a much larger volume, it is not beamed and the apparent luminosity is angle independent. Thus the ratio of the compact and extended radio fluxes is a measure of the orientation of the central radio source. This ratio (compact/extended) is found to be *smaller* in objects with *broader* emission lines. This may influence all line-width vs. continuum luminosity correlations in samples including radio-loud objects. It suggests that the emission lines in those radio-loud objects are emitted, preferentially, from material in a plane perpendicular to the direction of the radio beam, which may also be the plane of the central accretion disk.

10.4 The Baldwin Relationship

This relation, discovered by J. Baldwin in 1977 and confirmed in several later studies, is a strong correlation between the equivalent width (EW) of $C\,IV\,\lambda1549$ and the continuum luminosity. It is clearly observed in radio selected samples but seems to be weaker in optical samples. In particular, quasars discovered on objective prism plates show a weak, less significant correlation. This is, perhaps, not surprising given the fact that these objects are selected by the strength of their emission lines. It is also known that the correlation is different for different lines, in particular the EW of the optical lines is not well correlated with the optical continuum luminosity. Some of the uncertainty is due to the lack of well selected, bias free samples.

The original sample studied by Baldwin covered only a small range in continuum luminosity ($\sim 10^{1.5}$) and resulted in a well defined slope for EW($C\,IV\,\lambda1549$) vs. L_{1549}. Later studies extended the range to more than four orders of magnitude in continuum luminosity, and to a much larger number of objects. The correlation is still present, but its slope is very different. An example is shown in Fig. 31 where the $L\alpha$ EW of more than 300 AGNs is compared with the continuum luminosity at 1215Å. The best (harmonic mean) slope in this case is -0.3, i.e.

$$EW(L\alpha) \propto L_{1215}^{-0.3} \, . \tag{101}$$

A regression analysis for a sub-sample of the same data set, covering the range $10^{30} \leq L_{1215} \leq 10^{31.5} \, erg \, s^{-1} \, \mathring{A}^{-1}$, gives a much steeper slope, of 0.5, which is similar to the original slope found by Baldwin. This change of slope, as a function of the *luminosity range* of the sample, is a key to the understanding of the Baldwin relationship.

Several attempts have been made to explain the Baldwin relationship. The shape of the ionizing continuum may be luminosity dependent in such a way that the continuum is "softer" in more luminous objects. Because of that the $CIV\lambda1549$ line luminosity increases less than the continuum luminosity, resulting in smaller EW for brighter objects. This cannot be a large effect since high excitation lines, such as $NV\lambda1240$ and $OVI\lambda1035$, are strong in bright quasars. Alternatively, the ionization parameter in bright AGNs can be somewhat smaller than in fainter objects (see the discussion on the $L\alpha/CIV\lambda1549$ ratio in 10.2). This gives the right tendency but the difficulty with the high excitation lines is not resolved. Moreover, the Baldwin relationship for $L\alpha$ cannot be explained in this way, since there is no physical reason for a decrease in $EW(L\alpha)$ with increasing continuum luminosity. Photoionization calculations confirm most of these objections. They show that an increase in U can explain only a part of the effect, over a part of the observed luminosity range.

A third possibility is an inverse correlation between continuum luminosity and the covering factor. The tendency is consistent with the $L\alpha$ vs. contin-

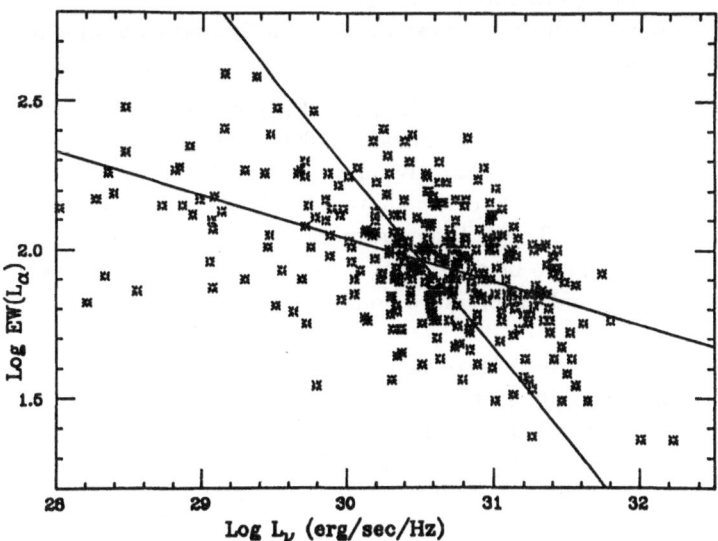

Fig. 31. $L\alpha$ equivalent width vs. continuum luminosity (the Baldwin Relationship) for 328 AGNs

uum relation shown in Fig. 28, but the deduced range in covering factor is not large enough to explain the Baldwin relationship. Also, the dependence on the luminosity range is not explained.

It has been suggested that large continuum variations, that are *not* associated with corresponding emission line variations, can produce the observed correlation. This is a plausible explanation for quasars, since emission line variability in them are small compared with the continuum variability. Some confirmation of this idea comes from the fact that the Baldwin diagram for *individual* Seyfert 1 galaxies, constructed from line and continuum measurements at different phases of activity, is not very different from the original relationship found for *a sample* of bright quasars.

Lately it has been realized that the presence of geometrically thin accretion disks may introduce an EW-continuum luminosity dependence. This is discussed in the following section.

10.5 AGN Accretion Disks

10.5.1 Theoretical disk models. Theoretical justifications for accretion disks and their properties are discussed by R. Blandford. Here we limit the discussion to the observational implications of geometrically thin, optically thick, accretion disks.

The calculations of the radiation emitted by thin accretion disks, around massive black holes, are extremely complex. There were several attemps to calculate the spectrum of "bare disks", i.e. those without a corona, but none combined a full atmospheric solution with a general relativistic treatment. The most sophisticated calculations, so far, combine a full relativistic treatment with a simplified transfer solution (all examples in this chapter are from this set of calculations). Differences between existing models are large, reflecting, most probably, the different approximation used.

The following is a short summary of those results that are most relevant to the observations of AGNs.

a: The spectrum of "bare disks" with $L/L_{\rm Edd} \sim 0.1$, gives a fair fit to the observations of many AGNs, in the spectral range 1000-6000Å. Disks around fast rotating black holes are hotter, and give a somewhat better fit to the observed spectra. The largest accretion rate allowed by the models, expressed in units of $L/L_{\rm Edd}$, is about 0.3. For larger accretion rate, the geometrical thickness of the disk is too large to be consistent with the "standard thin disk" assumption. The disks are radiation pressure dominated and their self-gravity radius is relatively small.

b: The polarization of AGN disks is considerably smaller than the polarization of Newtonian disks, with fully scattering atmospheres.

c: The soft X-ray continuum (0.3-2 keV) of many AGNs is too strong to be fitted by the disk models that fit the observed optical-ultraviolet continuum. The hard (2-20 keV) X-ray continuum of *all* AGNs is inconsistent with standard thin disk spectra.

d: Aspect dependent effects are very important. A simple, wavelength independent limb darkening law of the form

$$F(\theta) \propto \cos\theta(1 + a\cos\theta), \qquad (102)$$

with $a \sim 1.5$, gives a good approximation to the angle dependent flux at optical wavelengths. This limb darkening law fails at very short wavelengths, where relativistic effects dominate the emission. In particular, the continuum of an edge-on disk is much harder than that of a face-on disk. This is illustrated in Fig. 32 that shows the flux emitted by the disk at three inclination angles. The relativistic effects are important at soft X-ray energies for low luminosity AGNs, and at ultraviolet energies for high luminosity (i.e. more massive central black hole) objects.

e: The lack of a direct observational method to find the inclination angle in individual objects, implies that the fitting of AGN continua by disk models is not unique. It can amount to a factor of 10 uncertainty in the inferred central mass and accretion rate for a given object.

A fit of a theoretical disk model to the observation of a particular quasar is shown in Fig. 33. There are three major continuum components in this fit. The first is an infrared power-law, made to fit the 1-3μm range, and extended into shorter wavelengths with the same slope. Such a component is always needed since the disk emission cannot explain the observed infrared continuum. The assumption of a power-law shape is neither certain nor essential. It is a reasonable guess over the limited wavelength range in question, but other possibilities, such as thermal emission by dust and stars, can be made to fit

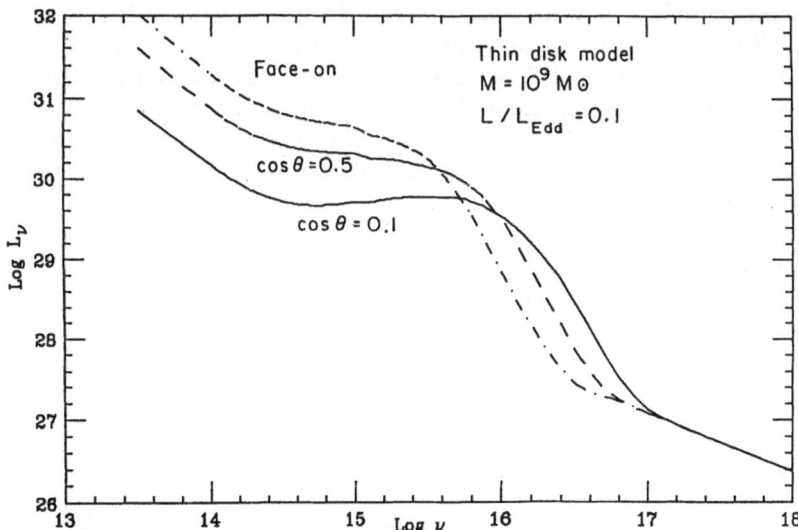

Fig. 32. Theoretical thin disk spectra at three different viewing angles. The infrared and the X-ray continua are approximated by power-laws, with spectral indices of 1.4 and 0.7, respectively (courtesy of A. Laor)

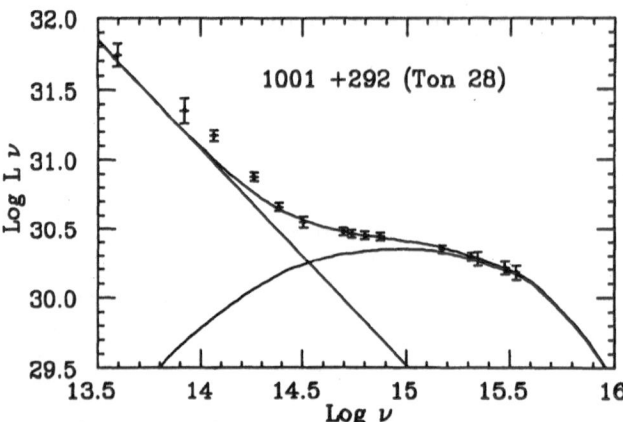

Fig. 33. A model fit to the spectrum of the quasar Ton 28, showing the infrared power-law component and a theoretical disk continuum (courtesy of A. Laor)

the data, at least in some cases. The second component is the thermal disk emission. It is about as intense as the infrared component at 1μ and produces most of the emission at shorter wavelengths. There are indications that the relative contribution of this component, compared with the infrared component, is larger in more luminous objects. Finally a third, X-ray component, is needed in all cases. It dominates the emission at energies larger than about 1 keV, the energy where all disk models are found to radiate well below the observed flux. It is important to note that despite the very good match to the observed spectrum, in this and other cases, some AGN spectra cannot be fitted by any disk model calculated so far.

Future, more realistic thin disk models, must incorporate several of the ingredients missing in present day calculations. The X-ray source must be given some thought. If this radiation is coming from the innermost part of the disk, then some of it may intersect the disk outer parts, changing the local energy balance and the surface temperature there. This can be due to direct illumination (flaring out of the disk or a thick central part) or via some scattering material above the disk. Hot corona is a component that can scatter the disk radiation back onto its surface, and can also be a local source of X-ray radiation. None of these have been treated in detail so far. Finally, better transfer calculations, employing full atmospheric solution, are greatly needed. They will provide the only theoretical way to answer the important question whether such disks are likely to have *emission and absorption lines and edges.* Such spectral features may turn out to be the best indications for the presence or absence of massive accretion disks in AGNs.

10.5.2 Thin accretion disks and the Baldwin relationship. The successful fitting of observed AGN continua by disk models does not give a conclusive evidence for the presence of such disks. There are other continuum emission processes that are consistent with the data and the model uncertainties are large. There are other ways to search for such disks, using viewing angle related effects. The observed disk flux depends on the inclination to the line of sight. On the other hand, the emission line flux originates, presumably, in isotropically emitting clouds, and is independent on the observer's viewing angle. Thus the line-to-continuum ratio, i.e. the line EW, is a measure of the disk inclination.

The intrinsic line/continuum ratio can differ from object to object and it is not possible to determine the disk inclination by a single EW measurement. The statistical properties of a large sample are more promising in this respect. The disk continuum flux is roughly proportional to $\cos\theta$ (102) and the line EW, assuming isotropic line emission is, therefore, proportional to $1/\cos\theta$. The probability distribution of equivalent widths, in a sample with random disk orientations, is thus

$$P(EW)d(EW) \propto (EW)^{-2}d(EW) \, . \tag{103}$$

The observed EW distributions in several AGN samples is consistent with that.

Testing the EW distribution, and other disk-related phenomena, is hampered by various selection effects. Some of those are well known and are related to the discovery methods and the incompleteness of magnitude limited samples. Thin disks, if they exist, may introduce yet another selection effect. Such objects are brighter when observed face-on, which may cause the drop-out of edge-on objects from flux limited samples. The effect is most difficult to estimate in objective-prism surveys, where the objects are detected due to some combination of their continuum brightness and line strength. Searching for thin disks in big AGN samples, by ways of their characteristic EW distribution, can only be performed if the discovery technique is independent of the disk properties. Radio selected samples, especially those found by their extended, steep radio spectrum, may prove to be most appropriate for that.

An example illustrating all this is shown below. Several theoretical thin disk models, with $10^8 \leq M \leq 10^{10}\,M_\odot$, $0.1 \leq \cos(\theta) \leq 1$ and two values of L/L_{Edd}, 0.1 and 0.3, were calculated to produce a set of theoretical continua. These were used in photoionization calculations to find theoretical $L\alpha$ intensities, like in chapter 4. The resulting theoretical $I(L\alpha)$ vs. $L_\nu(1215\text{Å})$ correlation is shown in Fig. 34. The horizontal lines in the diagram are due to the fact that *a given $L\alpha$ flux is associated with a range of continuum fluxes*, because of the disk inclination.

The same data, transformed into an $EW(L\alpha)$ vs. L_{1215} by assuming *a constant covering factor* of 0.1, is shown in Fig. 35. The previous horizontal lines are tranformed into diagonal lines and there is a tendency for $EW(L\alpha)$ to decrease with increasing L_{1215}. This is due to the combined effect of the disk inclination and the disk ionizing flux being softer in more luminous (larger mass) systems.

The theoretical distributions in Figs. 34 and 35 are similar to the observed distributions in Figs. 28 and 31, and can explain, therefore, the Baldwin relationship. They are also consistent with the observational finding that the slope of the Baldwin relation depends on the luminosity range of the sample. In this case, there is no need to assume a luminosity dependence of the covering factor or the ionization parameter. Note again the possible selection effects and the inhomogeneity of the sample used here.

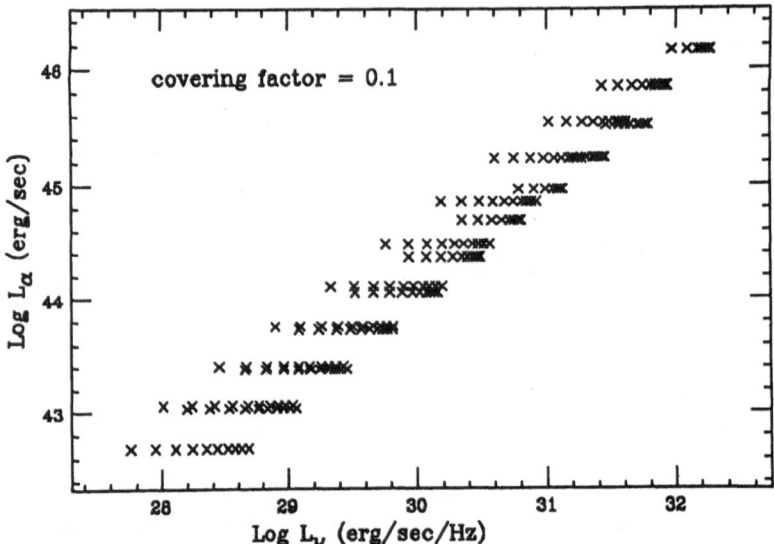

Fig. 34. Theoretical $L\alpha$ vs. $L_\nu(1215\text{Å})$ for disk continua

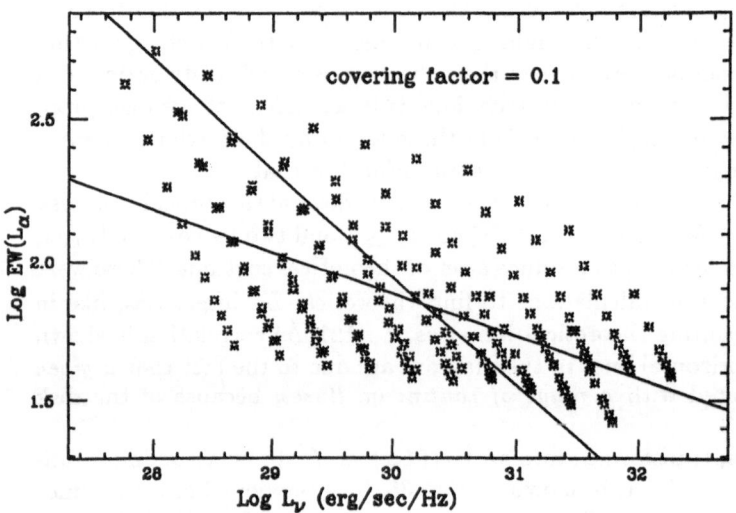

Fig. 35. Theoretical EW($L\alpha$) vs. $L_\nu(1215\text{Å})$ for AGN disks

10.6 Bibliography

Line-continuum correlations: There are many papers on this subject. See for example Baldwin, Gaskell and Wampler (1989), Joly (1987b) and references therein.

Line width vs. continuum luminosity: See Wandel and Yahil (1985), Joly (1987b), Mathews and Wampler (1985), Busko and Steiner (1989). The relation of the line width with the compact/extended radio luminosity ratio is discussed in Wills and Browne (1986). Some papers discussing M/L in AGNs, using line width and ionization parameters, are: Wandel and Yahil (1985), Joly et a. (1985), Wandel and Mushotzky (1986), Padovani et al. (1988) and Netzer (1989).

Thin disk models: For theoretical disk continua see Czerny and Elvis(1987), Laor and Netzer (1989), Sun and Malkan (1989), Laor (1990) and references therein. The calculations used here are from Laor and Netzer (1989) and Laor (1990). For the polarization properties of massive disks see Laor, Netzer and Piran (1990) and references therein. Recent papers on the fitting of AGN continua by disk models are Sun and Malkan (1989), Czerny and Elvis (1987) and Laor (1990).

The Baldwin relationship: The original paper is by Baldwin (1977). Some more observational studies are described in Wampler et al. (1984), Osmer and Smith (1980), Baldwin, Wampler and Gaskell (1989) and Kinney et al (1990). The Baldwin relation in individual objects is discussed in Wamsteker and Collina (1986), Kinney et al (1990) and Edelson et al (1990). Theoretical discussion can be found in Mushotzky and Ferland (1984, dependence on U and the covering factor), Murdoch (1983, line and continuum variability) and Netzer (1985, thin disks).

11 Quasars, Miniquasars and Microquasars

Much of the previous analysis focused on the broad emission lines. This limits the discussion to the more luminous AGNs, where these lines are strong and easy to measure. The aim in this chapter is to concentrate on fainter AGNs, those in which the broad lines are weak or absent. These are mostly low red-shift objects showing some, but not all the AGN characteristics. Many of their observed properties have already been discussed in chapters 1 and 2, and the emphasis here is on the physical interpretation. At the end of the chapter we use the observed narrow-line properties to suggest a classification scheme for the entire AGN family.

11.1 Narrow Line X-ray Galaxies

Many of these objects have been discovered in X-ray surveys and identified, subsequently, with emission line galaxies. Some of their observed characteristics have been discussed in chapter 1. They include strong narrow lines, mostly of high excitation, strong X-ray emission and weak ultraviolet continuum. There is a great similarity in the spectral distribution of the X-ray continuum of NLXGs and Seyfert 1 galaxies.

The key to the understanding of NLXGs is the faint, broad $H\alpha$ observed in many of them. In these objects the broad $H\beta$ is extremely weak, or absent, giving a strong lower limit to the broad $H\alpha/H\beta$ ratio of ~ 10. This can be understood if NLXGs are heavily reddened Seyfert 1 galaxies. Most observed differences between NLXGs and "normal" Seyfert 1s are consistent with this idea. The amount of dust is large, $A_V \sim 2$ mag. or so, but not large enough to absorb the hard X-ray radiation. This explains the weak ultraviolet continuum. There are some indications that the amount of reddening is correlated with the inclination of the host galaxy, with edge-on spirals showing the largest extinction. This suggests that reddening may be common in many AGNs, being less noticable in the so called "normal" Seyferts, which, in this scheme, are located in face-on host galaxies. The dust is associated, most probably, with the interstellar medium, and is not necessarily related to the dust suspected to cause some of the broad line reddening, closer to the center (chapter 7).

11.2 Seyfert 2 Galaxies

The narrow emission lines of Seyfert 2 galaxies cover a large range of ionization. Lines from $[OI]$ to [FeXI] are observed, and the spectrum is similar to the narrow line spectrum of Seyfert 1s. The equivalent width of the lines, relative to the nonstellar continuum, is systematically larger than in Seyfert 1s. The line width distribution is broad, from 250 to 900 $km\ s^{-1}$, with an average around 400 $km\ s^{-1}$. Blue asymmetry is observed in most lines, being stronger in lines of higher critical density. This again is similar to the observations of Seyfert 1 galaxies, and suggests that the higher density clouds move faster with respect to the central source.

The nonstellar optical, ultraviolet and X-ray continua of Seyfert 2 nuclei are very weak but their *ratios* are not too different from those in Seyfert 1s, thus the overall continuum *shape* is similar in the two sub-groups. The extrapolation of the observed ultraviolet continuum to high energies, does not seem to give enough ionizing flux to explain the observed emission lines (i.e. the required covering factor is larger than 1). Thus, there are several indications that much of the nonstellar continuum is not observed by us.

The key to the understanding of the difference between Seyfert 1 and Seyfert 2 galaxied is in recent polarization measurements of these objects. The degree of polarization is small, 1-3%, but when *the polarized flux* is plotted separately from the rest, as in Fig. 36, it shows, very clearly, a typical Seyfert 1 spectrum,

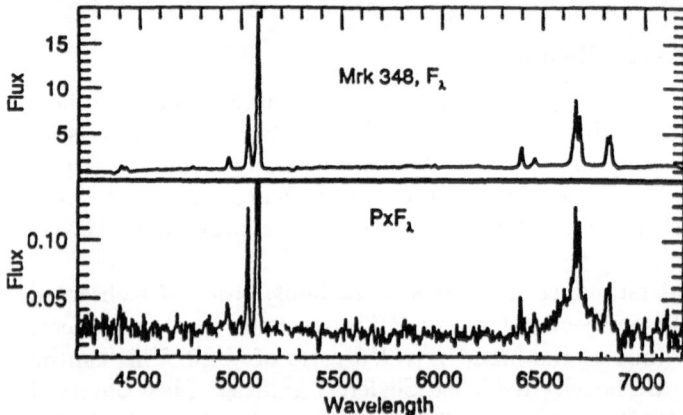

Fig. 36. Top: the spectrum of the Seyfert 2 galaxy Mkn 348. Botton: The polarized flux of Mkn 348, showing the typical broad emission lines (after Miller and Goodrich 1990).

with strong broad emission lines of hydrogen and FeII. The small fraction of polarized flux explains why this detection escaped the notice for many years. Currently there is a handfull of Seyfert 2 galaxies showing this pheonomenon.

The following physical model for Seyfert 2 galaxies has emerged (Fig. 37). A "normal" Seyfert 1 nucleus, surrounded by its BLR, is situated at the center of the system. A thick torus, of inner radius $\sim 1pc$ and similar thickness, is present too. Most of the torus material is in molecular clouds, that are shielded from the central radiation by dust and by free electrons that evaporate from the clouds. The central BLR is obscured from some observers by the molecular torus. Such observers can only see the small fraction of BLR light that is scattered in their direction. The intensity of the scattered light depends on the Compton depth of the medium, and the degree of polarization on the observer's viewing angle. Some viewing angles are not obscured, and a Seyfert 1 type spectrum is observed. The NLR size greatly exceeds the dimension of the torus, and the narrow emission lines are seen by all observers. However, the NLR illumination and ionization is not isotropic, because of the torus, and the *observed NLR* is likely to attain a jet-like structure. There are radio observations and narrow emission line maps that support the claim for a jet-like NLR in Seyfert 2 galaxies.

The above model provides a natural explanation for the difference between Seyfert 1 and Seyfert 2 galaxies. One suggestion is that the different viewing direction is the only distinction between the two groups of objets. In this case the thickness and dimension of the torus can be estimated from the relative number of Seyfert 1 and Seyfert 2 galaxies. There are difficulties too, such as several well studied Seyfert 2s where the polarized flux is extremely small and no broad lines are seen. The role of dust is not very clear and heavy reddening is likely to be present, at least in some directions. Light scattered by dust is polarized in a wavelength dependent way, and there are observational ways to test this idea. The Compton depth should be large enough to scatter part of the broad line radiation, but not too large to smear out the line and continuum

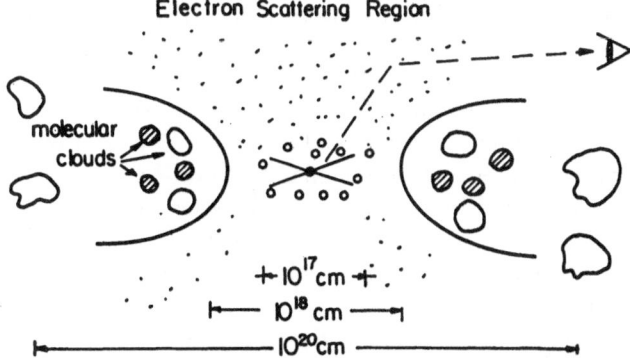

Fig. 37. A unified model for Seyfert galaxies. The central source and the BLR are surrounded by a thick torus of molecular gas. The Seyfert 1 galaxies are those objects observed from the pole direction. Seyfert 2 galaxies are those sources whose inner parts can only be seen through reflected radiation

variability in Seyfert 1 galaxies. There are theoretical uncertainties too, to do with the structure and stability of the torus.

An alternative explanation to the differences between the two groups of galaxies is related to their variability. The broad emission lines of some Seyfert 1 galaxies exhibit a large amplitude variations. In some objects the variability is so large that the spectrum at minimum light resembles a Seyfert 2 spectrum. It has been suggested that some Seyfert nuclei spend a large fraction of their active phase in a "turned-off" state, when the ionized flux is too weak to excite the gas. The recombination time of the broad line gas is short and the lines disappear several hours after the continuum decline. The recombination time of the low density gas is long enough to show strong narrow lines many years after the decline. Such an object, with strong narrow lines and weak continuum, will be classified as a Seyfert 2 galaxy. A possible cause for the drop in luminosity is a large decrease in the accretion rate. The big continuum bump observed in broad line AGNs has been attributed to emission from accretion disks. If this is indeed the case then the time scale for its fading is very long. It is hoped that future HST observations will help to decide whether the continuum of Seyfert 2 galaxies shows any sign of such a bump.

To summarize, Seyfert 2 galaxies may be Seyfert 1 nuclei that are hidden in space or in time. This may explain many, perhaps most observed properties of Seyfert 2s. However, we should not neglect the possibility of the presence of genuine narrow line objects, with no BLR at all.

11.3 LINERs

LINERs are the least luminous and the most common AGNs. At least 30% of all spirals show this phenomenon and the fraction may even be larger. LINERs show strong, low excitation emission lines, compared with the high excitation narrow lines of other AGNs. Their typical $[OII]\lambda3727/[OIII]\lambda5007$ line ratio is about 1, while in Seyfert 1s it is 0.5 or less. The $[OI]\lambda6300$ line in LINERs is very strong and lines of $[NeV]$ and $[FeVII]$ are not observed. There is very little information about the ultraviolet emission lines in LINERs, a situation that is likely to change with the successful operation of the HST. A LINER spectrum is shown in Fig. 38.

Typical emission line widths in LINERs are 200-400 $km\,s^{-1}$, smaller than in Seyfert galaxies and comparable to the stellar rotational velocity in the nucleus. The optical, nonstellar continuum luminosity is about 1% of the Seyfert 1 continuum luminosity, i.e. weak compared with the stellar background measured through a typical aperture of a ground-based telescope. There are clear indications, from the emission line spectrum, of ionization by a nonstellar continuum, but no direct observation of this continuum. X-ray emission has been detected in several LINERs but the optical/X-ray flux ratio of the nonstellar source is difficult to measure for the reasons explained above. For similar reasons, it is not yet clear whether the extrapolated ultraviolet continuum is intense enough to explain the observed emission lines. Thus, most of the information on LINERs comes from the analysis of their optical emission lines.

Two types of observations provide the key to the understanding of LINERs. The first is the detection, in many such objects, of faint, broad emission wings to $H\alpha$. This is demonstrated in Fig. 38 and is known to be quite common. Thus, a faint broad line region is probably present and many LINERs are faint, low excitation Seyfert 1 nuclei. Other broad emission lines are not observed and it remains to be found whether the broad lines are weak because of extinction, like in the NLXGs, obscuration or some other reason. The narrow line spectrum of LINERs is clearly distinguised from that of Seyfert galaxies, in having a much lower degree of ionization. The way to understand this is by a comparison of some emission line ratios with those of brighter AGNs. This is done by ways of constructing diagnostic diagrams, as explained below.

11.3.1 Diagnostic diagrams. Diagnostic diagrams are constructed from pairs of observed line ratios. They are very useful since different line ratios reveal different information on the ionizing continuum, the ionization parameter, the gas temperature etc. Such diagrams can help to separate blackbody photoionized nebulae from non-thermal photoionized ones and help to identify shock-excited gas. This is crucial for LINERs where the number of well measured emission lines is small, and their intensity is sometimes uncertain due to the presence of stellar absorption features. For example, the so called "Star-burst Galaxies", which are galaxies with giant nuclear HII regions, have often been confused with LINERs and line ratio diagrams can help to separate these groups.

Fig. 39 shows a diagnostic diagram that it is useful for separating HII regions, and other nebulae that are ionized by a blackbody continuum, from AGNs. The diagram compares the $[OIII]\lambda 5007/H\beta$ line ratio with the $[NII]\lambda 6563/H\alpha$ ratio, and is a good way of measuring the mean energy of an

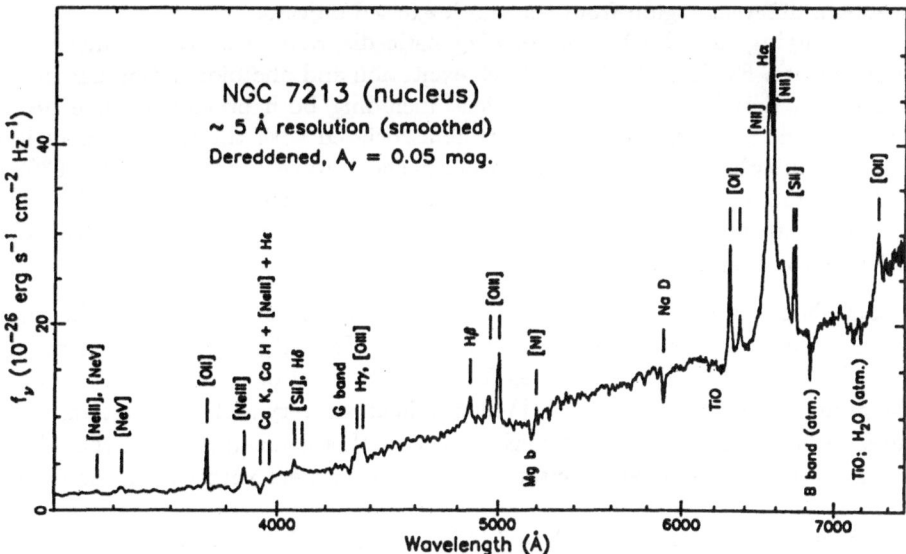

Fig. 38. The spectrum of the inner region of NGC 7213, showing the typical narrow line spectrum of a LINER and weak broad $H\alpha$ wings. (Filippenko and Halpern 1986)

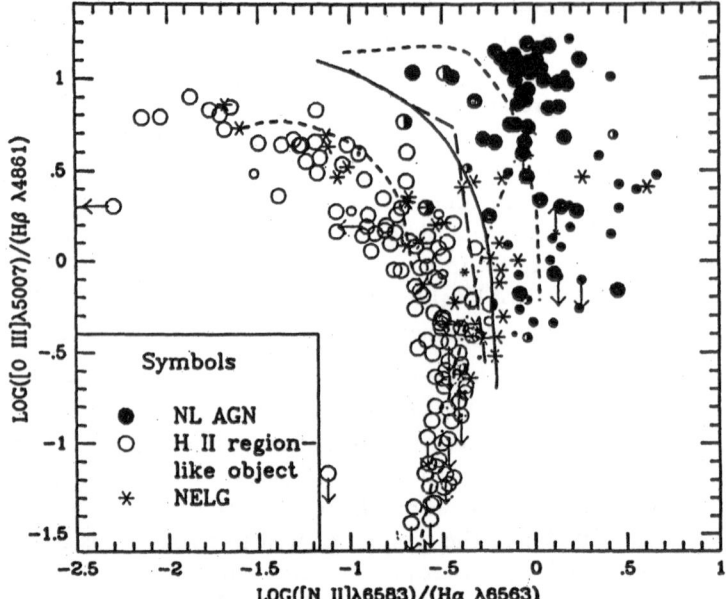

Fig. 39. A line ratio diagram for different emission line objects. The solid lines separate HII type objects from AGNs and the dashed lines are theoretical line ratios from several photoionization models (after Veilleux and Osterbrock, 1987)

ionizing photon. This can be used to separate LINERs from star-burst galaxies. Another useful diagram is $[OIII]\lambda5007/H\beta$ vs. $[OII]\lambda3727/[OIII]\lambda5007$. On this diagram the HII regions are well separated from planetary nebulae, and from object ionized by a power-law continuum, and all photoionized nebulae occupy a different region from the shock excited objects.

Classifying LINERs by way of diagnostic diagrams has been a subject of some debate. Evidence for both shock excitation and photoionization has been found and detailed theoretical models, including both processes, have been constructed. Some of the uncertainty is associated with the measurement itself. In particular, lines that are most useful shock-wave indicators, ($[OI]\lambda6300$, $[OIII]\lambda4363$, etc.) are weak in LINERs and confused with strong stellar absorption features. Below is a summary of this debate and an attempt to link LINERs to other classes of AGNs.

11.3.2 Shock-wave models for LINERs. The main motivation for considering shock-wave excitation, is the observation of $[OIII]\lambda4363$ in the spectrum of some LINERs. This line originates from the $^1S_0$ level of O^{+2}, at 5.3eV above the ground term. A strong $[OIII]\lambda4363$ indicates a very high kinetic temperature, $40,000\,K$ or so, which is inconsistent with photoionization, given the other observed lines. Shock-wave excitation is more consistent with such a temperature.

Shock-wave models applied to LINERs assume cloud motion through an interstellar medium, with velocities of the order of $200\ km\ s^{-1}$. The models are not entirely successful and several observed line ratios are not reproduced, despite the freedom exercised in choosing the velocity and density of the gas. The situation is not entirely clear since there are not enough constrains on such models. Future measurements of ultraviolet lines in LINERs will help to clarify the situation.

There are other uncertainties associated with $[OIII]\lambda4363$. First, the critical de-excitation density of the line is much higher than for $[OIII]\lambda5007$, and a large $[OIII]\lambda4363/[OIII]\lambda5007$ ratio does not necessarily mean high temperatures. High ($\geq 10^6 \ cm^{-3}$) densities could cause a similar effect. Second, measuring the line intensity is tricky because it falls in a spectral region full of stellar absorption features, and the recovery of its intrinsic intensity is uncertain. The intensity of other emission lines, such as $H\alpha$ and $H\beta$, are also affected by the stellar features.

11.3.3 Photoionization models for LINERs. Photoionization models for LINERs are motivated by the success of such models in explaining the spectrum of more luminous AGNs. The observational situation is, however, very different, since many of the important ultraviolet diagnostic lines have never been observed in LINERs. Another unknown is the shape of LINER ionizing continuum. Other uncertainties include the amount of dust and reddening, obscuration (like in Seyfert 2s) and the contamination of the spectrum by light from nuclear HII regions.

Present photoionization models for LINERs are simple and do not treat the density and covering factor in a very sophisticated way. A power-law ionizing continuum is assumed, with a rather steep slope, to account for the weak observed HeII and other high ionization lines. Despite their limitation, such models are quite successful in explaining many emission lines, and have already revealed an important relation between LINERs and other AGNs.

Fig. 40 shows one narrow line diagnostic diagram, $[OIII]\lambda5007/H\beta$ vs. $[OII]\lambda3727/[OIII]\lambda5007$, for a large number of AGNs. Shown also are calculated line ratios obtained in a simplified, single-cloud photoionization model, where the only variable is the ionization parameter. The density in these models is constant, $10^4 \ cm^{-3}$, and the ionizing spectrum as in Fig. 7. There are two representative abundances, "cosmic" (Table 2) and 0.1 cosmic. As evident from the diagram, all AGNs, from luminous quasars to the faint LINERs, form a continuous sequence of decreasing ionization parameter. The bright quasars and Seyfert 1s are at the top of the diagram, with $U \simeq 10^{-2}$, Seyfert 2 galaxies have somewhat smaller U, and LINERs are characterized by $U \simeq 10^{-3.5}$. This is a strong indication that LINERs are indeed active galactic nuclei. In this respect, the narrow emission lines are the most useful indicators of AGN activity.

There are several problems in LINERs photoionization models. The narrow $HeII\lambda4686$ line predicted by the models is stronger than observed in many LINERs (in the particular example discussed here $HeII\lambda4686/H\beta \simeq 0.09$), and other lines are only marginally consistent with the observations. More realistic models must therefore be computed. More diagnostic diagrams can help to clarify the situation. They should be compared with the theoretical calculations and a more detailed investigation of the parameter space (density,

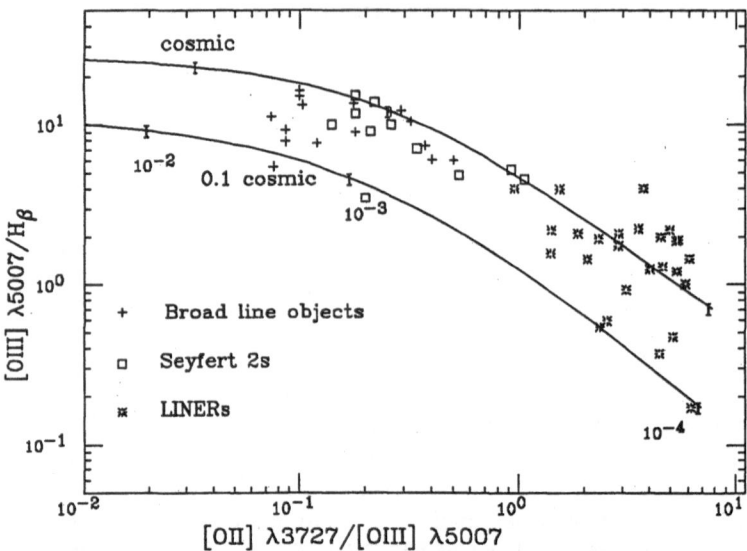

Fig. 40. A line ratio diagram for AGNs. Dereddened narrow line ratios of different types of objects are shown. The continuous curves are the results of two sets of photoionization calculations, with cosmic and 0.1 cosmic abundances. These are single cloud models, with a constant density of 10^4 cm^{-3} and variable ionization parameter, as marked. The ionizing spectrum is the standard continuum adopted in this work (Fig. 7). The narrow emission lines of all AGNs seem to form a continuous sequence of decreasing ionization parameter

covering factor etc.) must be perform. As for the comparison with shock-wave models, some line ratios are very promising in this respect, in particular lines of $[SII]$ and $[SIII]$ that provide a good distinction between the two suggested mechanisms. The few observations of these lines, already obtained, seem to agree better with photoionization model calculations.

11.3.4 Composite models for LINERs and Seyfert galaxies. There have been several attempts to combine shock-wave and photoionization excitation and to explain, in this way, the narrow line spectrum of LINERs and other AGNs. In this picture, the clouds are moving supersonically through the interstellar medium, being illuminated on one side by the central source. The shocked region can be on the illuminated (infall) or the back (outflow) face of the clouds, depending on the direction of motion. There are a large number of free parameters in the model, in particular the relative emission of the photoionized and the shocked gas must be specified. It means that the cloud thickness and the covering factor are inter-related. Such models can, by careful adjustment, produce almost all the observed emission lines. It is unfortunate that the observational limitations do not allow, at this stage, a serious test of this idea.

11.4 AGN Classification: QI, QII and QIII

Despite the large diversity, all AGNs seem to share some common properties. It is therefore useful to attempt to construct a detailed classification scheme, related to as many as possible observed characteristics. The main division I propose is to three luminosity classes: QI, QII and QIII. They overlap, in luminosity, with what is sometimes called Quasars, Miniquasars (Seyfert 1 and bright Seyfert 2 galaxies) and Microquasars (faint Seyferts and LINERs). The subdivision is into spectral types, using the intensity ratio of the broad and narrow component of the permitted lines. This follows a suggestion by Osterbrock and collaborators to assign a number between 1.0 and 2.0 to each Seyfert galaxy, according to its narrow/broad line ratio. Objects of type 1.0 do not show any narrow permitted lines and objects of type 2.0 have only narrow lines. The narrow hydrogen lines of many Seyfert 1 galaxies are 10-20% as intense as the broad lines. These have been classified as Seyfert 1.5 galaxies. As for NLXGs and LINERs that show faint broad $H\alpha$ wings, they were classified as Seyfert type 1.8 or 1.9. Thus a simple way to classify Seyfert galaxies, using the broad and narrow Balmer lines, is

$$spectral\ type\ = 1 + \Big[\frac{I(narrow\ line\ flux)}{I(total\ line\ flux)}\Big]^{0.4} . \qquad (104)$$

Other properties are self-explanatory and are summarized in Table 3.

Table 3. AGN classification

	QI	QII	QIII
Luminosity (X+ optical)	$10^{45-47}\ erg\ s^{-1}$	$10^{43-45}\ erg\ s^{-1}$	$10^{41-43}\ erg\ s^{-1}$
Spectral type	1.0 - 1.5	1.0 - 2.0	1.8 - 2.0
Line width $(km\ s^{-1})$	400 - 10,000	300 - 10,000	200 - 400
continuum variability	yes	yes	not observed
line variability	in some	in most	not observed
line or continuum reddening	?	in some	in most
$[OIII]\lambda5007/H\beta(narrow)$	6-12	6-12	1-2
$[OII]\lambda3727/[OIII]\lambda5007$	0.1-0.3	0.1-0.6	1-5
$[NII]\lambda6583/H\alpha(narrow)$	0.5-1	0.3-0.8	0.8-2
$[OI]\lambda6300/H\alpha(narrow)$	0.05-0.3	0.05-0.3	0.1-0.6
$[SII]\lambda6716+6731/H\alpha(narrow)$	0.3-0.7	0.3-0.7	0.5-1.5
EW($H\beta$)	~100Å	10 - 100 Å	1 -10 Å
Ionization parameter (BLR)	0.03 - 1	0.03 - 1	?
Ionization parameter (NLR)	$\sim 10^{-2}$	$10^{-2} - 10^{-3}$	$10^{-3} - 10^{-4}$

The table refers to *observed* properties, and some of the entries are still questionable. For example, the fraction of very bright objects (QI) showing any indication of reddening is not known, for reasons explained in chapter 7 (the entry in the table refers to the observed property of extinction and not to dust emission). Also, the intrinsic continuum luminosity of some QII objects (the Seyfert 2s and the NLXGs) may be much larger than indicated by their classification.

Some of the AGN properties seem to change, continuously, along the QI, QII, QIII sequence. The most obvious ones are:

The spectral type: The relative strength of the narrow emission lines increases along the sequence.

Reddening: Line and continuum reddening is more noticable in low luminosity AGNs.

The NLR ionization parameter: Accepting the idea of photoionization as the main excitation mechanism for all classes of AGNs, the narrow emission lines seem to form a single parameter sequence, with the ionization parameter decreasing from QI to QIII.

Variability: Much more common in QI, although the observations of the non-stellar continuum, and the broad lines in the QII and QIII classes leave much to be desired.

There are glaring omissions too, most notably the central mass and accretion rate. Adding these is a real challenge for future AGN studies.

Obviously, some of the above properties, and trends, are subject to ambiguity and selection effects. Nevertheless, the connection between the sub-classes starts to reveal itself, and we are perhaps justified in looking for a common physical mechanism in all AGNs.

11.5 Bibliography

Observations of narrow line AGNs: See chapter 2. The narrow line spectrum of Seyfert 1 galaxies is discussed by Cohen (1983).

NLXGs and dust: The relation between optical and X-ray properties, and the correlation with inclination and reddening, is discussed by Lawrence and Elvis (1982). More references can be found in a review article by Lawrence (1987).

Seyfert 2 galaxies: Polarization measurements: The first paper is by Antonucci and Miller (1985). More recent work is by Miller and Goodrich (1990). For models see Krolik and Begelman (1988).

LINERs: Useful review articles are Keel (1985) and Filippenko (1989). For basic observations see chapter 2. Evidence for broad $H\alpha$ and a large range of densities is given by Filippenko and Sargent (1985, 1988).

Diagnostic diagrams: See Baldwin et al. (1981), Veilleux and Osterbrock (1987) and Kirhakos and Phillips (1989). For a discussion on the use of $[SII]$ and $[SIII]$ lines see Diaz et al. (1985) and Kirhakos and Phillips (1989).

Shock wave models: See Shull and McKee (1979) and Binette et al. (1984).

Photoionization models: See Ferland and Netzer (1983), Halpern and Steiner (1983), Stasinska (1984, multi-component model) and Binnet (1985, abundances) for photoionization by a power-law continuum. See Pequignot (1984) for ionization by a blackbody source.

Composite models: The basic idea and most theoretical calculations are by Contini and Aldrovandi. See Viegas-Aldrovandi and Contini (1989) for references.

References

1. Alloin, D., Boisson, C., and Pelat, D., 1988, Astr.Ap., **200**, 17.
2. Antonucci, R. and Miller, J. 1985, Ap.J., **297**, 621.
3. Antonucci, R.R.J., Kinney, A.L., and Ford, H.C., 1989, Ap.J., **342**, 64.
4. Averett,E.H. and Loeser,R.,1988, Ap.J., **331**, 211.
5. Baldwin, J.A., 1975, Ap.J., **201**, 26.
6. Baldwin, J.A., 1977, Ap.J., **214**, 679.
7. Baldwin, J.A., and Netzer, H., 1978, Ap.J., **226**, 1.
8. Baldwin, J.A., Phillips, M.M., and Terlevich, R., 1981, PASP, **93**, 5.
9. Baldwin,J.A., Wampler,E.J. and Gaskell, G.M. 1989, Ap.J., **338**, 630.
10. Bahcall, J.N., Kozlovsky, B.Z., and Salpeter, E.E., 1972, Ap.J., **171**, 467.
11. Barvainis, R., 1987, Ap.J., **320**, 537.
12. Binette, L., 1985, Astr.Ap., **143**, 334.
13. Binette, L., Dopita, M.A., and Tuohy, I.R., 1984, Ap.J., **297**, 476.
14. Blandford, R.D., and McKee, C.F., 1982, Ap.J., **255**, 419.
15. Blumentha, G.R., and Mathews, W.G., 1975, Ap.J., **198**, 517.
16. Blumentha, G.R., and Mathews, W.G., 1979, Ap.J., **233**, 479.
17. Bonatto, C., Bica, E., and Alloin, D., 1989, Astr.Ap., **226**, 23.
18. Bradeley, S.E., and Puetter, R.C., 1986, Astr.Ap., **165**, 31.
19. Busko, I.C., and Steiner, J.E., 1989, MNRAS, **238**, 1479.
20. Capriotti, E.R., Foltz, C.B., and Byard, P., 1980, Ap.J., **241**, 903.
21. Capriotti, E.R., Foltz, C.B., and Byard, P., 1981, Ap.J., **245**, 396.
22. Carroll, T.J. 1985, MNRAS, **214**, 321.
23. Carroll, T.J., and Kwan, J. 1983, Ap.J., **274**, 479.
24. Carroll, T.J., and Kwan, J. 1985, Ap.J., **288**, 73.
25. Chen,K.,Halpern,J.P.,and Filippenko,A.V.,1989, Ap.J., **339**, 742.
26. Clavel, J., and Wamsteker, W., 1987, Ap.J.Lett., **320**, L9.
27. Clavel, J., Wamsteker, W., and Glass, I., 1989 Ap.J., **337**, 236.
28. Clavel et al. 1990, Ap.J., (in press)
29. Cohen, R.D., 1983, Ap.J., **273**, 489.
30. Collin-Souffrin,S.,1986, Astr.Ap., **166**, 115.
31. Collin-Souffrin,S. ,Dyson,J.E., McDowell,J.C, and Perry,J.J., 1988, MNRAS, **232**, 539.
32. Collin-Souffrin, S., and Dumont, A.M., 1986, Astr.Ap., **166**, 13.
33. Collin-Souffrin, S., and Dumont, A.M., 1989, Astr.Ap., **213**, 29.
34. Collin-Souffrin, S., and Dumont, A.M., 1990, Astr.Ap., **229**, 292.
35. Collin-Souffrin, S., Hameury, J.M. and Joly, M., 1988, Astr.Ap., **205**, 19.
36. Corbin, M.R., 1990, Ap.J., **357**, 346.
37. Cota, S.A., and Ferland, G.J., 1988, Ap.J., **326**, 889.
38. Czerny, B. and Elvis, M., 1987, Ap.J., **312**, 325.
39. Crenshaw, D.M., 1986, Ap.J.Suppl., **62**, 821.
40. Davidson, K., 1972, Ap.J., **171**, 213.
41. Davidson, K.,and Netzer, H. 1979, Rev.Mod.Phys., **51**, 715.
42. De Robertis, M., 1985, Ap.J., **289**, 67.
43. De Robertis, M., and Osterbrock, D.E., 1984, Ap.J., **286**, 171.
44. De Robertis, M., and Osterbrock, D.E., 1986, Ap.J., **301**, 727.
45. De Zotti, G., and Gaskell, C.M., 1985, Astr.Ap., **147**, 1.
46. Diaz, A.I., Pagel, B.E.J., and Wilson, R.G., 1985, MNRAS, **212**, 737.
47. Dumont, A.M., and Collin-Souffrin, S., 1990, Astr.Ap., **229**, 302.
48. Eastman, R.G., and MacAlpine, G.M., 1985, Ap.J., **299**, 785.
49. Edelson, R.A., and Krolik, J.H., 1989, Ap.J., **333**, 646.
50. Elitzur, M., 1984, Ap.J., **280**, 653.
51. Elitzur, M., and Ferland, G.J., 1986, Ap.J., **309**, 35.
52. Elitzur, M., and Netzer, H., 1984, Ap.J., **291**, 464.
53. Espey, B.R, Carswell, R.F., Bailey, .A., Smith, M.G., and Ward, M.J, 1989, Ap.J., **342**, 666.
54. Fabian, A.C., Guilbert, P.W., Arnud, K., Shafer, R.A., Tennant, A.F. and Ward, M.J., 1986, MNRAS, **218**, 457.
55. Ferland, G.J., and Mushotzky, R.F., 1984, Ap.J., **286**, 42.
56. Ferland, G.J., and Netzer, H., 1979, Ap.J., **229**, 274.

57. Ferland, G.J., and Osterbrock, D.E., 1986, Ap.J., **300**, 658.
58. Ferland, G.J., and Rees, M.J., 1988, Ap.J., **332**, 141.
59. Ferland, G.J., and Persson, S.E., 1989, Ap.J., **347**, 656.
60. Ferland ,G.J., and Shields, G.A., 1985 in *Astrophysics of Active Galaxies and Quasi-Stellar Objects* (J.Miller ed. p.157)
61. Ferland, G.J., Netzer, H., and Shields, G.A., 1979, Ap.J., **232**, 382.
62. Filippenko, A.V., in *Active Galactic Nuclei*, (IAU Symp. 134) p.495.
63. Filippenko, A.V. and Halpern, J.P. 1984, Ap.J., **285**, 485.
64. Fillipenko, A.V., and Sargent, W.L.W., 1985, Ap.J.Suppl., **57**, 503.
65. Fillipenko, A.V., and Sargent, W.L.W., 1988, Ap.J., **324**, 134.
66. Gaskell, M., 1982, Ap.J., **263**, 79.
67. Gaskell, M.,and Peterson, B.M., 1987, Ap.J.Suppl., **65**, 1.
68. Gaskell,M.,and Sparke,1986,Ap.J., **305**, 175.
69. Gaskell, C.M., Shields, G.A., and Wampler, E.J., 1981, Ap.J., **249**, 443.
70. Gondhalekar, P.M.,1990, MNRAS, **243**, 443.
71. Goodrich, R.W., 1989, Ap.J., **340**, 190.
72. Goodrich, R.W., 1990, Ap.J., **355**, 88.
73. Halpern, J.P., and Filippenko, 1988, Nature, **331**, 46.
74. Halpern, J.P. and Steiner, J., 1983, Ap.J., **269**, 137.
75. Hubbard, E.M.,and Puetter, R.C., 1985, Ap.J., **290**, 394.
76. Hummer, D.G., 1968, MNRAS, **138**, 73.
77. Hummer, D.G., and Kunasz, P.B., 1980, Ap.J., **236**, 609.
78. Joly, M.,1987a, Astr.Ap., **184**, 33.
79. Joly, M., 1987b, in *Emission Lines in Active Galactic Nuclei*, (P.Gondhalekar ed. p160)
80. Joly, M., Collin-Souffrin, S., Masnou, J.L., and Nottale, L, 1985, Astr.Ap., **152**, 282.
81. Kalkofen, W. (editor)1984, *Methods in Radiative Transfer* (Cambridge University Press).
82. Kallman, T. and Krolik, J., 1986, Ap.J., **308**, 805.
83. Kallman, T.,and McCray, R. 1982, Ap.J., suppl., **50**, 263.
84. Kawara, K., Nishida, M., and Gregory, B., 1990, Ap.J., **352**, 433.
85. Kazanas, D., 1989, Ap.J., **347**, 74.
86. Keel, W.C., 1983, Ap.J., **269**, 466.
87. Keel, W.C., 1985 in *Astrophysics of Active Galaxies and Quasi-Stellar Objects* (J.Miller ed. p.1)
88. Kinney, A.L., Rivolo, A.R., and Koratrar, A.P., 1990, Ap.J., **357**, 345.
89. Kirhakos, S., and Phillips, M.M., 1989, PASP, **101**, 949.
90. Korista, K.T., and Ferland, G.J., 1989, Ap.J., **343**, 678.
91. Koski, 1978, Ap.J., **223**, 56.
92. Krolik, J.H., 1988, Ap.J., **325**, 148.
93. Krolik, J,H., and Begelman, M.C., 1988, Ap.J., **329**, 702.
94. Krolik, J., and Kallman, T.A., 1988, Ap.J., **324**, 714.
95. Krolik, J., McKee, C.M. and Tarter, C.B., 1981, Ap.J., **249**, 422.
96. Kwan, J., 1984, Ap.J., **283**, 70.
97. Kwan, J., and Krolik, J.H., 1981, Ap.J., **250**, 478.
98. Kwan, J., and Carroll, T.J., 1982, Ap.J., **261**, 25.
99. Laor, A., 1990, MNRAS, (in press)
100. Laor, A., and Netzer, H., 1989, MNRAS, **238**, 897.
101. Laor, A., Netzer, H., and Piran, T., 1990, MNRAS, **242**, 560.
102. Lawrence, A., and Elvis, M., 1982, Ap.J., **256**, 410.
103. Lawrence, A., 1987, PASP, **99**, 30.
104. Lockett, P., and Elitzur, M., 1989, Ap.J., **344**, 525.
105. Maoz, D., and Netzer, H., 1989, MNRAS, **236**, 21.
106. Maoz, D., Netzer, H., Mazeh, T., Beck, S., Almoznino, E., Leibowitz, E., Brosch, N., Mendelson, H., and Laor, A., 1991, Ap.J., (in press)
107. MacAlpine, G.M., 1972, Ap.J., **175**, 11.
108. MacAlpine, G.M., 1985 in *Astrophysics of Active Galaxies and Quasi-Stellar Objects*, (J.Miller ed. p.259)
109. MacAlpine, G.M., Davidson, K., Gull, T.R., and Wu, C.C., 1985, Ap.J., **294**, 147.
110. Martin, P.G., and Ferland, G.J., 1980, Ap.J.Lett., **235**, L125.
111. Mathews, W.G., 1974, Ap.J., **183**, 23.
112. Mathews, W.G., 1982a, Ap.J., **252**, 39.

113. Mathews, W.G., 1982b, Ap.J., **258**, 425.
114. Mathews, W.G., and Wampler, J., 1985, PASP, **97**, 966.
115. Mathews, W.G., and Doane, J.S., 1990, Ap.J., **352**, 423.
116. Mathews, W.G., and Ferland, G.J., 1987, Ap.J., **323**, 456.
117. Mathews, W.G., and Capriotti, E.R., 1985, in *Astrophysics of Active Galaxies and Quas-Stellar Objects* (J.Miller ed. p.185)
118. Mihalas, D., 1978, *Stellar Atmospheres*, (San Francisco: Freeman)
119. Miley, G.K., and Miller, J.S., 1979, Ap.J.Lett., **228**, L55.
120. Miller, J.S., and Goodrich, B.F., 1990, Ap.J., **355**, 456.
121. Morris, S.L., and Ward, M.J., 1988, MNRAS, **230**, 639.
122. Murdoch, H.S., 1983, MNRAS, **202**, 987.
123. Mushotzky, R.F., and Ferland G,J., 1984, Ap.J., **278**, 558.
124. Netzer, H., 1985a, Ap.J., **289**, 451.
125. Netzer, H., 1985b, MNRAS, **216**, 63.
126. Netzer, H., 1987, MNRAS, **225**, 55.
127. Netzer, H., 1989, Comments on Astrophysics, **14**, 137.
128. Netzer, H., and Davidson, K. 1979, MNRAS, **187**, 871.
129. Netzer, H., Elitzur, M., and Ferland, G.J., 1985, Ap.J., **299**, 752.
130. Netzer, H., and Ferland, G.J., 1984, PASP, **96**, 593.
131. Osmer, P.S., 1980, Ap.J.Suppl., **42**, 523.
132. Osmer, P.S., and Smith, M.G., 1980, Ap.J.Suppl., **42**, 333.
133. Osterbrock, D,E., 1977, Ap.J., **215**, 733.
134. Osterbrock, D.E., 1984, Q.J.R.Asrt.Soc., **25**, 1.
135. Osterbrock, D.E., 1989, *Astrophysics of Gaseous Nebulae and Active Galactic Nuclei* (University Science Books)
136. Osterbrock,D.E., Koski, A.T., and Phillips, M.M, 1976, Ap.J., **206**, 898.
137. Osterbrock, D.E., and Mathews, W.G., 1986, Ann.Rev.Ast.Ap., **24**, 171.
138. Osterbrock, D.E., Shaw, R.A., and Veilleux, S., 1990, Ap.J., **352**, 561.
139. Osterbrock, D.E., and Shuder, J.M., 1982, Ap.J.Suppl., **49**, 149.
140. Padovani, P., Burg, R., and Edelson, R.A., 1990, Ap.J., **353**, 438.
141. Penston, M.V., 1987, MNRAS, **229**, 1p.
142. Penston, M.V., 1988, MNRAS, **233**, 601.
143. Penston, M.V., and Perez, E., 1984, MNRAS, **211**, 33p.
144. Penston, M.V., Croft, S., Basu, D., and Fuller, N., 1990, MNRAS, **244**, 357.
145. Perez, E., Penston, M.V., Tadhunter, C., Mediaville, E., and Moles, M., 1988, MNRAS, **230**, 353.
146. Perez, E., Penston, M.V., and Moles, M., 1989, MNRAS, **239**, 75.
147. Perez, E., Penston, M.V., Tadhunter, C., Mediavilla, E., and Moles, M, 1987, MNRAS, **230**, 353.
148. Perry, J.J., and Dyson, J.E., 1985, MNRAS, **213**, 665.
149. Persson, S.E., 1988, Ap.J., **330**, 751.
150. Peterson, B.M., 1988, PASP, **100**, 18.
151. Peterson, B.M., et al., 1991, Ap.J., in press.
152. Puetter, R.C., 1981, Ap.J., **251**, 446.
153. Puetter, R.C., and Hubbard, E.N., 1985, Ap.J., **295**, 394.
154. Puetter, R.C., and Hubbard, E.N., 1987, Ap.J., **320**, 85.
155. Rees, M., 1987, MNRAS, **228**, 47p.
156. Rees, M., Netzer,H. and Ferland, 1989, Ap.J., **347**, 640.
157. Reichert, G.A., Mushotzky, R.F., Petre, R., and Holt, S.S., 1985, Ap.J., **296**, 69.
158. Robinson, A., and Perez, E., 1990, MNRAS, **244**, 138.
159. Rudy, R.J., and Schmidt, G.D., 1988, Ap.J., **331**, 325.
160. Sanders, D.B., Phinney, E.S., Neugebauer, G., Soifer, B.T., and Matthews, K., 1989, Ap.J., **347**, 29.
161. Sargent, W.L.W., Steidel, C.C., and Boksenberg, A., 1988, Ap.J.Suppl., **69**, 703.
162. Schneider, D.P., Schmidt, M., and Gunn, J.E., 1989, A.J., **98**, 1507.
163. Scoville, N., and Norman, C., 1988, Ap.J., **332**, 163.
164. Shull, M.J.M, and McKee, C.E., 1979, Ap.J., **227**, 131.
165. Shull, J.M,, and Van Steenberg, M.E., 1985, Ap.J., **298**, 268.
166. Shields, G.A., 1976, Ap.J., **204**, 330.
167. Shields, G.A., 1978, Nature, **272**, 706.
168. Shields, G.A., 1989, in *Active Galactic Nuclei*, (IAU Symp. 134) p.577.

169. Shlosman, I., Vitello, P.A., and Shaviv, G., 1985, Ap.J., **294**, 96.
170. Smith, M.G., Carswell, R.F., Whelan, J.A.J., Wilkes, B.J., Boksenberg, A., Clowes, R.G., Savage, A., Cannon, R.D., and Wall, J.V., 1981, MNRAS, **195**, 437.
171. Smith, M.D., and Raine, D.J., 1985, MNRAS, **212**, 425.
172. Stasinska, G., 1984, Astr.Ap., **135**, 341.
173. Stirpe, G.M., van Groningen, E., and de Bruyn, A.G., 1989, Astr.Ap., **211** , 31.
174. Sulentic, J.W., 1989, Ap.J., **343**, 54.
175. Sun, W.H., and Malkan, M.A., 1989, Ap.J., **346**, 68.
176. Terlevich, R., and Melnick, J., 1988, Nature, **333**, 239.
177. Terlevich,R., 1989, in *Evolutionary Phenomena in Galaxies* (Beckman and Pagel ed. Cambridge Univ. Press)
178. Uomoto, A., 1984, Ap.J., **284**, 497.
179. Veilleux, S., and Osterbrock, D.E., 1987, Ap.J.Suppl., **63**, 295.
180. Viegas-Aldrovandi, S.M., and Gruenvald, R.B., 1988, Ap.J., **324**, 683.
181. Viegas-Aldrovandi, S.M., and Contini, M., 1989, Ap.J., **339**, 689.
182. Voit, G.M., and Shull, J.M., 1988, Ap.J., **331**, 197.
183. Wamsteker, W., and Colina, L., 1986, Ap.J., **311**, 617.
184. Wamsteker, W., Roriguez-Pascual, P, Wills, B.J., Netzer, H., Wills, D., Gilmozzi, R., Barylak, M., Talavera, A., Maoz, D., Barr, P., and Heck, A., 1990, Ap.J., **354**, 446.
185. Wampler, E.J., Gaskell, C.M., Burk, W.L. and Baldwin, J.A., 1984, Ap.J., **276**, 403.
186. Wandel, A.,and Yahil, A., 1985, Ap.J., **295**, L1.
187. Wandel, A., and Mushotzky, R.F., 1986, Ap.J.Lett., **306**, L61.
188. Ward, M.J., Geller, T., Smith, M., Wade, R., and Williams, P., 1987, Ap.J., **316**, 138.
189. Whittle, M. 1985a, MNRAS, **213**, 1.
190. Whittle, M. 1985b, MNRAS, **216**, 817.
191. Wilkes, B.J., 1986, MNRAS, **218**, 331.
192. Wilkes, B.J. and Carswell, R.F., 1982, MNRAS, **201**, 645.
193. Weymann, R.J.,and Williams, R.E. 1969, Ap.J., **157**, 1201.
194. Wills, B.J., Netzer, H., and Wills, D., 1985, Ap.J., **288**, 94.
195. Wills, B.J., and Browne, I.W., 1986, Ap.J., **302**, 56.
196. Wu, C.C., Boggess,A., and Gull, T.R., 1983, Ap.J., **266**, 28.
197. Zheng, W., 1988, Ap.J., **333**, 188.
198. Zheng, W., and O'Brien, P.T., 1990, Ap.J., **356**, 463.

List of Symbols

$A_c(r)$	Cloud cross section
A_{ji}	Radiative transition rate
a	Damping constant
B_{ij}	Absorption coefficient
b_i	Departure coefficient for level i
C_{ij}	Collisional excitation rate
$C(r)$	Cumulative covering factor (up to distance r)
c	Velocity of light
c_s	Sound speed
$E_l(r)$	Cumulative line flux (up to distance r)
E_λ	Line profile
$E(t)$	Emission line light curve
E_{ij}	Energy separation of levels ij
EW	Equivalent width
F_ν	Monochromatic continuum flux
f_d	Drag force
f_{ij}	Oscillator strength

G	Gravitational constant
g_{ff}	Gaunt factor for free-free transition
g_G	Gravitational acceleration
g_i	Statistical weight of level i
g_{rad}	Radiative acceleration
h	Planck's constant
I_ν	Radiation intensity
J	Mean radiation intensity
$j_c(r)$	Total cloud emission
k	Boltzmann constant
$k(\tau)$	Number of scattering factor
$k'(\tau)$	Path length increase factor
L	Integrated continuum luminosity
L_{46}	Luminosity in $10^{46}\ erg\ s^{-1}$
L_ν	Monochromatic continuum luminosity
$L(t)$	Continuum light curve
L_{Edd}	Eddington luminosity
M_c	Mass of a cloud
M_{BH}	Mass of the central black hole
M_9	$M_{\mathrm{BH}}/10^9 M_\odot$
$\dot{M}$	Accretion rate
$\dot{m}$	Accretion rate in units of L/L_{Edd}
m_e	Electron mass
m_p	Proton mass
m_λ	Magnitude of extinction
N	Particle number density
N_{10}	$N/10^{10}\ cm^{-3}$
N_e	Electron density
N_{HIM}	Density of the hot inter-cloud medium
N_{col}	Column density
$n_c(r)$	Number of clouds per unit volume at distance r
n_i	Population of level i
P	Pressure
$Q(\tau)$	Number of scattering before escape
R_c	Radius of a cloud
r	Distance of a cloud
r_{av}	Emissivity-weighted average radius
r_{in}	Inner radius of the emission line region
r_{out}	Outer radius of the emission line region
S	Source function
T_C	Compton temperature
T_e	Electron temperature
T_{ex}	Excitation temperature
T_{HIM}	Temperature of the hot inter-cloud medium
T_{rad}	Radiation temperature
t_{dyn}	Dynamical time

t_{rec}	Recombination time
t_{sc}	Sound-crossing time
U	Ionization parameter
V_c	Volume of a cloud
$v(r)$	Cloud velocity at a distance r
v_{Doppler}	Doppler velocity
v_{3000}	$v/3000\ km\ s^{-1}$
X_c	Relative continuum opacity
X_l	Relative line opacity
Z	Nuclear charge
z	Redshift
α	Recombination coefficient
α_T	Temperature averaged recombination coefficient
β	Escape probability
β_{eff}	Effective escape probability
γ	Continuum spectral index
$\epsilon_l(r)$	Line emission $(erg\ s^{-1}\ cm^{-2})$
ϵ_ν	Line emission coefficient
η_ν	Continuum occupation number
κ_ν	Line absorption coefficient
κ_c	Continuum absorption cross section at line center
κ_l	Line absorption cross section at line center
λ_{ij}	Wavelength of the ij transition
λ_0	Line center wavelength
μ	$\cos\theta$ in a spherical coordinate system
ν	Frequency
ν_0	Threshold ionization frequency
$\bar\nu$	Mean photon frequency or mean photon energy
ν_{cut}	Cut-off frequency
Ξ	(radiation pressure)/(gas pressure) $\propto U/T$
σ_ν	Continuum absorption cross section
σ_c	Effective Compton cooling cross section
σ_{es}	Electron scattering cross section
σ_h	Effective Compton heating cross section
τ	Optical depth
τ_{in}	Optical depth to the inner (illuminated) face of the cloud
τ_{out}	Optical depth to the outer face of the cloud
Φ_ν	Normalized line profile
$\Psi(t)$	Transfer function
Ω_{ij}	Effective collision strength
ω	$1/t$

Physical Processes in Active Galactic Nuclei

R.D. Blandford

With 44 Figures

1 Introduction

1.1 Overview of these Lectures

Active Galactic Nuclei (henceforth AGN) have been observed assiduously throughout the electromagnetic spectrum over the past 30 years and we know much about their collective and individual properties. Unfortunately, we are still a long way from being sure how they work.

My primary intention, in the lectures that follow, is to explain some of the most important mechanisms that are believed to operate in these objects to an audience with a graduate physics and general astronomical background, but little prior exposure to AGN. In this way, I hope to provide a "toolkit" useful for understanding the abundance of models that have already been proposed to interpret the observations that will be reviewed more systematically here by Drs. Netzer and Woltjer. I shall make no attempt to summarize the phenomenology of AGN; neither will I give a critique of the models, although some personal preferences will clearly emerge, especially in the final lecture in which I outline one possible unified interpretation of AGN. I shall, however, draw upon both observations and models selectively to illustrate these physical processes.

In the interests of brevity, I can only give heuristic derivations of most of the physical processes that I shall discuss. However, when possible, I will give more accurate formulae and references to their derivation.

1.2 The Basic Model

1.2.1 The scope of nuclear activity. There are many possible ways to organise a discussion of the properties of AGN. If we consider the totality of phenomena observed in all types of AGN, then we can imagine plotting them on a diagram of emitted frequency versus characteristic size. (See Fig.(1.1).) AGN are observed from $\lesssim 100$MHz radio frequencies, for example as low frequency radio sources, to $\gtrsim 100$MeV$\sim 2 \times 10^{22}$Hz gamma ray sources. The activity is made manifest directly on scales as large as the sizes of the giant radio sources $\lesssim 6$Mpc$\sim 2 \times 10^{25}$cm, and as small as the distance travelled by light in the shortest observed X-ray variability times $\sim 2 \times 10^{12}$cm. (Even larger scales are present if we include the effect of ionising radiation on the intergalactic medium.) Ignoring, for the moment, the actual shapes of the emitting regions,

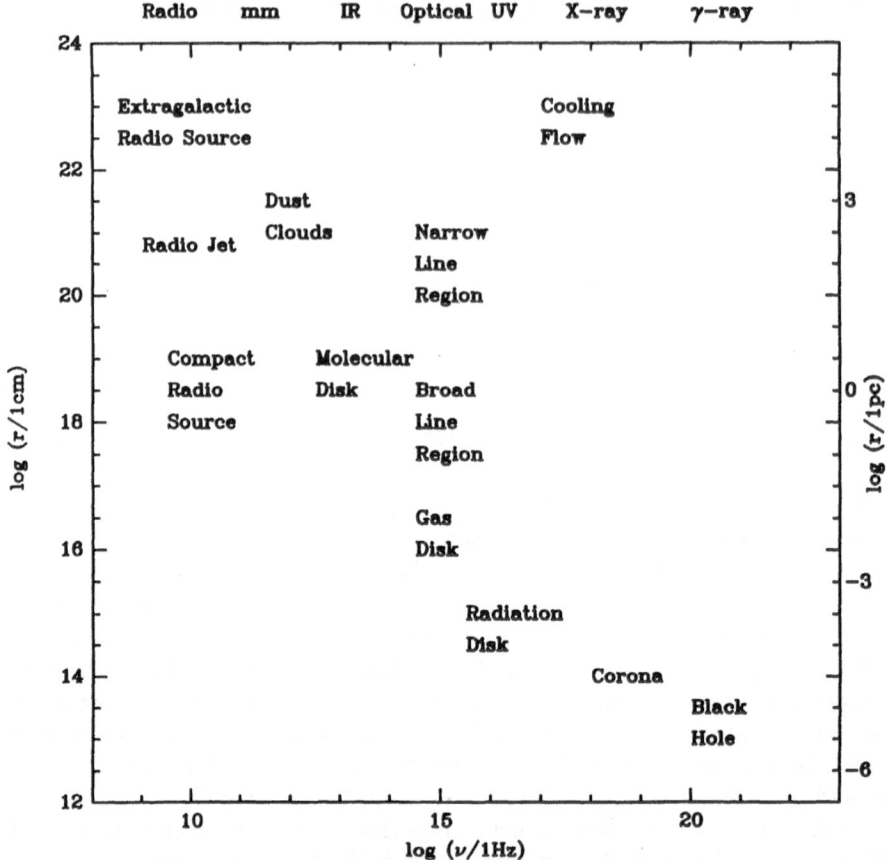

Fig. 1.1. Schematic illustration of the range of activity associated with the nuclei of galaxies. Phenomena have been observed spanning the full dynamic range of ~ 45 octaves of length scale from variable X-ray emission in Seyferts ($\lesssim 10^{12}$cm) to giant double radio sources ($\gtrsim 10^{25}$cm) and from metre wavelengths (low frequency variability) to 100MeV observations of some quasars and Seyfert galaxies (45 octaves of frequency). Some of these phenomena, *e.g.* radio jets, are observed directly. Others, *e.g.* accretion disk coronae and black holes are inferred indirectly and are of more questionable reality.

we can treat the size as essentially a radial distance from the centre of the AGN and the different features as being roughly nested with the UV ionising source lying within the optical continuum region which in turn lies inside the volume occupied by the emission line clouds and the compact radio sources and so on. (There is a crude analogy with sub-atomic physics where we see optical radiation emerging from scales $\sim$Å, X-rays from $\sim 10^{-2}$Å and γ-rays from nuclear scales $\sim 10^{-5}$Å. Indeed, the history of research into AGN has paralleled that in subatomic physics as smaller scales have become accessible with higher energy observations.)

There is really a third dimension to this diagram, power radiated in an octave of frequency over an octave of radius. This can range from $\sim 10^{39}$erg s^{-1}, the radio power of our galactic center, to $\sim 10^{48}$erg s^{-1}, the inferred UV power of the most powerful, high redshift quasars. (As we shall see, this may be a bit of an over-estimate as the power may not be emitted isotropically.) That AGN emit significant powers over such a broad frequency and radius range shows that they are far out of thermal equilibrium. This disequilibrium takes two forms. Firstly, smaller regions are typically hotter, just like in the interior of a star. However, unlike stars, the interiors of AGN are largely visible from the outside. Secondly, components of quite different temperature can co-exist, just like in the solar corona where photospheric radiation, $\sim 10^6$K gas, X-rays and relativistic particles are all found. Here, the difference is that, in an AGN, the powers associated with the different components are often within one or two orders of magnitude of each other.

Of course, different manifestations of nuclear activity are found in different objects; and very few objects are like 3C273 which is observed to exhibit essentially all of these features. However, as observations have become more sensitive and the full electromagnetic spectrum has been unveiled, "radio-silent" objects have become "radio-quiet", X-ray fluxes have been extensively measured and so on. The upcoming generation of ground- and space-based telescopes will no doubt continue this trend.

Another characteristic of the scope of nuclear activity is that the majority of *normal* galaxies appear to exhibit it at some level, usually through a broad emission line component. Furthermore, between ~ 1 and ~ 100 per cent of galaxies *were* quasars and an even larger fraction exhibited weaker activity. (Quasars are comparatively rare at present.)

1.2.2 Taxonomy. As will be apparent from the lectures of Dr. Netzer and Dr. Woltjer, the taxonomy of AGN is a very confused (and confusing) subject [1]. Part of the problem is observational because it is not possible to obtain full spectral coverage in all objects and it is therefore not easy to reconcile a classification based, say, on optical emission lines with one based on the X-ray properties of a quite different sample. Other difficulties are self-inflicted. Astronomers have been guilty in the past of spectral chauvinism. They have also been reluctant to discard historical baggage; the very phrase "quasi-stellar object" being a good example. However, the major impediment to serious taxonomy is more fundamental. AGN are quite heterogeneous objects, especially in their directly observed properties. It is one of the challenges of contemporary research to infer the underlying physical structure of AGN in the belief that this is simple, despite the confusing character of the observations.

1.2.3 Radiation Processes. Our most direct diagnostics of the conditions within AGN come from the radiation that we observe. This is produced by a wide variety of mechanisms, wider than encountered anywhere else in astronomy, almost certainly including high energy gamma ray processes, atmospheric re-processing, dust emission and synchrotron radiation. It is important to understand these processes very well in order to be able to infer when they operate and what can be deduced from the observed spectra, polarisation and variability.

1.2.4 Black holes. Physicists and astronomers reacted to the discovery of quasars in 1963 with a wide variety of suggestions ranging from the fairly prosaic (*e.g.*supernovae) [2], to the radical *e.g.*quarks [3], white holes [4], and a C-field [5]). Included among these early ideas was the prescient suggestion, developed notably by Zeldovich and Novikov [6], and Salpeter [7], that a quasar "prime mover" be a black hole. With the passage of nearly three decades, the arguments in favour of black holes (especially their efficiency, inevitability and stability) have only strengthened. In addition, various observational discoveries (central stellar cusps, rapid X-ray variability and small scale radio jets) are increasingly hard to interpret in terms of alternative models. Black holes have, in this manner, become the explanation of choice for the majority of researchers.

However, there is still no proof (even by the lax standards of astronomy) that massive black holes are present in AGN. Indeed, the evidence is notably weaker than that for stellar mass black holes in X-ray binaries, and will probably remain so until we have some direct, dynamical corroboration of their presence in the centres of nearby galaxies. Despite this lack of confirmation, I shall adopt the black hole interpretation uncritically in what follows.

1.2.5 Accretion disks. Accretion disks were initially a theoretical construct. They were hypothesised to form when gas, endowed with angular momentum [8], is accreted onto a central gravitating object. Accretion disks are believed to be present (with decreasing conviction) in white dwarf binaries, neutron star binaries, protostars and AGN [9].

Other modes of accretion onto a black hole are possible. In particular, spherical accretion has been widely discussed. In these lectures, I take the position that this is unlikely to be of widespread importance because it is hard to see how the gas can have so little angular momentum to make this a good approximation and, besides, there is abundant observational evidence for the breaking of spherical symmetry. Nevertheless many of the physical processes that I shall discuss are directly relevant to these alternative modes of accretion.

Disks are envisaged to play many roles in AGN. They are identified with the channels through which matter moves radially inward, and with the source of the continuum emission, which we observe directly, and which maintains the ionisation in the emission line clouds and the surrounding intergalactic medium. They are perhaps also the launch sites for bipolar winds and jets. Finally they are the sources of mass and spin responsible for building up the central black hole, whose gravitational field shapes the arena in which most of this activity takes place. In a sense, it is the disk, rather than the black hole, which is the prime mover of an AGN.

A very broad range of physics, including aspects of general relativity, gas dynamics, radiative transfer, magnetohydrodynamics and plasma astrophysics, must be mastered in order to begin to describe accretion disks. I shall discuss some of the physical considerations which enter into a theory of accretion disks, emphasizing the difficulties that underlie such a task. I shall try to keep this discussion focussed by relating to observations and models of AGN. However, I shall not maintain the pretense that there is one true model; such a course would totally misrepresent the diversity of interpretations of AGN and, almost surely, the diversity of real galactic nuclei.

1.2.6 Jets. I shall also discuss the origin, propagation and radiative properties of jets. I do this for two distinct reasons. Firstly, in many sources, the flux observed from jets, and the radio lobes that they feed, can dominate the direct emission from the disk. Although jets have been mainly observed at radio wavelengths, it seems likely that much of the non-thermal continuum seen at higher frequencies may be produced by jets in many AGN. This is largely due to *relativistic beaming*. Gas flowing at mildly relativistic speed will direct its emission preferentially along its direction of motion. Those sources, in whose beams we are located, will appear anomalously bright. They can also exhibit *superluminal motion* wherein small inhomogeneities moving with the jet appear to move across the plane of the sky faster than the speed of light.

Secondly, it is, by now, pretty clear that jets cannot be considered in isolation from disks; they appear to be a natural by product of the formation of a disk and may be the means by which the gas in the accretion disk loses angular momentum so that it can continue to accrete.

1.2.7 Fuelling of AGN. Consideration of accretion disks naturally leads to a discussion of the origin of the gas that flows through it. This is an issue that is directly accessible to observation as mm, infra-red and optical telescopes can all resolve structure associated with nearby AGN. It is clear that the galactic neighbourhood of a particular galaxy has a major bearing upon its nuclear activity, though the way in which this happens is not simple.

The interstellar medium of the host galaxy, as well as perhaps providing the gaseous fuel, can also be responsible for obscuring or reprocessing the central source. Much of the infra-red continuum may be UV radiation reprocessed by dust. In addition, if disks extend out to large radii, then they can be responsible for equatorial obscuration of the central continuum source. This is, in addition to relativistic beaming, a second way in which the orientation of the observer can be crucially important for our classification of a particular AGN.

1.2.8 Grand unification. A major thrust of current research is to try to understand the relationships between different manifestations of nuclear activity, quasars, radio galaxies, Seyfert galaxies and so on. It is also becoming irresistable to try to interpret the cosmological evolution of AGN, which is increasingly well determined, in these terms. In the spirit of demonstrating how to apply the preceding physical processes, I shall outline a particular unified theory, one out of several possibilities, in the final chapter.

1.2.9 Liners, starbursts and "normal" galaxies. As has already been mentioned and is discussed here in detail by Drs. Netzer and Woltjer, most galaxies exhibit some form of nuclear activity. Much of this activity may be unrelated to a central black hole and accretion disk, even if one is present. Liners (Low Ionisation Nuclear Emission Line Regions) [10], and starburst galaxies [11], are two prominent classes of such galaxies whose properties I shall mostly ignore.

1.2.10 Unbelievers. It should, in fairness, be remarked at this stage, that a significant minority of astronomers genuinely doubts the existence of massive black holes and accretion disks all together. Some doubt that quasar redshifts are cosmological [12]. I do not know how to help them because they also deny the conventional laws of physics. Others have accepted conventional estimates of the powers and sizes but have explored models featuring massive uncollapsed stars [13], spinars [14], clusters of stellar mass black holes or neutron stars [15], and "warmers" [16]. These models may be appropriate for some types of AGN; at the the very least by exploring alternative hypotheses, they are giving some measure of the overall plausibility of the black hole – accretion disk interpretation. Many of the physical processes that I shall discuss in the context of black holes and accretion disks are relevant to these alternative models.

1.3 References

1.3.1 General texts. Fortunately, many aspects of the physics of the black hole-accretion disk model of AGN have already been summarized in a few good textbooks and review articles. The most useful general references include Zeldovich and Novikov(1971) [17], Begelman, Blandford and Rees (1984) [18], Rees (1984) [19], Frank, King and Raine (1985) [20], Weedman (1986) [21], Osterbrock (1989) [22], and the conference proceedings [23], [24], [25], [26], [27], [28]. Additional, more specialized discussions will be cited below.

1.3.2 Bibliographic disclaimers. This is not the place to attempt to recount the tortuous route along which we have arrived at our present, incomplete understanding of AGN. As such, I shall not credit the originators of many of the discoveries that I shall describe. Furthermore, I will have to be highly selective and somewhat arbitrary in my choice of contemporary references, generally giving preference to those sources with which I happen to be more familiar and those which give a faster entry into the literature.

2 Radiation Processes

2.1 General Considerations

In this section, I will present short, heuristic discussions of the most commonly encountered radiation processes at a level sufficient for application to AGN. References to more complete treatments will be given where possible. In keeping with astronomical custom, I shall use a mix of Gaussian and astronomical units. The best general references for astrophysical radiation processes are Rybicki and Lightman (1979) [29]and Longair (1981) [30].

2.2 Synchrotron Radiation

2.2.1 Emission by a single particle. *Synchrotron radiation* is created by a relativistic electron spiraling in a static magnetic field. A *non-relativistic* electron orbits with a velocity-independent angular frequency, the *gyro-frequency*

$$\omega_G = \frac{eB}{m_e c} = 1.8 \times 10^7 B \text{rad s}^{-1}, \tag{2.1}$$

where B is measured henceforth in Gauss [31], [32]. At relativistic energy, m_e is replaced by γm_e, where γ is the Lorentz factor. (See Fig.(2.1).)

Now, the electric field at a distance r, radiated by a non-relativistic electron moving with velocity $c\beta$ and acceleration $c\dot{\beta}$ (a dot, henceforth, denotes differentiation with respect to coordinate time) is given by

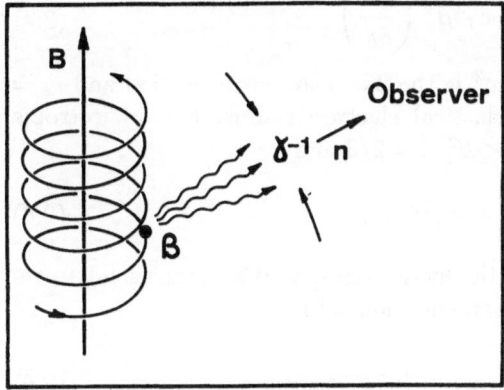

Fig. 2.1. Motion of a relativistic electron in a static magnetic field. The electron orbits the field with an angular frequency ω_G, independent of its motion parallel to the field. The accelerated electron radiates synchrotron radiation which is beamed parallel to the instantaneous direction of the electron motion.

$$E = \frac{e\dot{\beta}_\perp}{rc} \tag{2.2}$$

evaluated at the retarded time and resolved perpendicular to the line of sight. The power radiated is obtained by integrating the mean Poynting flux, $< E^2 > c/4\pi$ over a sphere of radius r.

$$P = \frac{2e^2 < \dot{\beta}^2 >}{3c} \tag{2.3}$$

where $<>$ denotes a time average.

In a frame in which the electron is at rest there will be no momentum radiated and so we can write a linear relation between two four vectors dp^μ, the change in 4-momentum of the electron and dx^μ, the associated change in position of the electron in spacetime

$$dp^\mu = P dx^\mu \tag{2.4}$$

As a consequence, P is a scalar invariant. We therefore seek a Lorentz invariant, relativistic generalisation of equation (2.3)for which the natural (and correct) choice is

$$P = \frac{2e^2}{3m_e^2 c} < \frac{dp^\mu}{d\tau} \frac{dp_\mu}{d\tau} > \tag{2.5}$$

where τ denotes the proper time. Rewriting this equation in terms of γ and $\beta = v/c$, we obtain

$$P = \frac{2e^2 \gamma^6}{3c} [\dot{\beta}^2 - (\beta \times \dot{\beta})^2] \tag{2.6}$$

For synchrotron radiation, the particle energy is constant and $\dot{\beta} = \omega_G \times \beta$. Equation(2.6)can be re-written in the convenient form

$$P = 2\sigma_T c \gamma^2 \beta_\perp^2 \left(\frac{B^2}{8\pi} \right) \tag{2.7}$$

where $\sigma_T = 8\pi r_e^2/3 = 6.6 \times 10^{-25} \mathrm{cm}^2$ is the Thomson cross section and $r_e = e^2/m_e c^2 = 2.8 \times 10^{-13}$cm is the classical electron radius. For an isotropic distribution of relativistic electrons,$< \beta_\perp^2 >= 2/3$ and

$$P = \frac{4}{3} \sigma_T c \gamma^2 U_{mag} \tag{2.8}$$

where $U_{mag} = B^2/8\pi$ is the magnetic energy density. It's more useful to remember the electron cooling time than equation(2.8)

$$t_{cool} = \frac{\gamma m_e c^2}{P} \simeq 5 \times 10^8 B^{-2} \gamma^{-1} \mathrm{s} \tag{2.9}$$

measuring B in G [29].

The synchrotron radiation spectrum emitted by a relativistic electron is broad band, in contrast to the narrow band cyclotron radiation spectrum radiated by a non-relativistic electron (Fig.(2.2)). The spectrum is centred on a *critical* frequency, ν_c. This can be estimated by observing that the time it takes the emission cone to sweep past the observer is $\Delta t_{em} \sim \gamma^{-1}(\omega_G/\gamma)^{-1} = m_e c/eB$. This is the time that would be measured by an observer watching an electron move by. An observer in the direction of the unit vector n will receive a short pulse of radiation in an interval $\Delta t_{ob} \sim (1 - n \cdot \beta)\Delta t_{em} \sim \Delta t_{em}/2\gamma^2$. Now if we take a Fourier transform of this pulse, the central frequency will be $\sim \Delta t_{ob}^{-1} \sim \gamma^2 \omega_G$. Reinstating numerical constants gives the approximation

$$\nu_c \sim \gamma^2 B \mathrm{MHz} \tag{2.10}$$

We can then use equation(2.10)to re-write equation(2.9)in the more useful form

$$t_{cool} \sim 6 \times 10^8 B^{-3/2} \nu_6^{-1/2} \mathrm{s} \tag{2.11}$$

again measuring field strength in G and frequency in MHz.

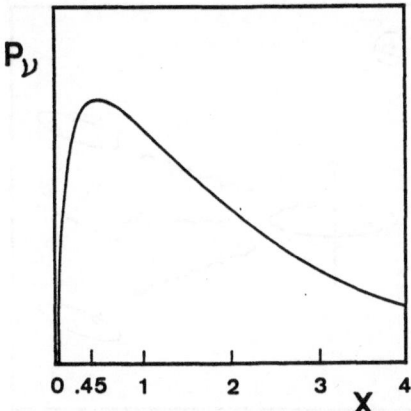

Fig. 2.2. Single particle synchrotron radiation spectrum emitted by an ultra-relativistic electron gyrating in a magnetic field with gyro frequency ω_G. The abscissa $x = 2\pi\nu m_e c / \gamma^2 eB$. The spectral power, P_ν is not normalised. At low frequency $P_\nu \propto \nu^{1/3}$; at high frequency, P_ν decays exponentially.

Some possible sites of synchrotron radiation associated with AGN are listed in Table(2.1). Observe that the synchrotron cooling times are typically shorter than the local dynamical timescales (*i.e.*the source sizes divided by their characteristic speeds).

2.2.2 Polarisation. Synchrotron radiation is linearly polarised. To see why, consider gyro radiation from a non-relativistic electron on a circular orbit (Fig.(2.3)). When viewed in the orbital plane, the radiation is 100 per cent linear polarised with the electric vector oscillating perpendicular to the magnetic field. When viewed from along the direction of the magnetic field, the radiation is 100 per cent circular polarised. Now let the electron move with a relativistic speed and beam this radiation in the direction of motion. The two components of circular polarisation will effectively cancel, whereas the linear polarisation will largely survive. The net effect is a typical degree of polarisation of $\sim$ 70 per cent. The radio emission from extragalactic radio sources is

Table (2.1). Possible synchrotron radiation sites and characteristic physical parameters.

Location	B G	ν Hz	γ	t_{cool} yr	t_{dyn} yr	p_{min} dyne cm^{-2}	U_{min} erg
Extended Radio Source	10^{-5}	10^9	10^4	10^7	10^8	10^{-11}	10^{59}
Radio Jet	10^{-3}	10^9	10^3	10^4	10^4	10^{-7}	10^{57}
Compact Radio Source	10^{-1}	10^9	10^2	10	10	10^{-3}	10^{54}
Outer Accretion Disk	10	10^{14}	$10^{3.5}$	10^{-4}	1	10	10^{49}
Inner Accretion Disk	10^3	10^{16}	$10^{3.5}$	10^{-8}	1	10^5	10^{47}
Black Hole Magnetosphere	10^4	10^{18}	10^4	10^{-10}	10^{-3}	10^7	10^{47}

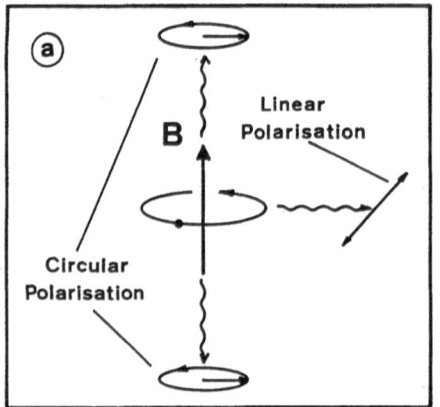
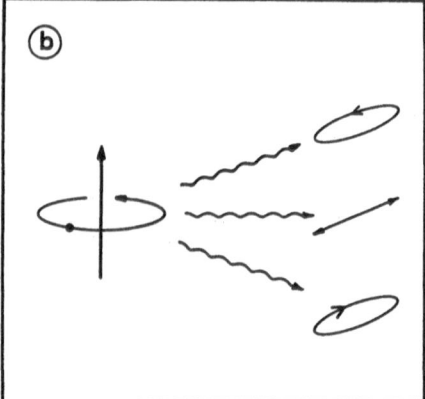

Fig. 2.3. a) Polarisation of cyclotron radiation from a non-relativistic electron. b) Polarisation of synchrotron radiation from a relativistic electron.

usually polarised to a degree that ranges from $\sim$ a few per cent to $\sim$ 60 per cent. Partly for this reason, it is thought to be synchrotron radiation.

2.2.3 Emission by a power-law electron distribution function. The electrons in cosmic radio sources have a wide range of energies. As is traditional in cosmic ray physics we assume that the electron distribution function is a power-law, *i.e.* the number density of relativistic electrons per unit Lorentz factor is given by

$$N_\gamma = K\gamma^{-s} \tag{2.12}$$

We can now integrate over this distribution to obtain the total emission. It is adequate for present purposes to treat the spectral emissivity by a single particle as a delta-function,

$$P_\nu \sim P\delta(\nu - \nu_c) \tag{2.13}$$

where P and ν_c are given by equations (2.8) and (2.10) respectively. The volume spectral emissivity j_ν is then given by integration over the particle distribution

$$
\begin{aligned}
j_\nu &\propto \int d\gamma N_\gamma P_\nu \\
&\propto KB^{1+\alpha}\nu^{-\alpha}
\end{aligned}
\tag{2.14}
$$

where we have integrated over the delta function in equation (2.13). The exponent α is known as the *spectral index* and is given by

$$\alpha = \frac{s-1}{2} \tag{2.15}$$

Spectral indices $\alpha \sim 0.7$ are observed in extended radio sources. The inferred electron distribution function slopes are therefore observed to be $s \sim 2.4$, somewhat smaller than the slope of the cosmic ray spectrum. The spectra of radio jets are somewhat flatter ($\alpha \sim 0.5$) than those from the extended radio components, which is perhaps indicative of more vigorous particle acceleration in jets.

A power-law distribution function can only be a good approximation to the true distribution function over a limited range of energy; at the lowest and highest energies, the particle density must diminish to zero. We can model this by introducing upper and lower cut-offs, γ_{min} and γ_{max} for the power law distribution. The exponent s then takes on a special significance. If we consider the pressure and energy density, $\propto \int d\gamma \gamma^{1-s}$ for relativistic electrons, then the integral is dominated by the upper cut-off at γ_{max} when $s < 2$. The synchrotron emissivity $\propto \int d\gamma \gamma^{2-s}$ is likewise dominated by the upper cut-off for $s < 3$. In many synchrotron sources, $0.5 \lesssim \alpha \lesssim 0.75$ so that $2 \lesssim s \lesssim 3$. Suppose that the source can only be observed over a limited frequency range, $\nu_1 < \nu < \nu_2$ and the associated range of electron energy is $\gamma_1 < \gamma < \gamma_2$. Now suppose that the complete power law electron distribution function extends from γ_{min} to γ_{max}. The electron energy contained in the source will then be larger than that inferred directly from the observed spectrum by a factor $\sim (\gamma_1/\gamma_{min})^{s-2}$ and the total power radiated will be larger than that associated with observed radiation by a factor $\sim (\gamma_{max}/\gamma_2)^{3-s}$. Both of these factors can be large and affect energetic arguments.

2.2.4 Minimum pressure of a synchrotron source.
In order to understand the dynamics of extragalactic sources, it is necessary to calculate how much energy they contain and, equivalently, how much pressure they exert. Both the energy and the pressure contain two essential components, associated with the relativistic electrons and the magnetic field. An important argument, due originally to Burbidge [33], allows us to put a lower bound on the combined electron and magnetic pressure. If we can resolve a source and know its redshift, then we can compute its distance and size and infer its volume spectral emissivity (conventionally expressed per steradian), $j_{\nu\Omega}$. We have argued that the pressure is dominated by that part of the spectrum where $s \sim 2$ and the spectral index $\alpha \sim 0.5$. In the case of a power law with $\alpha \neq 0.5$, this will either be the estimated upper or the lower cut off frequency. Call this frequency ν_p. The associated Lorentz factor is then $\gamma \propto (\nu_p/B)^{1/2}$. We use equation (2.14) to write $j_{\nu\Omega}(\nu_p) \propto \nu_p^{-1/2} p_e p_{mag}^{3/4}$ where $p_e = N\gamma m_e c^2/3$ is the relativistic electron pressure and $p_{mag} = B^2/8\pi$ is the magnetic pressure. Now the combined pressure $p = p_e + p_{mag}$ for a given $j_{\nu\Omega}$ is minimised when $p_e \sim p_{mag}$. Carrying out this procedure and scaling to convenient units, we obtain

$$p_{min} \simeq 10^{-9} \left(\frac{j_{\nu\Omega}(\nu_p)}{10^{-36}\text{erg Hz}^{-1}\text{ster}^{-1}\text{s}^{-1}\text{cm}^{-3}} \right)^{4/7} \nu_{p6}^{2/7} \text{dyne cm}^{-2} \quad (2.16)$$

where $\nu_{p6} = \nu_p/1\text{MHz}$. The numerical coefficient is slightly sensitive to the spectral shape; in particular, if $\alpha \sim 0.5$, then the pressure must be increased by a small logarithmic factor. The total energy in the source is roughly $3p_{min}V$ and the *equipartition magnetic field strength* is $\sim (12\pi p_{min})^{1/2}$ [34], [35]. Some representative minimum pressures and minimum energies are given in Table (2.1).

There may also be significant contributions to the pressure from relativistic protons and thermal plasma. (In the interstellar medium, the cosmic ray proton pressure exceeds the cosmic ray electron pressure by a factor of roughly 30.) A further possible complication is that a uniform magnetic field does not exert an isotropic pressure, but an anisotropic stress that is tensile along the field. For this reason, it is generally assumed that the magnetic field is sufficiently disordered that we can average the stress tensor over direction to define a mean magnetic pressure. Polarisation observations support this assumption because the degree of polarisation measured in most sources is much less than the maximum value possible from synchrotron radiation.

2.2.5 Synchrotron self-absorption. Some radio synchrotron sources are so compact that they self-absorb their own emission. Before seeing when this is important, let us first assume that the source is optically thin and define the *flux density* $S_\nu = \int dV j_\nu / 4\pi D^2$ where D is the distance to the source. (We ignore cosmological corrections for simplicity.) Flux density is conventionally measured in Jansky (Jy) where $1 \text{Jy} = 10^{-23} \text{erg cm}^{-2} \text{ s}^{-1} \text{ Hz}^{-1}$. The radiation in the source will generally be anisotropic. It is therefore also necessary to define the intensity I_ν which is measured in units of erg cm^{-2} s^{-1} Hz^{-1} sterad^{-1}. The intensity is a particularly useful quantity to define because, in the absence of scattering and absorption, it is conserved along a ray. In other words, the intensity measured by a radio telescope on earth is the same as that emerging from the surface of a distant source (again ignoring cosmological corrections). The flux density is related to the intensity by

$$S_\nu = \int d\Omega I_\nu \qquad (2.17)$$

where $d\Omega$ is an element of solid angle.

Radio astronomers frequently express the intensity in terms of an equivalent quantity, the *brightness temperature*, T_B defined by

$$T_B = \frac{c^2 I_\nu}{2 k_B \nu^2} \qquad (2.18)$$

where k_B is Boltzmann's constant. The brightness temperature is manifestly the temperature of a black body radiating intensity I_ν in the Rayleigh-Jeans (classical) part of the Planck spectrum. Thermodynamically, we would not expect the brightness temperature of the radiation at some frequency ν to exceed the kinetic temperature of the electrons radiating at that frequency. If all the emission from within the source were to contribute a brightness temperature larger than the electron temperature $\sim \gamma m_e c^2 / 3$, then the source will become self-absorbed and only its outer shell will be observable. In other words the brightness temperature is limited to

$$T_B \lesssim \frac{\gamma m_e c^2}{3 k_B}$$

$$\simeq 1 \times 10^9 \left(\frac{\nu_6}{B}\right)^{1/2} g(s) \mathrm{K}$$

(2.19)

where we have used equation(2.10)and measure the frequency in MHz, the field in G and $g(s)$ is a slowly varying function that increases smoothly from 0.5 to 2.0 as s increases from 1 to 3[34].

The flux density from a self-absorbed source satisfies $S_\nu \propto \nu^2 T_B \propto \nu^{5/2}$. Homogeneous, compact radio sources therefore exhibit rising spectra at low radio frequencies and falling spectra at high frequency where they are optically thin (Fig.(2.4)). As we shall see in section 5.3.1, more complex spectra can be created by inhomogeneous sources.

One application of this result is to the interpretation of the far infra-red cut off in radio-quiet quasars. It is observed that the flux density rises very rapidly between mm and infra red frequencies. In some sources it is asserted that $\alpha < -2.5$ [36]. This steep a spectrum cannot be created by a simple, homogeneous source (though a synchrotron radiation model can be saved with additional contrivance [37]).

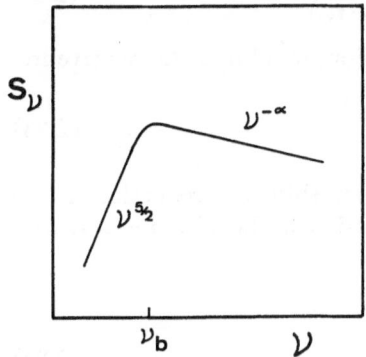

Fig. 2.4. Flux density from a homogeneous synchrotron source. The source is optically thick for $\nu \ll \nu_b$ and optically thin for $\nu \gg \nu_b$.

2.3 Compton Scattering

2.3.1 Thomson cross-section. An electron can also be accelerated by a classical electromagnetic wave. Using equation (2.2)it can be quickly shown that the differential cross section for scattering an electromagnetic wave propagating in direction with unit vector $\boldsymbol{n}$ polarised with electric vector in the direction of the unit vector $\boldsymbol{e}$ into a state specified by $\boldsymbol{n}'$ and $\boldsymbol{e}'$ is

$$\frac{d\sigma}{d\Omega} = r_e^2 (\boldsymbol{e} \cdot \boldsymbol{e}')^2$$

$$= \frac{1}{2} r_e^2 [1 + (\boldsymbol{n} \cdot \boldsymbol{n}')^2]$$

(2.20)

173

where the second line follows if we average over the incident polarisation states and sum over the scattered polarisation states [29],[31]. Note that the cross section is not strongly beamed in angle. If we now average over $n \cdot n'$, the average cross section is the familiar Thomson cross-section that we have already encountered in our discussion of synchrotron radiation (cf.equation(2.7)). Given a region containing free electrons with density n_e, it is often useful to define a *Thomson optical depth*

$$\tau_T = \int ds n_e \sigma_T \qquad (2.21)$$

along a line of sight. When $\tau_T \ll 1$, it measures the probability that a photon will be scattered. A source with $\tau_T \gg 1$ is said to be optically thick to Thomson scattering. Photons escape this source by diffusing out of it with a mean speed given roughly by c/τ_T.

2.3.2 Compton scattering. When the energy of the photon becomes comparable with the rest mass of an electron, the recoil of the electron can no longer be ignored and the scattered photon frequency is shifted from ν' to ν which can be calculated from the *Compton formula*, derived by conserving energy and momentum

$$\nu = \frac{m_e c^2 \nu'}{m_e c^2 + h\nu'(1 - n \cdot n')} \qquad (2.22)$$

[29]. Averaging over the scattered photon direction, we obtain the relative frequency shift

$$\left\langle \frac{\Delta \nu}{\nu} \right\rangle \simeq -\frac{h\nu}{m_e c^2} \qquad (2.23)$$

for $\nu \ll m_e c^2/h$. Associated with this frequency shift is a reduction in the total scattering cross section. The full cross section is the $Klein - Nishina$ cross section, for which two useful limits are

$$\begin{aligned} \sigma &\simeq \sigma_T \left(1 - \frac{2h\nu}{m_e c^2}\right), \qquad h\nu \ll m_e c^2 \\ &\simeq \frac{3}{8}\sigma_T \left(\frac{m_e c^2}{h\nu}\right) \left[\ln\left(\frac{2h\nu}{m_e c^2}\right) + \frac{1}{2}\right], \qquad h\nu \gg m_e c^2 \end{aligned} \qquad (2.24)$$

2.3.3 Comptonisation. We are now in a position to describe how electrons and photons approach equilibrium. Firstly consider photon heating of the electrons. Every scattering, the photons lose a fraction of their energy given by equation(2.23)(cf.Fig.(2.5)). Hence the mean heating rate of the electrons will be given by

$$W_+ = n_e \sigma_T c \int d\nu U_\nu \left(\frac{h\nu}{m_e c^2}\right) \qquad (2.25)$$

where U_ν is the spectral energy density of the radiation and we have assumed that $\nu \ll m_e c^2/h$. It is not necessary to assume that the radiation is isotropic because a small quantity of magnetic field will usually suffice to isotropise the electron distribution function.

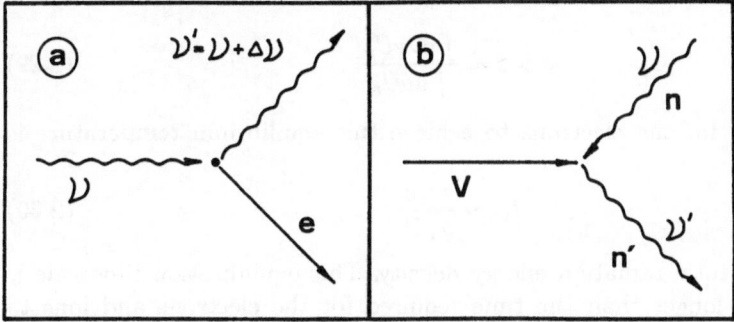

Fig. 2.5. a) Electrons are heated by the photons through Compton recoil. b) Photons are heated by the electrons by Doppler shifting their frequency.

There is a reverse energy transfer from the electrons to the photons. This can be most simply understood in terms of the Doppler shift. An electron moving with speed v will shift the photon frequency by $|\Delta\nu/\nu| \sim v/c$. To $O(v/c)$, blue shifts and redshifts are equally probable. However to $O(v/c)^2$, the rate of approaching (blueshifting) scatterings will exceed the rate of redshifting scatterings. Special relativistic corrections will also appear at this order. We can therefore write the *mean* frequency shift as

$$\left\langle \frac{\Delta\nu}{\nu} \right\rangle = x \frac{k_B T}{m_e c^2} \tag{2.26}$$

where T is the electron temperature and the constant x is to be determined. The electron cooling rate can be expressed as

$$W_- = n_e \sigma_T c \int d\nu U_\nu \left(\frac{x k_B T}{m_e c^2} \right) \tag{2.27}$$

Simple thermodynamic considerations fix the constant x. If the radiation field is dilute, then we can treat the photon gas as essentially classical and ignore induced scattering processes (see below). The spectral energy density of radiation in equilibrium with electrons of temperature T will then be given by $U_\nu \propto \nu^3 e^{-h\nu/k_B T}$. Substituting this energy density into equation(2.27)and then equating the answer to the heating rate, equation(2.25), from partial integration, we readily obtain $x = 4$.

We can now use equations(2.25)and (2.27)for a general radiation spectrum. In applications to AGN it is often true that the radiation energy density greatly exceeds the electron energy density. The electrons will then, given sufficient time, achieve a temperature dictated by the radiation spectrum. Balancing the heating with the cooling we find this equilibrium (or *Compton*) temperature to be

$$T_C = \frac{h <\nu>}{4 k_B} \tag{2.28}$$

where

$$<\nu> = \frac{\int d\nu\, \nu U_\nu}{\int d\nu\, U_\nu} \qquad (2.29)$$

The time taken for the electrons to achieve this equilibrium temperature is typically

$$t_C \sim \frac{m_e c}{\sigma_T U} \qquad (2.30)$$

where U is the total radiation energy density. This equilibration timescale is generally much longer than the time required for the electrons and ions to develop a Maxwellian distribution function, vindicating our assumption.

We can illustrate Comptonisation by considering the equilibrium of the tenuous gas surrounding the broad emission line region in Seyfert galaxies. We know that the hard X-ray spectrum from Seyfert galaxies is usually well described as a power law with spectral index $\alpha \sim 0.7$ [38]. If we suppose that it only extends from $\sim 1 keV$ up to $\sim 100 keV$, then we compute

$$<\nu> \sim \left(\frac{\alpha - 1}{2 - \alpha}\right)\left(\frac{\nu_{min}}{\nu_{max}}\right)^{\alpha - 1}\nu_{max} \qquad (2.31)$$

The equilibrium Compton temperature is then given by $T_C \sim 7 \times 10^7 \mathrm{K}$. (In deriving this temperature, we have ignored the cooling effect of the photons in the UV bump and collisional cooling processes like bremsstrahlung which might both be important.) This temperature can be used to place an upper limit on the pressure in the inter-cloud medium. This is generally found to be too small to confine the clouds (*cf.*[39], and Dr. Netzer's lectures).

2.3.4 Kompaneets equation. A more complete description of the frequency evolution of the radiation spectrum is possible when the radiation field can be regarded as isotropic and spatially homogeneous and $\nu << m_e c^2/h$. In this case, the photons gain and lose energy in small steps and their distribution function can be described by a modified diffusion (or more properly, *Fokker-Planck*) equation in frequency. The appropriate distribution function is the occupation number n defined as the mean number of photons in one polarisation per volume $4\pi\nu^2 d\nu dV/c^3$ of phase space. In thermal equilibrium at temperature T, n is given by the Planckian value

$$n = (e^{h\nu/k_B T} - 1)^{-1}. \qquad (2.32)$$

The relevant evolution equation for n due to Kompaneets [40], is given by

$$\frac{\partial n}{\partial t} = \frac{1}{\nu^2}\frac{\partial}{\partial \nu}\nu^4 \frac{h n_e \sigma_T}{m_e c}\left[\frac{k_B T}{h}\frac{\partial n}{\partial \nu} + n + n^2\right], \qquad (2.33)$$

(cf [29], [41]).

This equation can be understood from elementary considerations. The number of photons is conserved in Compton scattering and so their occupation

176

number must satisfy a conservation law of the form

$$\frac{\partial n}{\partial t} = -\frac{1}{\nu^2}\frac{\partial}{\partial \nu}\nu^2 j(\nu) \qquad (2.34)$$

where $j(\nu)$ is a radial "current" in $\mathbf{k}$- (or equivalently frequency) space. When Compton recoil dominates, there is a steady drift velocity of the photons in frequency given by equation(2.23), $d\nu/dt = -n_e\sigma_T h\nu^2/m_e c$, giving a contribution to the current of $nd\nu/dt$, the second term in equation(2.33). When the Doppler heating dominates, the photons diffuse in frequency space so that $j = -D_\nu \partial n/\partial \nu$ where the diffusion coefficient is given by $n_e\sigma_T c < \Delta\nu^2 > /3$ and the rms frequency shift experienced in a collision with an electron moving with speed v is $(\Delta\nu)_{rms} = \nu v/c$. This accounts for the first term in equation(2.33).

These two contributions to the current will suffice as long as $n \ll 1$. However, when $n \gg 1$, induced scattering becomes significant. In this case, the rate of frequency-reducing photon scattering is enhanced over the spontaneous rate by a factor $(1 + n)$, where n is the occupation number of the final state. This gives the third contribution to the current. Observe that the current vanishes when n has the Planckian form, equation(2.32). (In fact the current vanishes more generally when n satisfies a Bose-Einstein distribution. This is to be expected when we realise that photon number is conserved in pure Compton scattering and a second Lagrange multiplier is needed to solve for the equilibrium distribution function using statistical mechanics.)

Now let us use this equation to calculate the spectral distortion introduced in a dilute photon gas by hot electrons. We suppose, initially, that we can ignore the effects of Compton recoil and induced scattering. We model the radiative transfer in a simple manner by assuming that the photons are generated uniformly within some heating region of size R and propagate out of this region with a frequency-independent escape probability. If the heating region is optically thin to Thomson scattering, $i.e.\tau_T \ll 1$, then the estimate of the escape probability per unit time is c/R. (In fact we can use this as a definition of R.) However, if the source is optically thick, the mean escape speed of the photons as they random walk out of the heating region will be $\sim c/\tau$. We interpolate between these two limits by writing the escape rate as $nc/(1 + \tau)R$.

Under these assumptions, the modified Kompaneets equation can be written as

$$\frac{\partial n}{\partial t} = \frac{1}{\nu^2}\frac{\partial}{\partial \nu}\left(\nu^4\frac{n_e\sigma_T k_B T}{m_e c}\frac{\partial n}{\partial \nu}\right) - \frac{nc}{(1 + \tau)R} + S, \qquad (2.35)$$

where S is the photon source function. Now consider the stationary spectrum produced by Comptonising a monochromatic photon source of frequency ν_0 [42][43]. By inspecting the ordinary differential equation formed by setting the right hand side of equation (2.35) to zero, we see that the solution is a power-law,

$$n \propto \nu^{-(3+\alpha)}; \qquad \nu_0 < \nu < \nu_{rec} \qquad (2.36)$$

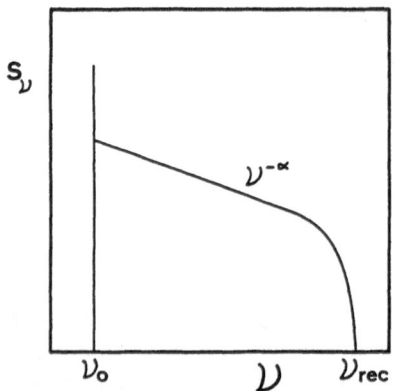

Fig. 2.6. Spectrum produced by a monochromatic photon source with frequency ν_0 embedded within a hot electron scattering region. The slope of the power law is dictated by the Comptonisation parameter y, given by equation(2.38). y is proportional to the ratio of the acceleration rate to the escape rate. The spectrum extends up to a maximum frequency, $\nu_{rec} = (3 + \alpha)k_B T/h$, where Compton recoil becomes significant.

where the spectral index α is given by the solution of the quadratic equation

$$\alpha(\alpha + 3) = \frac{4}{y} \tag{2.37}$$

The quantity

$$y = \frac{4k_B T\tau(1 + \tau)}{m_e c^2} \tag{2.38}$$

is known as the *Comptonisation parameter.* (Fig.(2.6).)

The neglect of Compton recoil is only permissible as long as $\nu << \nu_{rec} = (3+\alpha)k_B T/h$. We have shown that Comptonisation naturally produces a power-law spectrum over an extended frequency range. The spectrum that we have calculated is that within the scattering region. However, as the escape probability is frequency-independent, the spectrum observed from the source will have the same spectrum.

This mechanism for producing a power-law spectrum is essentially the same as that originally proposed by Fermi for accelerating Galactic cosmic rays and it suffers similar drawbacks. Firstly, it depends upon both the rate of frequency boosting and the escape probability being energy-independent. This is not the case above energies ~40keV and the increase in the escape probability will cause some steepening of the spectrum. Secondly, it requires that the ratio of the heating and escape rates have some fixed value in order to generate a preferred value of the spectral index. In order to reproduce the observed Seyfert X-ray spectral index $\alpha = 0.7$, requires $y = 1.5$. However if, for example, the source changes so that y is increased to 2.3, then α is reduced to 0.3. This fine tuning is unlikely to arise in practice unless there is some feedback by which the electron temperature changes in response to the photon heating. Curved spectra, like those seen is blazars, can also be reproduced by Compton scattering (Loeb, McKee and Lahav, preprint).

2.3.5 Inverse Compton scattering. Compton scattering by relativistic electrons is usually called *inverse Compton scattering*, although it really only involves the classical Thomson cross section. Using equation(2.5)and the electrodynamic equation of motion we can express the radiated power in the form

$$P = \frac{\gamma^2 \sigma_T c}{4\pi} \left\langle (E + \beta \times B)^2 - (E \cdot \beta)^2 \right\rangle$$
$$= \frac{4}{3} \gamma^2 \sigma_T c U_{rad} \tag{2.39}$$

where U_{rad} is the radiation energy density and we have assumed that the radiation is isotropic and that $\gamma \gg 1$. Now although this is a classical process, it useful to think in terms of photons. The photon scattering rate per unit volume of **k** space and per polarisation is $< n\sigma_T c(1 - \beta \cdot n) > = n\sigma_T c$ for an isotropic distribution. Comparing with (2.39), we deduce that the mean scattered photon frequency is

$$< \nu > = \frac{4}{3} \gamma^2 \nu' \tag{2.40}$$

where ν' is the incident frequency. The two powers of γ can be traced to two relativistic Doppler shifts; one from the the radiation frame to the electron rest frame, the other back again to the radiation frame [41], [29]. (See Fig.(2.7).)

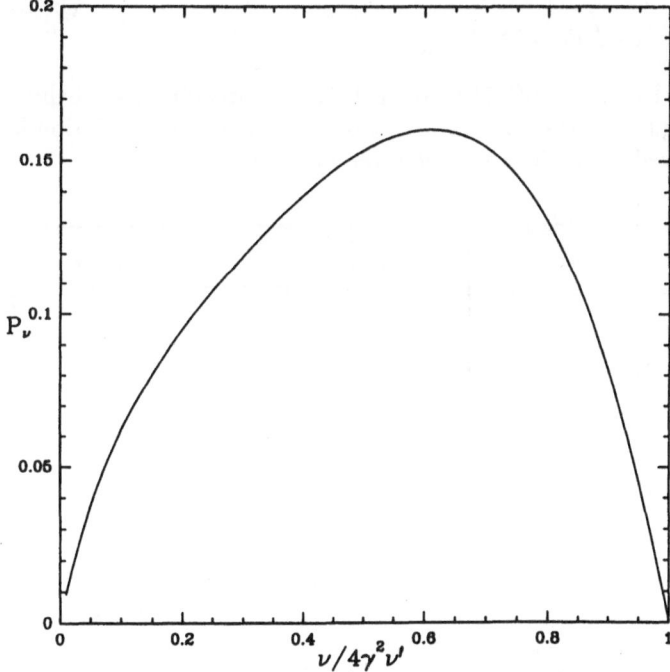

Fig. 2.7. Radiation spectrum emitted by monoenergetic electrons inverse Compton scattering monochromatic radiation. In the ultrarelativistic limit, $\gamma \gg 1$, the mean scattered frequency is $4\gamma^2\nu'/3$ and the peak scattered frequency is $4\gamma^2\nu'$, where ν' is the incident frequency.

Equations(2.39)and(2.40)are similar to equations(2.8)and (2.10)for synchrotron radiation. We can therefore appropriate the argument from synchrotron radiation theory for the slope of the spectrum radiated by a power law electron distribution function. Therefore, if s is the slope of the electron distribution function, the spectral index of the scattered monochromatic radiation is $\alpha = (s-1)/2$.

So far we have ignored Compton recoil. This is permissible as long as the incident photon energy, Lorentz transformed into the electron rest frame is less than the electron rest mass. in other words, we require that

$$\gamma\nu' << m_e c^2/h \qquad (2.41)$$

When this condition is violated, the Klein-Nishina cross section, equation(2.24), must be used and the radiated power is reduced.

2.3.6 Synchrotron-self Compton radiation. When a synchrotron source is sufficiently compact that the synchrotron radiation photons are inverse Compton scattered by the relativistic electrons, then the emergent spectrum is known as *Synchrotron-self Compton radiation* [44]. The scattered flux density can be calculated approximately using the delta function approximation and integrating over both the synchrotron radiation spectrum and the electron distribution.

$$S_c(\nu) \propto \int d\nu'\nu'^{-\alpha} \int d\gamma\gamma^{-s}\delta(\nu - 4\gamma^2\nu'/3)$$
$$\propto \int d\nu'\nu'^{-\alpha-1} \qquad (2.42)$$

where s, α are related by equation(2.15)and the limits of integration are exhibited graphically in Fig.(2.8). The spectral index of the synchrotron radiation is approximately preserved in the inverse Compton emission.

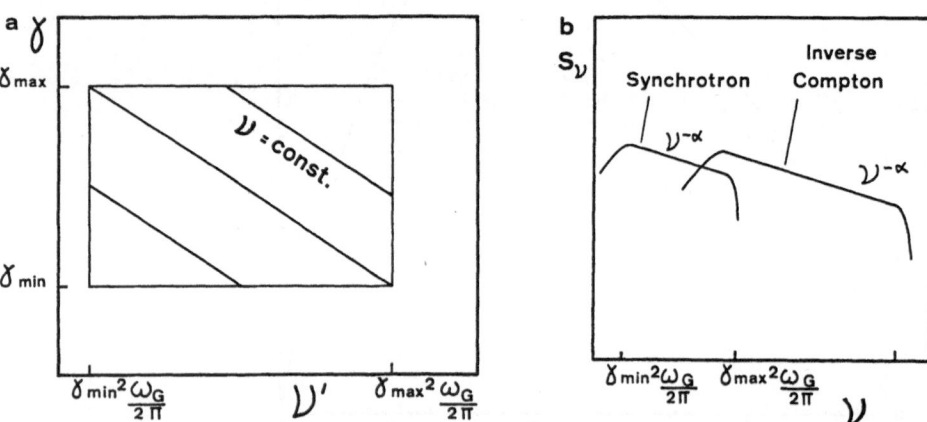

Fig. 2.8. a) Limits of integration for equation(2.42). The electron distribution function is supposed to extend as a power law from γ_{min} to γ_{max}. Combinations of synchrotron photon frequency and electron energy radiating a given Compton frequency are shown. b) The Compton spectrum extends as a rough power law (ignoring weakly-varying logarithmic factors) from $\sim \gamma_{min}^4\omega_G/2\pi$ to $\sim \gamma_{max}^4\omega_G/2\pi$.

2.3.7 Inverse Compton limit. Equation(2.42)applies to the once scattered Compton radiation. If the energy density of this radiation exceeds that of the magnetic field, then the twice-scattered Compton radiation will greatly exceed the once-scattered radiation and so on. In fact the Compton scattering will extend up to the γ—ray region when condition(2.41)becomes violated. The requirement that the once-scattered radiation energy density not exceed the synchrotron radiation energy density is known as the *inverse Compton limit*[35].

This limit can be converted into a convenient rule of thumb. When the synchrotron radiation spectrum peaks at frequency ν_m and the brightness temperature is T_m, the ratio of the once-scattered Compton power to the synchrotron power is

$$\frac{S_C}{S_S} \simeq \frac{U_S}{U_{mag}}$$
$$\propto \frac{T_m \nu_m^3}{B^2} \qquad (2.43)$$
$$\propto T_m^5 \nu_m$$

If we evaluate the constants of proportionality, then the condition that the Compton power be less than the synchroton radiation power can be written as a limit on the brightness temperature

$$T_m \lesssim 10^{12}(1-\alpha)^{4/5}\nu_{max9}^{(\alpha-1)/5}\nu_{m9}^{-\alpha/5}\text{K} \qquad (2.44)$$

where the frequencies are now measured in GHz. This formula must be modified by a spectral factor if $\alpha \gtrsim 1$. It also ignores the effects of relativistic beaming (see section 4 below) which can be significant. However, in a non-relativistic source, the brightness temperature should be limited by the conventional inverse Compton limit of $\sim 10^{12}$K. In contemporary invocations of the inverse Compton limit it is usual to use the measured X-ray flux or upper limit to set a slightly stronger upper limit on the radio brightness temperature for a non-relativistic source and hence use this a predictor of relativistic motion when this bound is violated [45].

One method for measuring the brightness temperatures in compact radio sources involves using the variability timescale (times c) to estimate the linear size. The brightness temperatures computed in this manner frequently exceed the inverse Compton limit of $\sim 10^{12}$K and, in one instance, exceeds $\sim 10^{18}$K [46].

2.3.8 Induced Compton scattering. Let us now return to the non-linear term $\propto n^2$ in the Kompaneets equation. As we have mentioned, this comes from induced scattering. Consider, for simplicity, two photon states, primed and unprimed, in the electron rest frame. In the Thomson limit, when we ignore electron recoil, the two states will have the same frequency and the induced scattering rate into the unprimed state $\propto n'n$ will be cancelled exactly by the induced scattering out of the unprimed state $\propto nn'$. However, when we include the effect

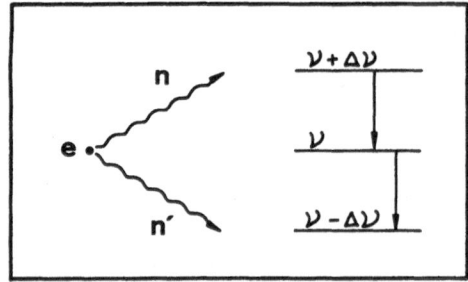

Fig. 2.9. Induced scattering into and out of an unprimed state with frequency ν.

of Compton recoil we find that the photons scattered into the unprimed state at frequency ν start with frequency $\nu - h\nu^2(1 - \boldsymbol{n} \cdot \boldsymbol{n}')/m_e c^2$ and those that are scattered out of the unprimed state have frequency $\nu + h\nu^2(1 - \boldsymbol{n} \cdot \boldsymbol{n}')/m_e c^2$ (Fig.(2.9)). The net induced scattering rate into the unprimed state is given by the difference between the rate of scattering into a photon state and the rate of scattering out of that state

$$\frac{\partial n}{\partial t} = \frac{3nn_e\sigma_T h}{8\pi m_e c} \int d\Omega'[1 + (\boldsymbol{n} \cdot \boldsymbol{n}')^2](1 - \boldsymbol{n} \cdot \boldsymbol{n}')\frac{\partial(n'\nu^2)}{\partial \nu} \qquad (2.45)$$

[47].

If the radiation field were isotropic, and in general it won't be, then equation(2.45)simplifies to

$$\frac{\partial y}{\partial \tilde{t}} = \frac{\partial y^2}{\partial \nu} \qquad (2.46)$$

where $y = h\nu^2 n/m_e c^2$ and $\tilde{t} = n_e \sigma_T ct$. Equation(2.46)can be easily solved by integrating along its characteristics, $d\nu/dt = 2y$. As shown in Fig.(2.10), induced Compton scattering causes high brightness radiation to drift to lower

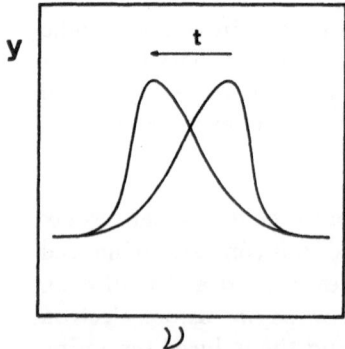

Fig. 2.10. Spectral evolution of $y \propto I_\nu/\nu$ in an isotropic radiation field under the influence of induced Compton scattering. After a sufficient length of time, shocks will develop in frequency space.

frequency, steepening the rising portions of the spectrum y and flattening the falling portions. After a sufficient interval, the low frequency spectrum will become formally double-valued though, just as in fluid dynamics, a shock will actually develop in frequency space, with thickness determined by the electron velocity dispersion. The main characteristic of induced Compton scattering is its strong dependence on frequency. It should also create large, variable degrees of linear polarisation. These are the two main signatures that ought to allow induced Compton scattering to be recognisable observationally.

For the more general case, it is convenient to introduce an optical depth to induced scattering

$$\tau_I \sim \tau_T n \frac{\Delta \nu}{\nu}$$
$$\sim \frac{k_B T}{m_e c^2} \tau_T \qquad (2.47)$$
$$\sim 2 T_{10} \tau_T$$

where we have used the approximation $n \sim k_B T/h\nu$, cf.equation(2.18). (T_{10} is the brightness temperature measured in units of 10^{10}K).

There is, as yet, no evidence from the variability or the polarisation of the radiation that induced Compton scattering occurs in *any* source. If we take the maximum reported variability brightness temperature of $10^{18} K$ at face value, the failure to observe the effects of induced Compton scattering limits the line of sight Thomson optical depth to $\lesssim 10^{-8}$, an unreasonably small value. This provides a second argument, additional to the inverse Compton scattering limitation that compact radio sources are expanding relativistically.

2.4 Pair Production and Annihilation

2.4.1 Motivation. There are several reasons for taking seriously the notion that the central regions of AGN contain large densities of positrons as well as electrons. The first is that a few gamma ray sources are positionally (though in most instances uncertainly) identified with AGN [48]. Gamma rays more energetic than the electron rest mass will react with lower energy photons to create electron positron pairs. Secondly, it has recently been shown that X-ray sources are rapidly variable. Roughly a third of type 1 Seyferts galaxies are already known to vary[38], and the most dramatic example, NGC6814, changed by a factor 5 in 1 minute [49]. The measured variability times and inferred source sizes are so small that co-extensive γ rays would not be able to escape and would instead create electron-positron pairs.

A third argument is theoretical. If the X-rays from AGN are made by relativistic electrons alone then, as we have emphasized, the radiative cooling times are inevitably very short compared with the source dynamical times. This implies that cooled electrons will accumulate in the source and make it optically thick to Thomson scattering. It will then take a long time for the photons to random walk out of the source and the variation will be slow. However, if electrons and positrons are accelerated and cool, they can annihilate, alleviating this constraint [50].

For these reasons we will consider the properties of a simple source model containing relativistic and thermal electrons and positrons as well as high energy photons. The location of the source is most naturally an active corona above the innermost parts of an accretion disk. We model this source in a similar fashion to the way cosmic ray physicists model cosmic ray transport in the Galaxy, as a homogeneous box in which a series of elementary reactions proceed and from which photons can escape with a certain probability. Let us consider some of these elementary processes.

2.4.2 Pair production. Electrons and positrons are produced most efficiently under these circumstances by γ ray-X-ray collisions. The threshold for pair production is readily shown to be

$$\nu_\gamma > \left(\frac{m_e c^2}{h}\right)^2 \frac{2}{\nu_X(1 - \boldsymbol{n}_\gamma \cdot \boldsymbol{n}_X)} \tag{2.48}$$

so that 100 MeV γ rays can react with ~ 3keV X-rays. The cross section near threshold is roughly $\sim 0.2\sigma_T$, declining $\propto \nu_\gamma^{-1}$ at higher energy [51][52].

It is convenient to define a *compactness parameter*

$$\ell_i = \frac{L_i \sigma_T}{m_e c^3 R} \tag{2.49}$$

for radiation component i, where R is the source size, estimated, perhaps, by the variability [53]. ℓ_i is most simply interpreted for ~ 1MeV γ-rays. The number density of ~ 1MeV photons in the source can be estimated by $\sim L_{1MeV}\sigma_T/m_e c^3 R^2$. The optical depth to two photon pair production is then $\sim \ell_{1MeV}$. (An extra factor of 4π is sometimes included in the definition of ℓ_i.) Unfortunately, it is often usually hard to measure this quantity directly and its value must be extrapolated from measurements of ℓ_X. Measurements of ℓ_X do, however, measure the optical depth of the source to pair production of photons of energy $\sim m_e^2 c^4/h\nu_X$. When the X-ray compactness $\ell_X \gg h\nu_X/m_e c^2$, it will not be possible for γ rays to escape the source without creating pairs.

2.4.3 Inverse Compton scattering. The dominant contribution to the radiation energy density in AGN is probably the UV emission from the inner parts of an accretion disk. Accelerated electrons will inverse Compton scatter UV photons losing energy at a rate $\dot\gamma \propto \gamma^2$ (as long as $\gamma \lesssim 10^4$ and Klein-Nishina suppression does not set in). Now, in a steady state, the flow of electrons (and positrons) in energy space must be constant so that the electron distribution function satisfies

$$N_\gamma \propto \dot\gamma^{-1} \propto \gamma^{-2} \tag{2.50}$$

[54], [55], [56]. The associated spectral index is $\alpha = 0.5$ (*cf.*equation(2.15)).

In this calculation, we have assumed that there is just one generation of accelerated electrons. However, the escaping gamma rays pair produce secondary electrons which themselves create more gamma rays and so on. In the opposite limit of many electron generations, it is not the flow of particles in energy space which is conserved, but the flow of energy. Therefore,

$$N_\gamma \propto \gamma^{-1}\dot{\gamma}^{-1} \propto \gamma^{-3} \qquad (2.51)$$

with associated spectral index of $\alpha = 1$ [57].

Now, as we have described, there does seem to be a preferred X-ray spectral index in type 1 Seyfert galaxies, $\alpha = 0.7$ intermediate between these two limiting cases and so it was hoped that some natural feedback would establish this spectral slope under a broad range of physical conditions. Unfortunately, this has not yet been demonstrated [58][59].

2.4.4 Annihilation. The inverse process to pair production is pair annihilation. In non-relativistic plasmas, with electron speeds in the range $e^2/\hbar \lesssim v \lesssim c$, this proceeds at a rate $\sim 0.4 n_e^2 \sigma_T c$ per unit volume, roughly independent of temperature with the decrease in particle speeds being compensated by an increase in cross section as the particles cool [60][52]. n_e is the total electron plus positron density and we assume that both constituents have comparable density in a steady state. (As the radiative cooling is so efficient, annihilation of relativistic electrons is unimportant.)

The net result is that in a steady state, electron creation by γ-rays is balanced by annihilation so that

$$\frac{m_e c^3 \ell_\gamma}{4\pi \sigma_T R^2 h\nu_\gamma} \sim 0.4 n_e^2 \sigma_T c \qquad (2.52)$$

where ℓ_γ refers to those γ rays for which the source is optically thick to pair production and ν_γ is a typical γ-ray frequency. The left hand side of this equation is simply the pair production rate per unit volume. Rewriting this equation, we can solve for the mean Thomson depth in the source

$$\tau_T \sim 0.3 \left(\frac{\ell_\gamma m_e c^2}{h\nu_\gamma} \right)^{1/2} \qquad (2.53)$$

For the conditions envisaged, $\tau \sim 1-10$ and a significant population of electrons and positrons accumulates. These electrons and positrons should thermalise quickly because their energy exchange times are short compared with their annihilation times.

2.4.5 Coulomb scattering. Energy exchange between thermal electrons and positrons is mediated by Coulomb collisions. The impact parameter for two electrons to undergo a large angle collision is $b_{\pi/2} \sim e^2/k_B T$. Therefore the non-relativistic energy exchange timescale is given by

$$t_{cool} \sim (n_e b_{\pi/2}^2 v_e)^{-1}$$
$$\sim \frac{1}{n_e \sigma_T c} \left(\frac{k_B T}{m_e c^2} \right)^{3/2} \qquad (2.54)$$

where v_e is the electron thermal velocity and we have ignored the influence of distant encounters which reduce this line by a logarithmic factor. Numerically,

$$t_{cool} \sim T_8^{3/2} n_e^{-1} \text{s} \tag{2.55}$$

where $T_8 = T/10^8 \text{K}$ and n_e is measured in cm^{-3} [61][59]. This is typically shorter than the annihilation time. The electron temperature is then determined by Compton scattering and, in full numerical simulations of pair plasmas, is $T \sim 10^9 \text{K}$ [58], [59].

By contrast, relativistic Coulomb scattering is significantly slower than radiative cooling and is relatively unimportant[59].

2.4.6 Compton scattering of the emergent radiation. Having established that there may be a hot, Thomson thick pair plasma, with Comptonisation parameter, defined by equation(2.38)of order unity, we must consider that the UV radiation envisaged to be emitted by the accretion disk may emerge from the corona with a significantly hardened spectrum [62]. Time variability of UV and X-ray spectra may provide a diagnostic of this type of reprocessing[58], [59].

2.4.7 Bremsstrahlung. Bremsstrahlung radiation, is emitted when electrons collide with ions or positrons. The radiated spectrum can be estimated using simple considerations. When an electron passes by an ion it will be accelerated by the Coulomb force and radiate. The radiation electric field is given by equation(2.2). The spectrum of this radiation can be found from its Fourier transform. At low frequency, this is given approximately by

$$\tilde{E} \sim \int dt e^{-i\omega t} E \sim \frac{e\Delta v_\perp}{2\pi r c^2} \tag{2.56}$$

where $\Delta v_\perp$ is the change in perpendicular velocity. We can then use Parseval's theorem and integrate over the sphere at infinity to compute the energy radiated per unit frequency in a single encounter $\sim \tilde{E}^2 r^2 c$. Most power will be radiated by collisions in which $\Delta v \sim v$ or $b \sim b_{\pi/2} \sim e^2/m_e v^2 \sim e^2/k_B T$. The maximum energy photon that can be radiated has frequency $\nu \sim m_e v^2/2h \sim k_B T/h$. Therefore the spectral emissivity is the encounter rate multiplied by the energy radiated per unit frequency in an encounter, *i.e.*

$$j_{\nu\Omega} \sim n^2 \pi b_{\pi/2}^2 v \tilde{E}^2 r^2 c$$

$$\sim n^2 e^2 r_e^2 \left(\frac{k_B T}{m_e c^2}\right)^{-1/2}; \quad \nu \lesssim k_B T/h \tag{2.57}$$

A more careful calculation gives

$$j_{\nu\Omega} = 1 \times 10^{-22} n_{e10}^2 T_8^{-1/2} g e^{-h\nu/k_B T} \quad \text{erg cm}^{-3}\text{s}^{-1}\text{Hz}^{-1}\text{sterad}^{-1} \tag{2.58}$$

where we have assumed an electron-ion plasma with cosmic abundance. g is a slowly varying function of temperature and frequency known as the Gaunt factor, which is given in [29] and can be approximated as unity for simple estimates.

Integrating equation(2.58)over frequency and solid angle, we obtain the total emissivity

$$j \sim n^2 \left(\frac{e^2}{\hbar}\right) r_e^2 (k_B T m_e c^2)^{1/2}$$

$$= 15 n_{e10}^2 T_8^{1/2} \text{erg cm}^{-3} \text{s}^{-1}$$

(2.59)

where we again assume an average Gaunt factor ~ 1 and cosmic abundance.

Free-free radiation is also radiated in electron-positron collisions at roughly the same rate [63]. (Non-relativistic electron-electron and positron-positron collisions radiate electric quadrupole as opposed to electric dipole radiation at a slower rate.) Free-free radiation is usually of minor importance in electron-positron pair plasmas.

2.4.8 Other processes. There are other radiative processes which could be important in relativistic plasmas under conditions different from the radiation-rich conditions envisaged here. They include synchrotron emission of the soft photons[59], pair production by proton-photon [64], proton-proton (via pions) [65], and proton-proton (via neutrons) [66]scattering. Neutrons, which travel a distance $\sim 2 \times 10^{13}\gamma$cm before decaying, may even escape the AGN altogether before producing relativistic electrons.

2.5 Particle Acceleration

2.5.1 Particle acceleration sites in AGN. There are three distinct environments within an AGN where efficient acceleration of relativistic particles may occur. The first is associated with a rapid outflow such as a relativistic jet. The relativistic electrons responsible for the non-thermal synchrotron emission appear to be created locally. The second site is associated with the inflowing accretion disk, or more probably with an active corona above it. The third site is above the black hole itself where the gas density might be extremely low.

Several general particle accleration mechanisms have been proposed [67]. It should be noted at first that there are special considerations absent from better studied particle acceleration sites like the solar corona, the interplanetary medium and the interstellar medium. In particular, the high radiation energy density ensures that only very powerful, electron acceleration can overcome radiative energy loss near to the black hole.

2.5.2 Shock acceleration. The inflowing gas near a black hole moving either quasi-radially at high latitudes or through a disk at small latitude, is probably in highly supersonic motion. Shock waves will readily form and these are known to be efficient particle accelerators. Let us describe one mechanism - first order Fermi acceleration - by which shock waves can efficiently accelerate relativistic particles [68].

We idealise a shock front as a stationary plane discontinuity across which the background fluid speed decelerates from u to u/r where r is the shock compression ratio. The region where the gas flow is supersonic before encountering the shock is usually referred to as being *upstream* from the shock front and the post-shock, subsonic part of the flow is designated *downstream*. Now, in the interplanetary and interstellar media, we observe that particles are scattered by magnetic disturbances called *Alfvén waves*. Alfvén waves are transverse modes driven by magnetic tension that propagate in a conducting medium with the *Alfvén speed*, $a = \mathbf{B} \cdot \hat{\mathbf{k}}/(4\pi\rho)^{1/2}$ where ρ is the density of the background medium and $\hat{\mathbf{k}}$ is a unit wave vector. Usually the Alfvén speed is very slow and the waves can be regarded as stationary magnetic disturbances. High energy particles will interact with these waves when their relativistic gyro radii match the reciprocal of the wave vector. If the rms wave amplitude is δB then each Larmor orbit a particle will be scattered in pitch angle by an amount $\sim \delta B/B$. These scatterings will add stochastically so that the effective "collision frequency" will be

$$\nu_{coll} \sim \left(\frac{\delta B}{B}\right)^2_{k=\omega_G/v} \omega_G \qquad (2.60)$$

and the associated spatial diffusion coefficient will be

$$D = \frac{v^2}{3\nu_{coll}} \qquad (2.61)$$

If these scatterers are present in the vicinity of the shock front, (and a variety of instabilities should result in their being copiously generated there), then they cause high energy particles moving with the fluid to cross the shock front many times before finally being transmitted downstream. Every time a particle crosses the shock front it experiences a Doppler shift in energy associated with the change in reference frame. The average change in energy for a relativistic particle can be calculated to be $< \Delta\gamma/\gamma >= u/c$ per forward and backward shock crossing. Note that the particles always gain energy at a rate that is linear in the speed of the scatterers rather than quadratic as with traditional Fermi acceleration. (Traditional Fermi acceleration is a second order acceleration scheme in which high energy particles are accelerated to form a power law distribution function by colliding elastically with randomly moving magnetic clouds. We have already encountered a variant on this mechanism in our treatment of Comptonisation *cf.*equation(2.36).)

Individual particles will therefore gain energy at an exponential rate until they are transmitted downstream. However, as the particle motions are stochastic, the number of shock crossings is variable and a distribution of particle energies will be transmitted downstream. The easiest way to compute this distribution is to observe that the probability that a given particle, once downstream, does not return to the shock front is the escape probability $P_{esc} = u/c$. The fraction of the particles remaining after m scatterings is therefore $e^{-mu/c}$. However, the particles gain energy according to $\gamma = \gamma_0 e^{mu/c}$, where γ_0 is the initial energy. Therefore the fraction of accelerated particles with energy in excess of $\gamma m_e c^2$ will vary $\propto \gamma^{-1}$ and the particle distribution function will vary as the derivative of this, *i.e.*

$$N(\gamma) \propto \gamma^{-2} \qquad (2.62)$$

When this calculation is performed more generally, the result is a power law in momentum for the transmitted distribution function

$$f(p) \propto p^{-3r/(r-1)} \qquad (2.63)$$

dependent solely upon the shock compression ratio [68].

Shocks provide an efficient means of accelerating charged particles as the particle energy is drawn directly from the bulk kinetic energy of the flow. In this regard, shock acceleration is preferrable to more gradual acceleration schemes, for example those relying upon the absorption of hydromagnetic wave energy. The acceleration can be quite rapid, as is necessary to overcome radiative loss. Furthermore the primary accelerating agent, shocks, will form readily in the highly supersonic flows around a black hole.

In fact the flow speeds are not just supersonic but are likely to be mildly or even ultra-relativistic. The argument for the shape of the transmitted cosmic ray spectrum must then be modified. The acceleration becomes relatively more efficient and the spectrum is hardened [69].

There is, however, a limit to the energy to which a relativistic electron can be accelerated. In the limiting case, when wave scattering is maximally efficient, the particles will cross the shock front once per Larmor period. If the shock is relativistic, then the particle can, at best, double its energy every shock crossing. Balancing this acceleration against synchrotron loss, we obtain

$$\frac{dE}{dt} \sim eBc - 2\gamma^2 \sigma_T c U_{mag} \qquad (2.64)$$

Setting the net energy gain to zero and using equation(2.10), we obtain the result that the maximum energy that can be emitted as synchrotron radiation by shock-accelerated electrons is $\sim c/r_e \sim 10^{23}$Hz [70]. Higher energy photons can be radiated as inverse Compton radiation though.

2.5.3 Proton acceleration. The severity of the radiative losses in AGN has prompted some authors to argue that protons rather than electrons are the primary accelerated species, perhaps also by shock waves[64], [65], [66]. Much higher particle energies can be achieved in this manner. The radiating relativistic electrons can be created as secondaries through pion production when these protons collide with either thermal protons or photons (cf.section 2.4.8).

2.5.4 Electrostatic acceleration. As we shall discuss briefly below, strong, variable magnetic fields in the vicinity of a black hole will induce electric fields of comparable strength. These can accelerate electrons directly. If the radiation energy density is comparable with the magnetic energy density then the limiting energy is given by equation(2.64). Taking fiducial values, an electric field of $\sim 10^6$V cm^{-1} can be induced by a $\sim 10^3$G magnetic field. The total potential difference in the vicinity of a black hole (of radius $\sim 10^{14}$cm can therefore be as large as 10^{20}V [71]. However, using equation(2.38), we discover that inverse

Compton losses in an Eddington-limited radiation field will limit the particle energy to $\sim 10^6$. This is still more than adequate to account for the observed radiation.

2.6 Distortion of X-ray Spectra by Reflection and Transmission

2.6.1 Spectral Curvature. Recent reports that Seyfert X-ray spectra are not simple power laws but, instead, exhibit negative spectral curvature (humps) [72], have a simple interpretation in terms of Compton scattering [73][74]. Suppose that photons are incident upon a cold, plane atmosphere (*e.g.*the surface of an accretion disk), that is optically thick to Thomson scattering. Let us also suppose, somewhat arbitrarily, that the incident spectrum is a power law with the standard spectral index $\alpha \sim 0.9$ (*cf.*equation(2.51). A purely elastic scattering atmosphere, would eventually reflect all the incident radiation without absorption and leave the power law shape unchanged. However, at low energy $\lesssim 10\mathrm{keV}$ the high Z ions in the plasma will absorb the X-rays with an absorption cross section that exceeds σ_T. Averaging over the common elements a rough approximation to the absorptive opacity of partially ionised gas is

$$\kappa_{abs} \simeq \left(\frac{Z}{Z_\odot}\right)\left(\frac{E}{10\mathrm{keV}}\right)^{-2.5}\kappa_T \qquad (2.65)$$

As a typical photon must be scattered several times before being reflected, there will be partial absorption and suppression of the spectrum below $\sim 10\mathrm{keV}$ (*cf.*Fig.(2.11)). A combination of direct power law and reflected spectra may account for the observations. The cold material need not be confined to a disk and might, instead, comprise many clouds [75].

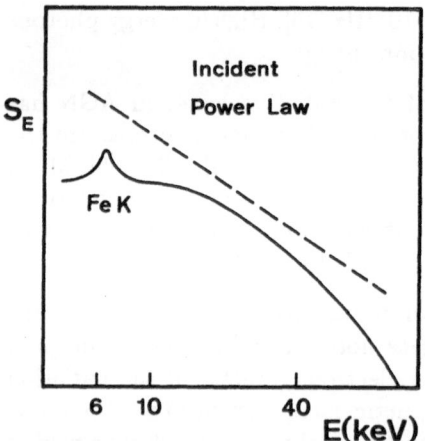

Fig. 2.11. Distortion induced in the reflection spectrum when a power law X-ray spectrum is incident upon a cold, thick plasma slab. Absorption edges of heavy elements are responsible for the flux reduction below 10keV. Energy loss due to Compton recoil is responsible for the curvature of the spectrum above $\sim 40\mathrm{keV}$.

Above 40 keV, the Compton shift of the incident photon becomes important. Photons lose energy by ~ 10percent per scattering at ~ 40 keV and so a few scatterings are all that is needed to produce a significant distortion.

2.6.2 Iron fluorescence. Another signature of the coexistence of cold plasma with the X-ray source is the report of iron fluorescence lines [76]. The most prominent lines are K lines mainly produced by recombination after the iron atoms are photoionised. The photoionisation threshold varies between $7-9$ keV as the ionisation state varies. The associated Kα photon energy varies similarly between 6.1-6.5keV.

It is conventional to measure the strength of the Kα emission line by its equivalent width. In order to compute this, we need to know the yield Y of Kα photons per recombination and the incident photon spectrum. A straightforward calculation gives

$$EW = \frac{\int d\Omega \int_{\nu_0} d\nu I_\nu Y (1 - e^{-\tau_0})}{4\pi I_\nu(\nu_\alpha)} \qquad (2.66)$$

where ν_0 denotes the relevant edge frequency and the ν_α refers to the Kα line. Evaluating this expression, we obtain

$$EW \simeq 300 \left(\frac{\Delta\Omega}{4\pi}\right) \left(\frac{Z}{Z_\odot}\right) \tau_T \mathrm{eV} \qquad (2.67)$$

The reported equivalent widths of iron lines in Seyfert galaxies are in the range $100 - 300$eV, which suggests that most lines of sight are Thomson thick or the metal abundance is significantly in excess of solar.

2.7 Re-radiation by Dust Grains

In recent years, it has been discovered that nearby AGN contain surprisingly large quantities of molecular gas [77]. Presumably this gas, like molecular gas in the solar neighbourhood, contains dust which can absorb ultra-violet and optical photons and re-radiate it as infra-red radiation [78][79]. Interstellar dust is found to comprise grains with sizes a in the range $30 - 3000$ Å containing predominantly silicates, graphite, and in cooler regions, ice. In recent years, it has become clear that there is a continuity at small a to macromolecules like *polycyclic aromatic hydrocarbons*. Grains absorb and scatter light, both processes effectively removing photons from a beam. The effective cross section for a beam of light is related to the geometrical cross section by two factors Q_a, Q_s which are both functions of wavelength. The number of magnitudes of visual extinction along the line of sight to a star is therefore

$$A_V = 1.1 N_g \sigma_g (Q_a^V + Q_s^V) \qquad (2.68)$$

where N_g is the column density of grains and σ_g is their geometrical cross section [80]. (The numerical coefficient is just $2.5/\ln 10$.) When there is a solar

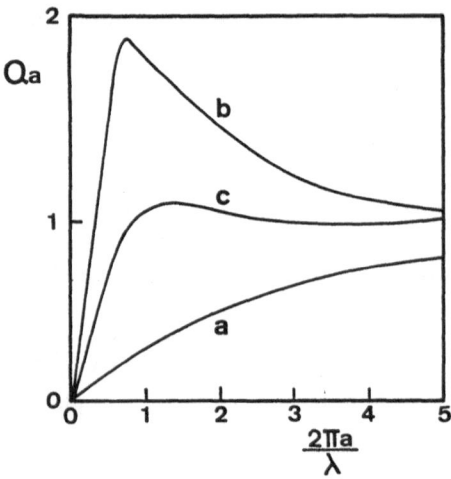

Fig. 2.12. Approximate absorptive extinction efficiency for spheres of a) dirty ice[80], b) iron[80] c) graphite [82]. Note that $Q_a \propto \lambda^{-1}$ at long wavelengths. Q_a is quite sensitive to the assumed grain shape.

metal abundance, the majority of the heavy elements are confined in grains which therefore have a mass density $\rho_g \sim 6 \times 10^{-3}\rho$ where ρ is the total density. Equation(2.68) can then be converted into two equivalent useful forms

$$A_V \sim \left(\frac{N_H}{2 \times 10^{21}\text{cm}^{-2}} \right)$$
$$\kappa_V \sim 200\text{cm}^2\text{g}^{-1}$$

(2.69)

Usually, dusty regions in an AGN are unresolved. We are then only interested in the absorptive extinction efficiency, Q_a. For short wavelengths, $\lambda \lesssim 2\pi a$, this is roughly unity. For $\lambda \gtrsim 2\pi a$, $Q_a \propto 2\pi a/\lambda$ ((2.12)) where the coefficient of proportionality depends upon the composition but is generally of order unity [81].

3 Black Holes

3.1 Motivation for Black Holes

3.1.1 Discovery of extraglactic radio sources and quasars. Soon after the discovery of powerful radio galaxies and quasars, it was argued that their power was ultimately gravitational in origin; nuclear and atomic processes were just too inefficient [83]. Early studies of relativistic stars [13], proved unpromising [84], and attention focussed on massive black holes [6], [7], spinars [14] and star clusters [2]. Evidence against stellar processes mounted as some individual outbursts were found to have energies in excess of a stellar rest masses and variability timescales shorter than those normally associated with supernovae [85]. It also proved hard to account for compact radio jets in the context of the star cluster model [86], [87], [88].

The argument against supermassive stars and spinars was essentially theoretical. They were argued to be either dynamically unstable or at least relatively short lived [89], Although they may possibly be identified with certain classes of AGN, they were thought unlikely to account for the majority.

3.1.2 Current evidence. The most compelling arguments in favour of the black holes include:

(i) Rapid variability, with timescales as short as 1 min in low power Seyfert galaxies [49]. The associated light travel time across the source can be as small as the Schwarzschild radius of a $\sim 10^7 M_\odot$ black hole. The characteristic variability timescale increases roughly with the luminosity of the AGN.

(ii) High efficiency of conversion of rest mass into radiant energy [90], [70]. The way that this argument works is that we sum the power of all quasars observed at earth and find that the associated radiation energy density of their *emitted* radiation is, in appropriate units, $\sim 10^5 h^{-3} M_\odot c^2$ per bright galaxy, (where h is the Hubble constant measured in units of 100 km s^{-1} Mpc^{-1}) (*cf.*section 7.5.1). Observations of the nebulosity around modest redshift quasars indicate that most quasars are indeed identified with bright galaxies. The nuclear masses have, presumably, only increased since the epoch of maximal quasar activity and so by limiting the central masses in nearby bright galaxies we can place a lower bound on the efficiency ϵ equal to the ratio of the radiation energy emitted to the nuclear mass. It is found that $\epsilon \gtrsim 0.01 h^{-3}$. Despite the uncertainty in several key factors, it is hard to see how a sufficiently high efficiency of conversion of mass into radiant energy can be brought about by nuclear or atomic processes. This leaves gravitational energy as the likely origin. The variability strongly suggests that a single coherent object is present, rather than, for example, a cluster of stellar mass black holes or neutron stars. This leaves massive black holes or relativistic stars as the likely energy source. It is hard to rule out massive relativistic stars on observational grounds, although no convincing, persistent periodicity, which might have been associated with rotation, has been reported from any AGN. The theoretical argument against relativistic stars is that they cannot avoid catastrophic dynamical instability [91]. The structure of spinars and magnetoids is less well defined but they too are believed to be no more than an ephemeral stage in the life cycle of an AGN [89].

(iii) Many of the brighter flat spectrum radio sources have been shown to exhibit superluminal expansion, that is to say, they display radio features which separate with kinematic speeds in excess of that of light [92]. Although the detailed interpretation of this phenomenon is still a matter for debate, superluminal expansion is to be expected when there are relativistic fluid motions in a source region and relativistic speeds are to be expected if there is a relativistically deep gravitational potential well.

(iv) Another piece of radio astronomical evidence is the persistence of radio source axes fixed in space for times no shorter than 10 million years in some examples [93]. A compact gyroscope, like a spinning black hole, is a natural way to accomplish this.

(v) As the central velocity dispersion of stars within the nucleus of a galaxy is typically $\sim 10^{-3}c$, a black hole of mass M, should dominate the gravitational potential out to $\sim 10^6$ Schwarzschild radii $\sim 10 M_8 \mathrm{pc}$, (where $M_8 = M/10^8 \mathrm{M_\odot}$ henceforth). Small increases in the central velocity dispersion have been reported within nearby galaxies, [94], [95], [96], including our own [97]. Dynamical determinations of the mass give values ranging from $\sim 3 \times 10^6 \mathrm{M_\odot}$ to $\sim 5 \times 10^8 \mathrm{M_\odot}$. Relatively contrived stellar distributions involving anisotropic velocity dispersions can mimic the effect of a central mass [98]. Even if there is a central mass it could be a compact star cluster, though, again specialised conditions must again be invoked to ensure its survival for a Hubble time [99].

(vi) Careful spectroscopy has revealed that most galactic nuclei have faint broad emission lines, suggesting that most galaxies have deep central potential wells [100].

These observations, taken together and discussed more critically elsewhere [101], strongly motivate the *black hole hypothesis*, namely that essentially all galactic nuclei contain $\sim 10^6 - 10^9 \mathrm{M_\odot}$ black holes in their nuclei and that these objects together with their orbiting accretion disks are the prime movers for most of the powerful activity. Note that this does not preclude a significant contribution to the radiation in lower power AGN, notably starbursts, coming from massive stars, supernovae and compact stellar remnants. It hardly needs saying that a well-motivated hypothesis is not the same thing as a proof.

Discussions of the properties of massive black holes in the environment of an AGN include [18], [19], [20], [101]. A conference proceedings organised around this topic is [102].

3.2 Spinning Black Holes

3.2.1 Numerical values. For the purposes of making numerical estimates, we note that the effective size of a black hole is of order the gravitational radius

$$m = \frac{GM}{c^2} \sim 1.5 \times 10^{13} M_8 \mathrm{cm} \tag{3.1}$$

where $M_8 = M/10^8 \mathrm{M_\odot}$. This is far too small to be resolved by any telescope. The associated light crossing time is

$$t_g = \frac{GM}{c^3} \sim 500 M_8 \mathrm{s} \tag{3.2}$$

The spin angular momentum, S, per unit mass, M, of a black hole is conveniently written

$$S/M = ac$$

$$= 5 \times 10^{23} \left(\frac{a}{m}\right) M_8 \text{cm}^2 \text{s}^{-1} \tag{3.3}$$

where $a < m$ for a black hole. This is much smaller than the specific angular momentum of interstellar gas within the surrounding galaxy $\sim 10^{30} \text{cm}^2 \text{s}^{-1}$. For this reason, accretion onto the black hole is believed to proceed via an accretion disk.

Henceforth, in this section, it will be convenient to set $G = c = 1$ so that mass, time and specific angular momentum are all measured in units of length.

3.2.2 Black holes. Black holes are solutions of the general relativistic field equations for an asymptotically flat spacetime possessing an *event horizon* [103], [104], [105]. An event horizon is a 3D hypersurface that separates events that can be seen from infinity from those that cannot be seen. Black holes are usually regarded as stationary and it is consequently allowed to treat the event horizon as a two dimensional surface. The event horizon surrounds a singularity where the laws of classical physics must break down. The *cosmic censorship conjecture* implies that this singularity will always be unobservable classically. It appears to be true.

The simplest black holes to analyse do not spin and are spherically symmetric. Their spacetime geometry is described by the Schwarzschild metric, which is commonly expressed in Schwarzschild coordinates. In these coordinates, the radius of the event horizon is $r = 2m$.

3.2.3 Kerr metric. A surprising result of classical general relativity is that there is essentially only one two parameter family of uncharged black holes. (The charge is actually a third parameter, but it can be shown that naturally occuring black holes cannot contain a gravitationally significant quantity of charge; they would discharge in less than a light crossing time. We will therefore ignore charge.) The space-time associated with this family of black holes is known as the Kerr metric. A convenient choice of coordinates for many astrophysical purposes are *Boyer-Lindquist* coordinates (t, r, θ, ϕ). The line element is given by

$$ds^2 = -\left(1 - \frac{2mr}{\Sigma}\right) dt^2 - \frac{4amr \sin^2 \theta}{\Sigma} dt d\phi + \frac{\Sigma}{\Delta} dr^2$$
$$+ \Sigma d\theta^2 + \left(r^2 + a^2 + \frac{2mra^2 \sin^2 \theta}{\Sigma}\right) \sin^2 \theta d\phi^2 \tag{3.4}$$

where

$$\Sigma = r^2 + a^2 \cos^2 \theta$$
$$\Delta = r^2 - 2mr + a^2 \tag{3.5}$$

[103], [91]. We can read off the metric coefficients $g_{\alpha\beta}$ from equation (3.5).

Let us interpret these various parameters operationally. The mass m can be measured by timing a satellite in a circular orbit of radius $r >> m$ and using Kepler's laws. The specific angular momentum a is required not to exceed m and can be measured by allowing a gyroscope to fall freely in a circular orbit in the equatorial plane with radius $r >> m$ and precess with respect to a non-rotating frame at infinity at a rate $-am/r^3$.

The radius coordinate r must also be interpreted. Consider, for simplicity, the Schwarzschild metric with $a = 0$. It can be measured in the equatorial plane by forming a circle with fixed radius and dividing the proper length of the circumference by 2π. Note that increasing the radius by a proper distance $d\rho$ produces an increase in the circumference $dC < 2\pi d\rho$. This is a consequence of the curvature of the spatial hypersurfaces around the black hole. This naturally generalises to the case of the Kerr metric.

Next, let us interpret the time coordinate t. Once again, just consider the Schwarzschild metric. If we measure the proper time interval $d\tau$ between two events occuring at fixed (r, θ, ϕ), we find that the change in the coordinate time $dt = (1 - 2m/r)^{-1/2} d\tau$. This is the time that would be measured by an observer watching at a large distance from the hole and it becomes infinite as $r \rightarrow 2m$. However, if an observer were to hover just above the horizon, his acceleration with respect to an observer freely-falling into the black hole would become infinite as $r \rightarrow 2m$.

3.2.4 Gravitational orbits in a Schwarzschild metric.
We can illustrate some of the key properties of black holes by analysing gravitational orbits in a Schwarzschild spacetime ($a = 0$). As the Schwarzschild metric is independent of the two coordinates t, ϕ, there are two constants of the motion, the energy and the angular momentum. These are defined in terms of the covariant components of the proper velocity

$$e = -u_0 = (1 - 2m/r)\dot{t}$$
$$h = u_\phi = r^2 \dot{\phi} \tag{3.6}$$

[103], [91]. If we use the invariant length of the proper velocity, $u^\alpha u_\alpha = -1$, then we obtain

$$\dot{r}^2 = e^2 - \left(1 - \frac{2m}{r}\right)\left(1 + \frac{h^2}{r^2}\right) \tag{3.7}$$

It is easiest to think about the solutions to equation(3.7)by treating $(1 - 2m/r)(1 + h^2/r^2)/2$ as an effective potential $V(r)$ and the constant $e^2/2$ as a Newtonian energy. The problem is then equivalent to that of the motion of a particle in a one dimensional potential well (Fig. (3.1)).

Stable circular orbits have radii given by minima in this potential

$$r_{min} = \frac{h^2}{2m}[1 + (1 - 12m^2/h^2)^{1/2}]; \quad h > 12^{1/2} m \tag{3.8}$$

If the particles have energy $e > [2V(r_{min})]^{1/2}$, then they can oscillate in radius about this minimum. In other words they will describe non-circular orbits.

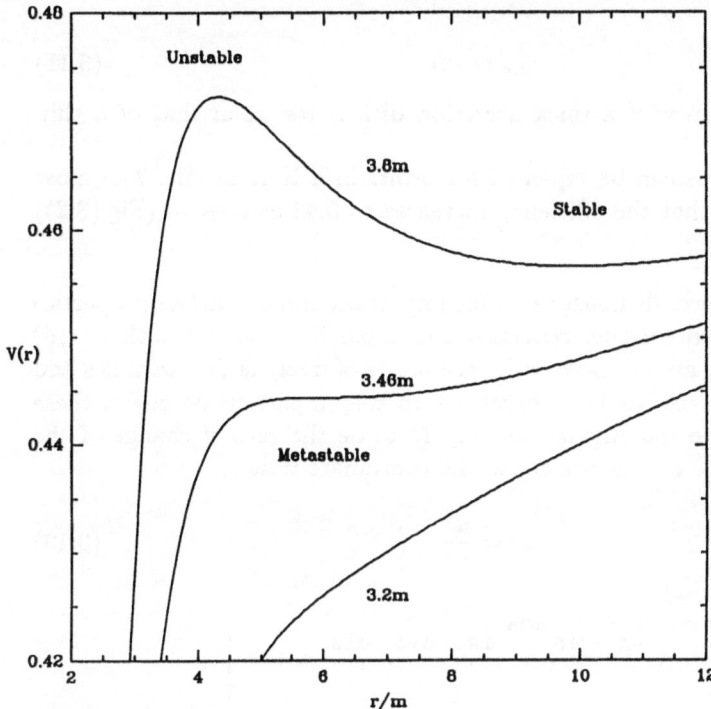

Fig. 3.1. Effective potential $V(r)$ for a particle in a gravitational orbit in a Schwarzschild metric.

However, unlike Keplerian orbits in a Newtonian potential. these orbits do not close. Orbits with radii given by maxima in the potential will be unstable. The smallest stable circular orbit has radius

$$r_{ms} = 6m \tag{3.9}$$

and the associated energy is

$$e_{ms} = \frac{8^{1/2}}{3} \tag{3.10}$$

and so the binding energy of the least stable circular orbit is $1 - e_{ms} = 0.057$.

This quantity is of interest in the theory of accretion disks. Specifically if gas spirals slowly inward through a thin accretion disk toward a non-rotating black hole then at any instant it will follow a circular gravitational orbit to a good approximation. However, the disk can only exist as long as these orbits are stable; thereafter it will spiral quickly into the hole. So at its inner edge, the gas will only have liberated $0.057c^2$ of energy per unit mass. This is the efficiency of the accretion disk.

This argument must be modified if the disk is thick and pressure forces cannot be neglected. In this case, orbits can be stabilised within r_{ms} and can extend all the way to the radius where $V(r)$ is maximised. If we allow the angular momentum to vary but require the orbit to be bound, then the marginally bound orbit is

$$r_{mb} = 4m \qquad (3.11)$$

Of course the efficiency of a thick accretion disk is less than that of a thin accretion disk.

These calculations can be repeated for orbits in a Kerr metric. The most important result is that the efficiency increases to 0.42 as $a \rightarrow m$ (Fig.(3.2)) [103], [91].

3.2.5 Orbits. In order to demonstrate some important and distinctive properties of Kerr holes, we next consider general orbits of particles moving so that (r, θ) are constant. (These are not necessarily the orbits of freely falling particles and non-gravitational forces would be necessary to keep a particle on one of these orbits.) Let us define the angular velocity Ω to be the rate of change of the azimuthal coordinate ϕ with respect to the coordinate time t,

$$\Omega = \frac{u^\phi}{u^0} \qquad (3.12)$$

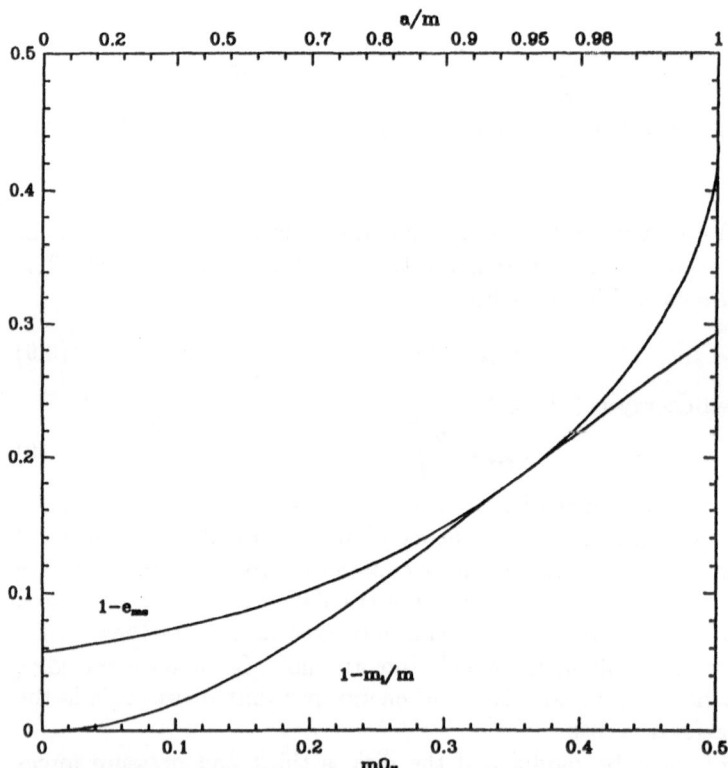

Fig. 3.2. Efficiency of energy release by gas accreting through a thin accretion disk onto a spinning black hole. The quantity plotted is $1 - e_{ms}$ as a function of the hole angular frequency Ω_H, and the specific angular momentum a. Also plotted is the extractable fraction of the hole's mass.

where u^α is the proper velocity of the particle. Using the invariant length of the proper velocity, $u^\alpha u_\alpha = -1$ we obtain a quadratic for Ω,

$$\Omega^2 g_{\phi\phi} + 2\Omega g_{0\phi} + g_{00} + (u^0)^{-2} = 0 \qquad (3.13)$$

As u^0 is real, we obtain the inequalities

$$\Omega_{min} = \frac{-g_{0\phi} - (g_{0\phi}^2 - g_{00}g_{\phi\phi})^{1/2}}{g_{\phi\phi}} < \Omega < \frac{-g_{0\phi} + (g_{0\phi}^2 - g_{00}g_{\phi\phi})^{1/2}}{g_{\phi\phi}} = \Omega_{max} \qquad (3.14)$$

In flat space, $g_{00} = -1$, $g_{0\phi} = 0$ and these conditions simply reduce to the requirement that the space velocity in an inertial frame be less than that of light. However, around a Kerr black hole, with $a > 0$ and $g_{0\phi} < 0$, we discover that $\Omega_{min} > 0$ when $g_{00} > 0$, that is to say when

$$r < r_E = m + (m^2 - a^2 \cos^2 \theta)^{1/2} \qquad (3.15)$$

When this happens, physical particles on time-like geodesics at fixed r, θ, ϕ must orbit with respect to non-rotating distant observers. r_E is known as the *static limit*. Its existence is a consequence of the dragging of inertial frames by the spinning black hole. A second critical surface is the *event horizon* where photons (and all material particles) are trapped and can no longer increase their radius r. Equivalently, it is a surface of infinite redshift when seen from large distance. It is located where $\Delta \to 0$. Using equation(3.5), we find that its radius is given by

$$r_+ = m + (m^2 - a^2)^{1/2} \qquad (3.16)$$

independent of θ.

At the event horizon, the upper and lower bounds on Ω approach a common value

$$\Omega_H = \frac{a}{2mr_+} = \frac{a}{r_+^2 + a^2} \qquad (3.17)$$

independent of the latitude. This is defined to be the angular velocity of the hole[103], [91].

The region between the event horizon and the static limit is sometimes called the *ergosphere*. It has the remarkable property that particles moving freely across the event horizon can have negative total energy (rest mass plus gravitational plus kinetic in the Newtonian limit). These orbits do not extend outside the static limit but can cross the event horizon. So, if we imagine a particle placed on such a negative energy orbit and allow it to cross the event horizon with this energy, then the mass of the hole will *decrease*. This thought experiment, known as the *Penrose process* [106], demonstrates that it is possible to extract spin energy from a Kerr black hole.

By analysing the behaviour of null geodesics at an event horizon, Hawking [107], was able to show that the area A of the horizon

$$A = \int d\theta d\phi (g_{\theta\theta}g_{\phi\phi})^{1/2} = 8\pi m r_+ \qquad (3.18)$$

cannot decrease. This result is reminiscent of the law of increase of entropy, (an analogy which turns out to be exploitable [108], and which leads ultimately to the phenomenon of *Hawking radiation* [109]). It is convenient to define an *irreducible mass*

$$m_i = \left(\frac{A}{16\pi}\right)^{1/2} \tag{3.19}$$

which cannot decrease as mass (including other black holes) is added to the black hole.

An interesting way to rewrite equation(3.19)is to express the total mass in terms of the irreducible mass,

$$m = \frac{m_i}{(1 - 4m_i^2\Omega_H^2)^{1/2}}; \quad \Omega_H < \frac{1}{8^{1/2}m_i} \tag{3.20}$$

where the upper bound on Ω_H is the requirement that $a < m$. (It can be shown that whenever the spin of a black hole is increased, there must be an associated mass increase so that a remains less than m.) Equation(3.20)is reminiscent of the relativistic mass formula from special relativity. From it, we obtain a bound on the *reducible* mass.

$$m - m_i < 0.29m \tag{3.21}$$

Therefore up to 29 per cent of the mass of a spinning black hole can be extracted, either by a Penrose process, or, as we shall see, electromagnetically.

3.2.6 Lense-Thirring precession. The spin of a black hole can also be reponsible for a variety of precessional effects which might have observational consequences. It is easiest to treat these effects in the post-Newtonian approximation using a formulation of weak-field gravity that is modelled on classical electromagnetic theory [110], [101]. In this formulation, the gravitational field is split into a *gravito-electric* field, $g = -\nabla\Phi$ and a *gravito-magnetic* field $h = \nabla \times a$, where

$$\Phi = \frac{(1 - g_{00})}{2} \tag{3.22}$$

$$a_i = g_{0i}$$

If we expand in powers of $v = O(r/m)^{-1/2}$ and suppose that $|h| = O|vg/c|$, then it can be shown that g, h satisfy four Maxwell-like equations.

$$\nabla \cdot g = -4\pi\rho$$
$$\nabla \cdot h = 0$$
$$\nabla \times g = 0 \tag{3.23}$$
$$\nabla \times h = -4\left[4\pi\rho v - \frac{\partial g}{\partial t}\right]$$

The corresponding weak field equation of motion is

$$\frac{\partial v}{\partial t} = g + v \times h \tag{3.24}$$

Note the different numerical factors from electromagnetic theory which are a consequence of the need to describe the gravitational field using tensors and the sign changes which follow from the attractive nature of gravity. Note the absence of a $\partial h/\partial t$ term in the third of equations(3.23), and the presence of a displacement current, consistent with our ordering, but different from the ordering normally encountered in electromagnetic theory.

We can use this formalism to compute the precession rate for a ring of matter orbiting about a slowly spinning hole. Let us first compute the gravito-magnetic field exploiting the well known calculation of the magnetic field from a magnetic dipole. We find that

$$h = \frac{3(S \cdot r)r - Sr^2}{r^5} \tag{3.25}$$

where $S = am$ is the spin of the hole. The gravito-magnetic torque that the hole exerts on the ring can be computed by integrating the torque per unit mass $r \times (v \times h)$ around the ring and is given by

$$G = \frac{2S \times l}{r^3} \tag{3.26}$$

where l is the orbital angular momentum of the ring. (The torque that the ring exerts on the hole is the negative of this.) We can extract the precessional frequency of the ring from equation(3.26)

$$\Omega_P = \frac{2S}{r^3} \tag{3.27}$$

This *Lense-Thirring* precession[105], is important for understanding how an inclined accretion disk will behave. It has been supposed that gas will settle into the equatorial plane when its inflow timescale equals the precessional period [111]. The inner parts of an accretion disk should then lie in the equatorial plane of the hole, and if jets are launched perpendicular to the disk, this might explain the observed stability of radio jet axes (see below). However, recent studies suggest that the dynamics of twisted accretion disks is more complicated than this [112].

3.2.7 Binary black holes and geodetic precession. A second application of this formalism is to the dynamical interaction of two black holes. When treated as point masses, they will orbit each other with circular orbits gradually losing energy by their frictional interaction with stars and gas in the nucleus. The Keplerian orbital period when their separation is r pc is

$$P \sim 10^4 r^{3/2} M_8^{-1/2} \text{yr} \tag{3.28}$$

where M is interpreted as the mass of the heavier hole. This is generally too long to be observed directly because when $r \lesssim 0.003$pc, gravitational radiation becomes effective (see below). However, if both black holes possess spin, they will also interact through their gravito-magnetic fields. Their spins and the orbital angular momentum will precess about the direction of the total angular momentum. This is called *geodetic precession*[105]. (If the central star cluster is sufficiently non-spherical, then the orbit will be subject to Newtonian precession as well.) We can again use our weak field formalism to compute the rate of precession.

Consider mass m_2 moving through the gravito-electric field g of a mass $m_1 >> m_2$. When we transform into an inertial frame moving with mass m_2, we find that there will be a gravito-magnetic field $h = -v \times g$ which will contribute a spin-orbit precession with angular frequency $\Omega_{SO} = -h/2 = m\ell/2r^3$, where ℓ is the orbital angular momentum. However, this is not the whole story. As space is curved, there will be an additional contribution to the precession associated with the space curvature of $2\Omega_{SO}$ (see Fig.(3.3).)

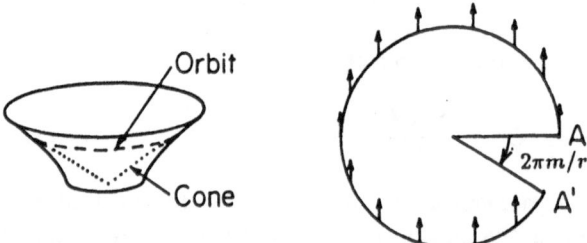

Fig. 3.3. Origin of space curvature precession. If we adopt the trumpet horn shape obtained by embedding the curved equatorial plane in three dimensional space, we find that we can approximate the space locally as being flat and equivalent to the the space formed by a cone tangential to the curved shape. If we now unfold this cone, we find that there is an angle deficit, $2\pi m/r$, so that each time an angular velocity vector is transported (technically parallel transported) around an orbit, it appears to rotate through an angle $2\pi m/r$.

If we generalise to the case $m_1 \sim m_2$, the final result is

$$\Omega_{geo} = \left[\frac{3(m_1 + m_2)^2 + m_1 m_2}{m_1 + m_2} \right] \frac{l}{2r^3} \qquad (3.29)$$

The incidence of binary black holes might be quite high (*cf.*section 6.4.2). The nucleus of the ingested galaxy will be denser than the outer parts of the capturing galaxy and will therefore sink toward the bottom of the potential well in a few orbital periods, gradually losing its peripheral stars through tidal forces as the density in the capturing galaxy increases [113].

Several observed phenomena have been cited as evidence for black hole precession (Fig.(3.4).) These include:

(i) Double nuclei, most recently that observed at infra red wavelengths in the core of the infra red luminous galaxy Arp 220 [114].

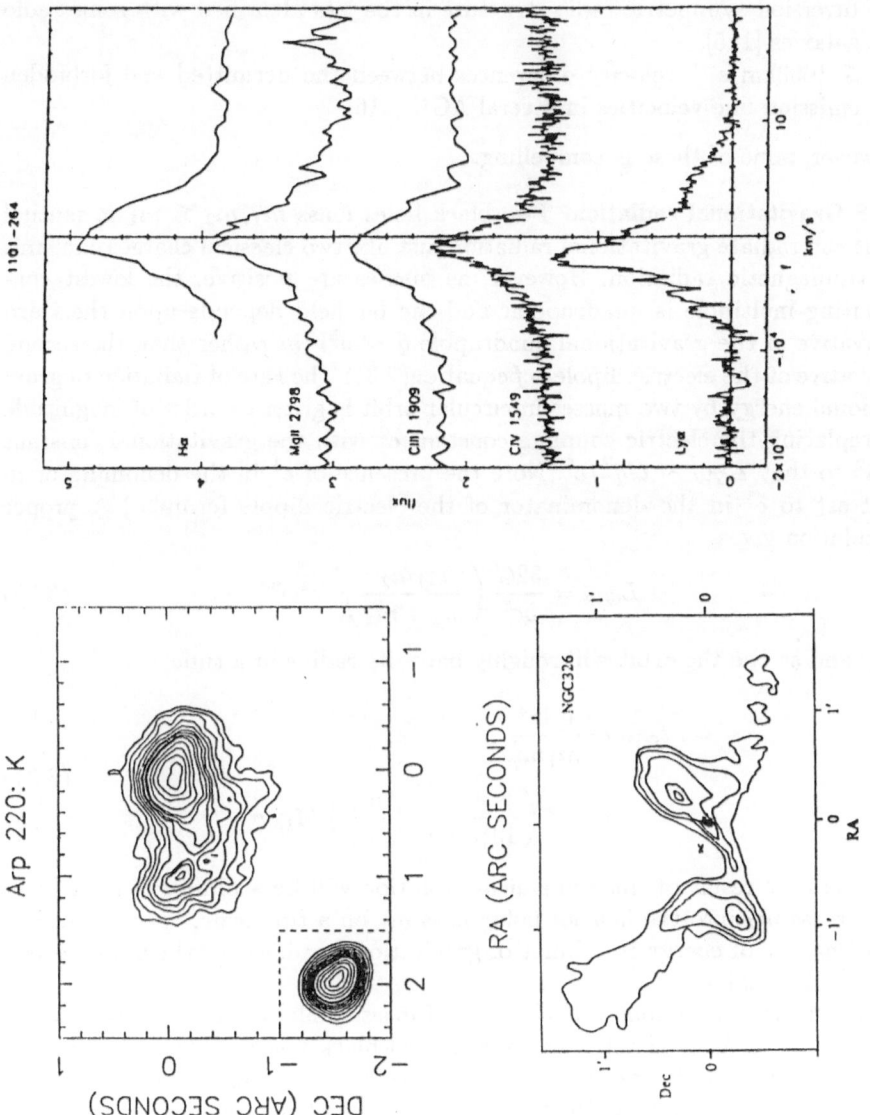

Fig. 3.4. Three types of evidence that have been cited for the existence of binary black holes in AGN. a) Double nucleus in the core of Arp220[114]. b) Inversion symmetric radio structure in the radio galaxy NGC326[115]. c) Velocity shift between the permitted and forbidden lines in the low redshift quasar Q1101-264[116].

(ii) Inversion symmetric radio structure in the jets identified with some radio galaxies [115].

(iii) $\lesssim 1000$km s^{-1} velocity differences between the permitted and forbidden emission line velocities in several AGN [116].

However, none of these is compelling.

3.2.8 Gravitational radiation. Two black holes mass $m_1, m_2 \lesssim m_1$ in mutual orbit can radiate gravitational radiation just like two classical charges can emit electromagnetic radiation. However, as masses are positive, the lowest contributing multipole is quadrupolar and the far field depends upon the third derivative of the gravitational quadrupole $\dddot{q} \sim \omega^3 r^2 m$ rather than the second derivative of the electric dipole (*cf*.equation(2.3)) The rate of radiation of gravitational energy by two masses in circular orbit is given to order of magnitude by replacing the electric coupling constant e^2 with the gravitational constant Gm^2 so that $L_{GW} \sim G\dddot{q}^2/c^5$. Note the presence of c^5 in the denominator in contrast to c^3 in the denominator of the electric dipole formula.) A proper calculation gives

$$L_{GW} = \frac{32G}{5c^5} \left(\frac{m_1 m_2}{m_1 + m_2} \right)^2 r^4 \omega^6 \tag{3.30}$$

[32], and so the the orbit will roughly halve its radius in a time

$$
t_{GW} \sim \frac{0.1 r^4}{m_1 m_2^2}
$$
$$
\sim 10^6 \left(\frac{r}{10^3 m_1} \right)^4 \left(\frac{m_1}{m_2} \right) M_{18} \text{yr}.
\tag{3.31}
$$

The final result of this dynamical evolution will be a plunge of the lighter hole, mass m_2 into the heavier hole, mass m_1 on a timescale, $\sim m_1$, releasing $\sim 0.1 m_2^2/m_1$ of energy in a burst of gravitational radiation, which might, one day, be detectable.

There are two reasons why a cluster of massive black holes is unlikely to be built up in an AGN. Firstly, newly captured black holes might undergo three body interactions with pre-existing black hole binaries, which result in ejection of all three holes with more than the escape velocity from the nucleus. Secondly, the final plunge of a low mass hole into a high mass hole can produce an asymmetric burst of gravitational radiation which carries away *linear* momentum and leaves the black hole remnant with a significant recoil velocity, sufficient to escape the nuclear potential well [117]. There might well be as many such black holes roaming intergalactic space as there are galaxies, although they will be extremely hard to detect.

3.3 Magnetic Effects

3.3.1 Electromagnetism in a curved spacetime.

We have shown that up to 29 per cent of the mass of a black hole is, in principle, extractable. One possible way in which this may happen is through the application of large scale electromagnetic torques. As this involves some unfamiliar and still controversial features, we will illustrate some properties of electromagnetic fields around a spinning black hole with the aid of four thought experiments[105], [101].

First let us place a small black hole, in a uniform electric field and ask how the electric field lines are affected (Fig.(3.5).) This question can be answered by

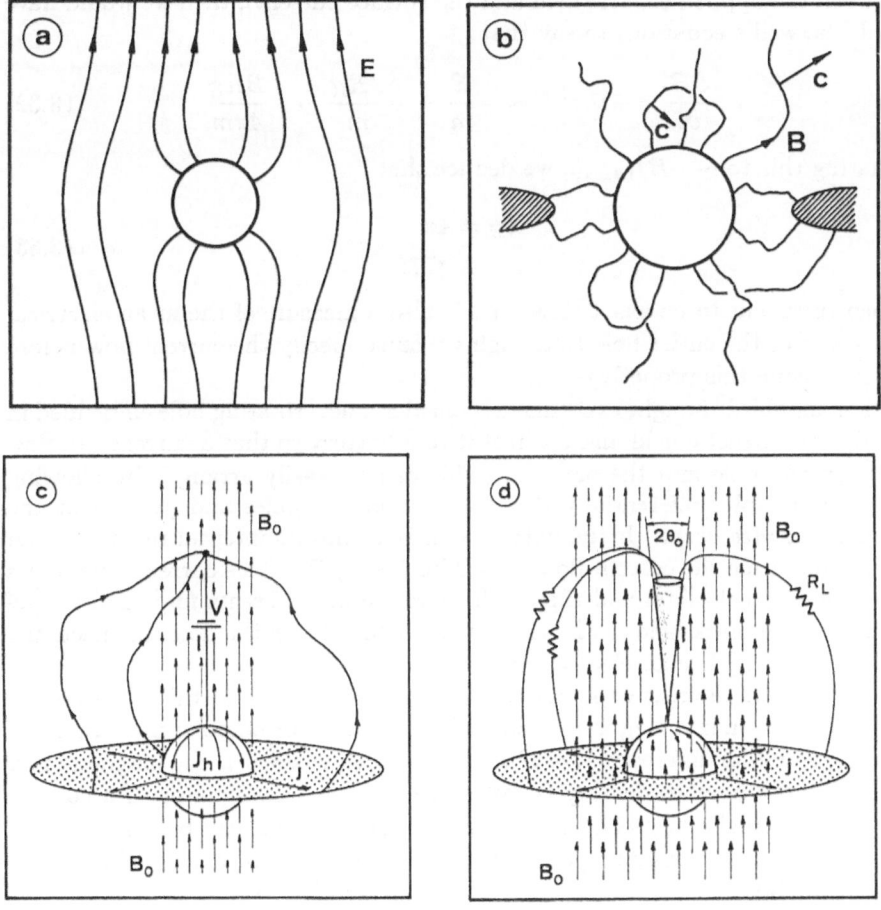

Fig. 3.5. Four thought experiments that illustrate the electromagnetic properties of a black hole. a) Hole in a uniform electric field. b) Decay of transient electromagnetic field. c) A Schwarzschild black hole, in a uniform magnetic field, starts to spin when a current flows through it. d) When a hole spins in a magnetic field, a potential difference is induced between the pole and the equator. If these are connected to an external resistor through wires, then a current will flow and power can be dissipated in the resistor. This power is derived from the spin of the hole. (Adapted from [105].)

solving the general relativistic version of Maxwell's equations in a Schwarzschild metric. This is a straightforward exercise and we find that the hole's event horizon becomes an equipotential surface. Suitably defined electric field lines cross the horizon normally. In other words, the horizon behaves in much the same way as an electrical conductor.

Our second thought experiment demonstrates that it is not a perfect conductor. Suppose that a magnetised cloud of plasma is engulfed by the hole. After the matter has crossed the horizon and no external currents can flow, there will be rapidly varying and decaying electromagnetic fields around the horizon. The decay timescale t_{decay} is typically a few light crossing times. Now if we had endowed the horizon with a surface electrical resistance R_H (the ratio between the applied electric field and the surface current), then we would have used Maxwell's equations to say that

$$\frac{\partial B}{\partial t} = -\nabla \times E \sim -\frac{E}{m} \sim -\frac{JR_H}{m} \sim -\frac{BR_H}{4\pi m} \tag{3.32}$$

equating this to $\sim -B/t_{decay}$, we deduce that

$$\begin{aligned} R_H &= 4\pi \\ &\equiv 377\Omega \end{aligned} \tag{3.33}$$

which turns out to be exact [118]. R_H is also a measure of the mean electrical resistance of the entire hole (although we must specify the current flow before we can define this properly).

For the third thought experiment, consider a non-rotating hole embedded in a uniform magnetic field and connect it to a battery so that a current can flow between the pole and the equator. (This can be easily arranged by allowing electrons to fall preferentially into the hole at the poles and protons at the equator, or *vice versa*.) If the battery emf is V and its internal resistance can be ignored, then the current flowing will be $I \sim V/R_H$. This current must cross the magnetic field lines and there will therefore be a Lorentz force and torque acting on the black hole $\sim IB$ which will act to spin up the hole. In effect, the hole is behaving like an electric motor.

The fourth thought experiment is really the converse of this. The hole is supposed to spin in a magnetic field. As it is a conductor unipolar induction will cause a potential difference $V \sim \Omega_H m^2 B \sim \Omega_H \Phi$, where Φ is the magnetic flux threading the hole, to develop between the pole and the equator. If a current can flow externally between the pole and equator, then it can be dissipated or do work on the surrounding gas. Either way, energy will be extracted from the spin of the hole.

We can quantify this by considering two adjacent field surfaces in an axisymmetric field distribution (*cf.*(3.6)). We can decompose the poloidal current flow into superposed circuits as shown. If the current flowing around the circuit is I and the flux between the magnetic surfaces $\Delta\Phi$, then we can compute the potential difference between the two magnetic surfaces at the horizon and across the load.

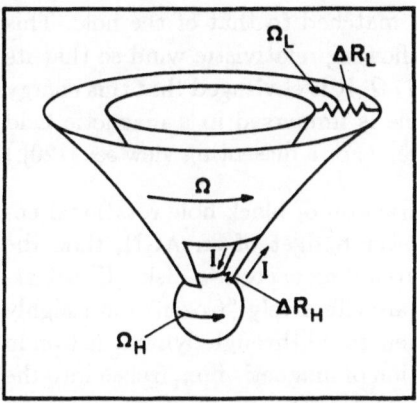

Fig. 3.6. Electrical circuit in which current flows along adjacent magnetic surfaces and through the black hole and an external load resistance, rotating with angular frequency Ω_L. The hole spins with angular frequency Ω_H.

$$\Delta V_H = \frac{(\Omega_H - \Omega)\Delta\Phi}{2\pi} = I\Delta R_H$$

$$\Delta V_L = \frac{(\Omega - \Omega_L)\Delta\Phi}{2\pi} = I\Delta R_L \tag{3.34}$$

where Ω is the angular velocity of the magnetic field lines, defined so that an observer orbiting with this angular velocity would not see any electric field. Ω is constant along a given magnetic field line. Eliminating I between equations(3.34), we solve for Ω.

$$\Omega = \frac{\Omega_H \Delta R_L + \Omega_L \Delta R_H}{\Delta R_L + \Delta R_H} \tag{3.35}$$

Ω is an intermediate angular velocity between that of the hole and the load.

As a consequence of this electromagnetic force, a torque ΔG will act on the hole, and remove angular momentum at a rate

$$\Delta G = \frac{I\Delta\Phi}{2\pi} = \frac{I^2(\Delta R_H + \Delta R_L)}{(\Omega_H - \Omega_L)} \tag{3.36}$$

This torque will be transmitted along the magnetic field, which has no moment of inertia and will supply angular momentum to the load.

However, it is the energy equation which is of most importance. This is just a rearrangement of equation(3.36). The torque will do work on the load and the hole and the sum of these two powers plus the power dissipated in the hole and load must vanish as there is no external energy source.

$$-\Delta G\Omega_H + \Delta G\Omega_L + I^2\Delta R_H + I^2\Delta R_L = 0 \tag{3.37}$$

These equations, which are correct relativistically under the force-free approximation, show that a significant energy can be extracted from the spin of the hole when the resistance of the load is matched to that of the hole. This happens automatically if the load is an outflowing relativistic wind so that its impedance is roughly that of free space $\sim 377\Omega$. It is envisaged that this energy extraction occur when a spinning black hole is immersed in a magnetic field generated by external currents [119][70], [105]. (For a dissenting view see [120].)

3.3.2 Radio galaxies. If electromagnetic extraction of black hole rotational energy is a significant contributor to the power budget of an AGN, then the magnetic field must be anchored in the surrounding accretion disk. (Under astrophysical conditions, magnetic field lines are effectively "frozen" onto highly conducting plasma [121]. The field may be generated through dynamo action in the disk or simply reflect the inward convection of magnetic flux, frozen into the accreting plasma. To order of magnitude accuracy, the electromagnetic power extracted from a rapidly spinning hole is given by combining equations(3.34), (3.37)

$$
\begin{aligned}
L_{EM} &\sim \frac{\Delta\Phi^2\Omega_H^2}{32\pi} \\
&\sim 10^{45}\left(\frac{a}{m}\right)^2 B_4^2 M_8^2 \mathrm{erg\ s}^{-1}; \quad a << m
\end{aligned}
\tag{3.38}
$$

where $B \sim 10^4 B_4$G. The total amount of energy stored in a rapidly spinning hole is given roughly from equation(3.20)as

$$
\begin{aligned}
m - m_{irr} &\sim 4m^3\Omega_H^2; \quad a < m \\
&\sim 5\times 10^{61}\left(\frac{a}{m}\right)^2 M_8 \mathrm{erg}; \quad a << m
\end{aligned}
\tag{3.39}
$$

If $M_8 \gtrsim 1$ and $a \sim 0.5m$, this is ample to fuel the most energetic of radio sources. Note that the energy may be extracted electromagnetically even when the accretion rate is small and so the radio power of a radio galaxy may exceed the UV luminosity of its associated galactic nucleus [122].

4 Accretion Disks

4.1 Introduction

4.1.1 General remarks. As we have already remarked, the specific angular momentum of gas in orbit in the surrounding galaxy is roughly $\sim 10^{30}\mathrm{cm}^2\mathrm{s}^{-1}$, far larger than the maximal spin angular momentum of the hole $\sim 5\times 10^{23}M_8\mathrm{cm}^2\mathrm{s}^{-1}$. Gas that accretes onto the black hole should form an accretion disk. Despite this, there have been extensive analyses of spherical or quasi-spherical accretion flows onto black holes. These highlight many general principles of accretion theory. Good introductory references to the theory of accretion disks include Shakura and Sunyaev(1973) [123], Novikov and Thorne(1973) [124], Pringle(1981) [125], Shapiro and Teukolsky(1983) [91],

Frank, King and Raine(1985) [20], Treves, Maraschi and Abramowicz(1988) [126], and Shields(1989) [127]. Conference proceedings specifically devoted to accretion disks include [9], [128], [129],

We shall continue to set $G = c = 1$ and regard the hole mass m as a length in physical equations.

4.1.2 Fiducial numbers. There is a critical luminosity associated with quasi-spherical accretion called the *Eddington luminosity* which exerts a radiation pressure on an electron-proton pair that just balances the attractive force of gravity.

$$L_E = \frac{4\pi M m_P}{\sigma_T} \simeq 10^{46} M_8 \text{erg s}^{-1} \tag{4.1}$$

Associated with this is an Eddington accretion rate, that would be able to sustain an Eddington luminosity with unit efficiency for conversion of mass into radiant energy,

$$\dot{M}_E = L_E \simeq 10^{25} M_8 \text{g s}^{-1} \simeq 0.2 M_8 M_\odot \text{yr}^{-1} \tag{4.2}$$

a (mass-independent) Eddington time, (the e-folding time for the mass of a black hole accreting at the Eddington rate)

$$t_E = \frac{M}{\dot{M}_E} \sim 4 \times 10^8 \text{yr} \tag{4.3}$$

and an associated Eddington particle density (characteristic of the particle density near the horizon when the hole accretes at the Eddington rate)

$$n_E = \frac{\dot{M}_E}{4\pi m^2 m_P} \sim 10^{-13} M_8^{-1} \text{cm}^{-3} \tag{4.4}$$

Note that the characteristic Thomson optical depth under these conditions is $\tau_T \sim n_E \sigma_T m = 1$. For completeness we can also define an Eddington temperature

$$T_E = \left(\frac{L_E}{4\pi m^2 \sigma_{SB}} \right)^{1/4} = 5 \times 10^5 M_8^{-1/4} \text{K} \tag{4.5}$$

and an Eddington magnetic field strength

$$B_E = \left(\frac{L_E}{m^2} \right)^{1/2} = 4 \times 10^4 M_8^{-1/2} \text{G} \tag{4.6}$$

4.2 Steady Newtonian Viscous Accretion Disks

4.2.1 Conservation laws. When mass accretes onto a central massive black hole under these conditions it is supposed to form an axisymmetric accretion disk of gas orbiting the central mass. For simplicity, we shall assume that the disk lies strictly in the equatorial plane of the black hole and that its self-gravity can be neglected. We shall assume that the disk is sufficiently cool that its thickness is a small fraction of its radius and that pressure gradients are only a small perturbation to Newtonian gravity. We shall also assume that its physical properties do not change with time except for a slow and negligibly small secular change in the mass of the central black hole. We can develop an elementary theory of steady Newtonian accretion disks by conserving mass, angular momentum and energy[91], [20].

Mass conservation takes the form

$$\dot{M} = 2\pi r \Sigma v = \text{const} \tag{4.7}$$

where Σ is the surface density in the disk and v is now the inward radial velocity. This equation expresses the assumption that the flow of mass per unit time crossing some radius r is independent of r.

Similarly angular momentum conservation can be treated by supposing that the quantity of angular momentum carried inward across a circle of radius r by the accreting gas minus the angular momentum transported outward by the torque G that the disk within r exerts on the exterior disk is constant. If the radial velocity v is small compared with the Keplerian speed then the specific angular momentum is $(mr)^{1/2}$. If we further assume that the torque vanishes at the inner radius of the disk, r_{min}, then we can identify the constant

$$G = \dot{M}[(mr)^{1/2} - (mr_{min})^{1/2}] \tag{4.8}$$

(Were we dealing with accretion onto a neutron star, we would have to modify the inner boundary condition). In either case, $G \sim \dot{M}(mr)^{1/2}$ for $r >> r_{min}$.

Next we turn to energy conservation. As gas moves radially inward through the disk, it releases its gravitational binding energy, $(m/2r)$. Consider a thin, annular ring of radial thickness dr. The binding energy released per unit time by accreting gas within this ring is $-\dot{M}d(m/2r)$. However, the viscous torque G at radius r does work on the exterior disk at a rate $-G\Omega$ where $\Omega = (m/r^3)^{1/2}$ is the Keplerian angular velocity of the disk. The difference of the work done on the inner and outer radii of the ring is $-d(G\Omega)$. The sum of the binding energy release and the net work is the increment in luminosity dL.

$$dL = -\dot{M}d\left(\frac{m}{2r}\right) - d(G\Omega) \tag{4.9}$$

Evaluating the differentials, we obtain

$$\frac{dL}{dr} = \frac{3\dot{M}m}{2r^2}\left[1 - \left(\frac{r_{min}}{r}\right)^{1/2}\right] \tag{4.10}$$

If we integrate the luminosity over radius we find that

$$L = \int_{r_{min}}^{\infty} dr \frac{dL}{dr} = \frac{\dot{M}m}{2r_{min}} \tag{4.11}$$

In other words we have global energy conservation in the sense that the total binding energy released by the time an element of gas has reached r_{min} is the sum of all the luminosity that it has radiated on this journey. The internal viscous torques simply ensure that more of it is radiated early on at large radius. In fact, for $r >> r_{min}$, three times as much power is released as a consequence of these viscous torques as is liberated locally. If we set $r_{min} = 6m$, which is a good approximation for a Schwarzschild black hole, then the half power radius is $32m$ (ignoring relativistic corrections). This has the observable consequence that accretion disks radiate with lower effective temperatures than we might otherwise guess [127].

4.2.2 Angular momentum release. In a conventional accretion disk, angular momentum is continuously transported radially outward. It is of interest to inquire about the eventual fate of this angular momentum. One possibility is that the disk has an ever expanding outer rim. This would require the disk mass to exceed $\sim (m/r)^{1/2}$ times the mass of the hole which, as we shall see, has interesting implications [130]. Another possibility is that the excess angular momentum is carried off through the weak gravitational force the moving disk exerts on passing stars [131]. Alternatively, it has been envisaged that there is a non-rotating halo that exerts a frictional torque on the disk [132].

However, the most promising external torque is of magnetic origin [133], [134]. There are good precedents for magnetic torques operating in solar type stars [121], in cataclysmic variable stars [135], and during star formation [136]. As we shall see in the following section, the very existence of jets in these environments furnishes an argument that magnetic torques are intimately involved in the global evolution of accretion disks.

The behaviour of large scale magnetic fields around accretion disks is usually described under the MHD or magnetohydrodynamic approximation [121]. The plasma, whether fully or only partially ionised, has such a high electrical conductivity and the length scales involved are so large that the magnetic field lines can be thought of as being frozen into the moving fluid. In other words, an observer moving with the fluid would measure no electrical field, just a magnetic field. As magnetic flux must be conserved, the magnetic field strength must intensify when the flow converges and weaken when it diverges.

The second key physical effect in MHD is that magnetic field exerts a stress. The force per unit volume is

$$f = \frac{j \times B}{c}. \tag{4.12}$$

(The electrical force is negligible under non-relativistic conditions.) It is convenient to express this force as the divergence of the *Maxwell stress tensor*, using the Maxwell relation $j = c\nabla \times B/4\pi$.

$$f = -\nabla \cdot T \tag{4.13}$$

where

$$T_{ij} = \frac{B^2 \delta_{ij}}{8\pi} - \frac{B_i B_j}{4\pi} \tag{4.14}$$

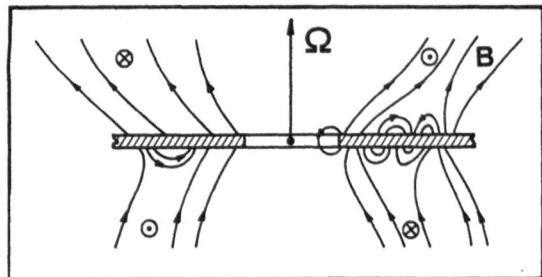

Fig. 4.1. Accretion disk evolving under the action of external magnetic torque.

The Maxwell stress tensor is anisotropic and can be decomposed into an isotropic magnetic pressure plus a magnetic tension acting along the field.

Let us make a simple model by supposing that magnetic flux threads an accretion disk as shown in Fig.(4.1). From equation (4.14), the magnetic torque acting on unit area of the disk will be $\mathcal{G} = <B_z B_\phi> r/2\pi$, taking into account the top and the bottom surfaces. If this were the only torque, the law of angular momentum conservation would read

$$\mathcal{G} = \frac{1}{2\pi r}\frac{\partial}{\partial r}\dot{M}(mr)^{1/2} = \frac{\dot{M}\Omega}{4\pi} \tag{4.15}$$

In this case there is no dissipation in the disk and mechanical energy is extracted by the magnetic Maxwell stress as a Poynting flux at a rate

$$\frac{dL}{dr} = 2\pi r\mathcal{G}\Omega = \frac{1}{2}\dot{M}\Omega^2 r \tag{4.16}$$

With an external torque, the half power radius is $2r_{min}$, significantly smaller than the value obtained above for a viscous accretion disk. A real accretion disk probably evolves under a combination of internal and external torques.

4.2.3 Vertical structure of a thin disk. A section of an accretion disk is like a miniature star because energy is released in the interior of the disk and has to be transported to the surface where it is radiated from an atmosphere[91]. Each of these features can be modelled. Let us look first at the disk structure. The vertical equation of hydrostatic support can be obtained by balancing the vertical pressure gradient with the vertical gravitational force. For a thin accretion disk, this becomes

$$\frac{dp}{dz} = -\rho\Omega^2 z \tag{4.17}$$

We need an equation of state to solve this equation. A simple, and not unreasonable possibility is to suppose that the gas is isothermal, $p = \rho s^2$, where s is the sound speed. In this case equation(4.17)is easily integrated to give a Gaussian density profile of width

$$H = s/\Omega. \tag{4.18}$$

212

Note that the azimuthal motion of the gas in the disk is highly supersonic and has a formal Mach number

$$\mathcal{M} = r/H \qquad (4.19)$$

It should not be surprising if shocks develop and these have been proposed as a means of transporting angular momentum internally. [137].

More generally, the pressure will be a combination of gas and radiation pressure familiar from the theory of stellar structure.

$$p = \frac{1}{3}aT^4 + \frac{\rho kT}{\mu m_P} \qquad (4.20)$$

where $a = 7.5 \times 10^{-15}$erg cm^{-3}K^{-1} is the radiation constant, μ is the mean molecular weight and the temperature T is determined by heat transport. Actually, there is no guarantee that it will be possible to achieve thermal equilibrium and this assumption must be checked for self-consistency in complete models. In some parts of accretion disks, the pressure is radiation-dominated, whereas in other parts, it is gas-dominated.

4.2.4 Viscous Torque. Let us suppose that the viscosity is Newtonian in the sense that it is local and the shear stress is proportional to the velocity gradient. It can be measured by a coefficient of kinematical viscosity ν. The torque G can then be expressed in terms of an integral through the disk.

$$\begin{aligned} G &= - \int 2\pi r dz \rho r \nu r \frac{d\Omega}{dr} \\ &= \dot{M}(mr)^{1/2}; \quad r >> r_{min} \end{aligned} \qquad (4.21)$$

where we have used equation(4.8). (Note that the azimuthal component of the shear viscous stress is proportional to the angular velocity gradient consistently with its vanishing when $\Omega =$const.) As $\Omega \propto r^{-3/2}$ in a Keplerian disk, we obtain

$$\dot{M} = 3\pi\nu\Sigma \qquad (4.22)$$

or using the equation of mass conservation,

$$v = \frac{3\nu}{2r} \qquad (4.23)$$

We must now face the controversial issue of specifying the viscosity mechanism. On one point all are agreed. Conventional "molecular" viscosity is inadequate to drive mass transfer in accretion disks. Shakura and Sunyaev[123] therefore proposed that the viscosity be turbulent and that it scale with the natural length scale, the thickness of the disk H and the natural speed, the sound speed s. *i.e.*

$$\nu = \alpha' s H \qquad (4.24)$$

where $\alpha' \lesssim 1$. (In fact they proposed an equivalent formula, that the shear stress in an accretion disk be proportional to the pressure, *i.e.*stress$\sim 3\nu\rho\Omega/2 = \alpha p$, where $\alpha = 3\alpha'/2$.) Yet another way of parametrising this *ansatz* is in terms of the Reynolds' number $Re = r^2 \Omega / \nu$. In the intervening eighteen years, no one has been able to provide a compelling justification for this prescription; neither has anyone been able to come up with a significantly better idea. The "α" prescription, as it is known, has some observational support from the modelling of the conditions under which dwarf novae undergo outburst which seem to indicate that $\alpha \sim 0.1 - 1$ [138].

Internal magnetic viscosity can also be described in a similar manner. Suppose that there are horizontal loops of magnetic field in a differentially rotating accretion disk. These will be convected by the conducting gas so that a non zero Maxwell shear stress develops (see (4.2)). However the toroidal field cannot build up indefinitely. It will be limited by both buoyant escape [139], and reconnection [140]. Let us suppose that in a steady state, the mean stress is given by

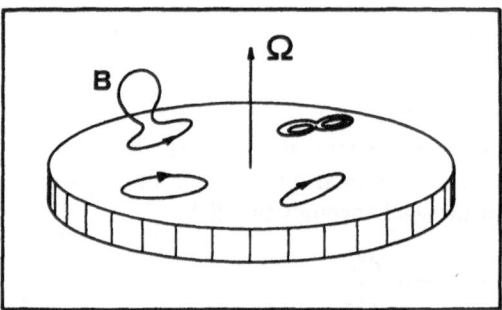

Fig. 4.2. Internal magnetic torques created by shearing of horizontal loops of magnetic flux.

$$\left\langle \frac{B_r B_\phi}{4\pi} \right\rangle \propto \left\langle \frac{B^2}{8\pi} \right\rangle \propto p \qquad (4.25)$$

and let the constant of proportionality be α once again. This prescription, just like its turbulent viscosity counterpart is plausible, though unproven.

An interesting point of principle arises when the pressure has to be specified. Most authors have implicitly assumed that the appropriate pressure p is the total pressure. This is probably appropriate under conditions of infinite optical depth. However, accretion disks are not infinitely thick and when the magnetic stresses become comparable with the gas pressure, the magnetic flux will probably escape buoyantly. For this reason, it has been argued that the proper prescription for the shear stress is αp_{gas}. This makes a big difference in the inner parts of AGN accretion disks [141].

4.2.5 Internal vs external torques. Now that we have given prescriptions for the viscous stress, we should compare its magnitude with that of an external stress. If we just confine our attention to magnetic stress, then the ratio of the external surface to the internal viscous torque is given approximately by

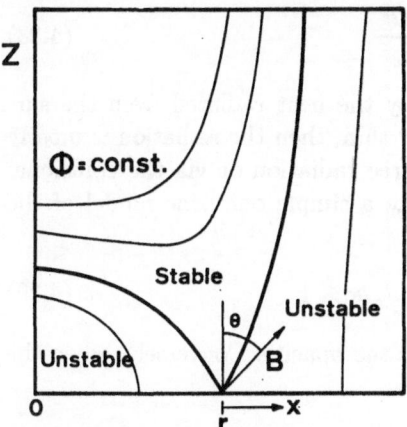

Fig. 4.3. Combined gravitational and centrifugal potential for an element of gas frozen onto a magnetic field line tied to an orbiting accretion disk. (Adapted from [142].)

$$\frac{\pi r^2 \mathcal{G}}{G} = \frac{<B_z B_\phi> r}{4 <B_r B_\phi> H} \qquad (4.26)$$

External torques dominate if the vertical field projecting out of the disk is sufficiently strong that $|B_z| \gtrsim 4H|B_r|/r$.

If magnetic torques are significant then they may drive an external wind. To see if this is likely, consider the effective potential for an element of ionised gas frozen onto a magnetic field line protruding from the disk at radius r and making an angle θ with the vertical (Fig. (4.3)). Under the conditions of perfect MHD, the field line can be thought of as orbiting with a fixed angular velocity $\Omega = (m/r^3)^{1/2}$. The potential is formed from the sum of the gravitational and centrifugal potential [142].

$$\begin{aligned}
\Phi &= \frac{-m}{[(r+x)^2 + z^2]^{1/2}} - \frac{m(r+x)^2}{r^3} \\
&\simeq -\frac{\Omega^2(r)}{2}[r^2 + 3x^2 - z^2 + \ldots]
\end{aligned} \qquad (4.27)$$

where x is the radial displacement and z the vertical displacement from the foot point of the magnetic field. From equation(4.27), we see that Φ decreases away from the disk if the angle $\theta > 30°$. An element of ionised gas frozen onto a magnetic field line at the surface of the accretion disk will therefore be unstable if $\theta > 30°$ and will be flung outward along the field line by centrifugal force. We shall consider the consequences of initiating centrifugal winds from disks below.

4.2.6 Radiative transport. Let us now return to the transport of radiation in a disk in which heat is generated by internal viscous torques and for which we can ignore irradiation of the surface. From equation(4.10), we find that the heat generated per unit area of disk when $r \gg r_{min}$ is given by

215

$$Q^+ = \frac{3\dot{M}\Omega^2}{4\pi} \tag{4.28}$$

In a steady state, this must be balanced by the heat radiated from the surfaces of the disk Q^-. If the disk is optically thin, then the radiation is mostly produced by two body processes like free-free radiation or via the collisional excitation of metal ions. We can then make a simple one zone model of the radiative transport and write

$$Q^- \sim \rho^2 \Lambda(T) H; \quad \Sigma\kappa \lesssim 1 \tag{4.29}$$

where $\Lambda(T)$ is the cooling function and κ is the opacity. Conversely, when the disk is optically thick, we have

$$Q^- \sim \frac{8acT^4}{3\kappa\Sigma}; \quad \Sigma\kappa \gtrsim 1 \tag{4.30}$$

The dominant opacity depends upon the temperature and density. At high temperature and/or low density, the Thomson opacity dominates,

$$\kappa_T \sim 0.4 \text{cm}^2\text{g}^{-1} \tag{4.31}$$

At lower temperature or higher density, the Rosseland opacity which is an approximate expression combining bound-free and free-free opacity for solar abundance (probably a reasonable approximation in AGN), is appropriate.

$$\kappa_R \sim 7 \times 10^{22} \rho T^{-3.5} \text{cm}^2\text{g}^{-1} \tag{4.32}$$

Interestingly, when we consider both the dominant opacity and the dominant pressure, there are really only three possibilities (Fig.(4.4)). These are conventionally known as the inner disk ($\kappa_T > \kappa_R, p_{rad} > p_{gas}$), the middle disk ($\kappa_T > \kappa_R, p_{rad} < p_{gas}$) and the outer disk ($\kappa_T < \kappa_R, p_{rad} < p_{gas}$)[91], [20].

4.2.7 Stationary accretion disk structure. We now have sufficient equations to solve for the structure of a stationary accretion disk. Our equations are essentially algebraic and it only requires fortitude to solve them[126]. However, the choices are beginning to multiply. Let us restrict our attention to one illuminating example, a radiation pressure and Thompson opacity dominated optically thick inner disk. In this case, we equate Q^+ to Q^-. We identify the pressure as $p = aT^4/3$ and use equations (4.28), (4.18)to obtain

$$\begin{aligned}\frac{3\dot{M}\Omega^2\mathcal{J}}{4\pi} &= \frac{8p}{\kappa_T\Sigma} \\ &= \frac{4\Omega^2 H}{\kappa_T}\end{aligned} \tag{4.33}$$

where $\mathcal{J} = 1 - (r_{min}/r)^{1/2}$ is the correction factor introduced to take account of the suppression of viscous dissipation near the inner edge of the accretion disk at r_{min}, cf.equation(4.10). The disk thickness is given by

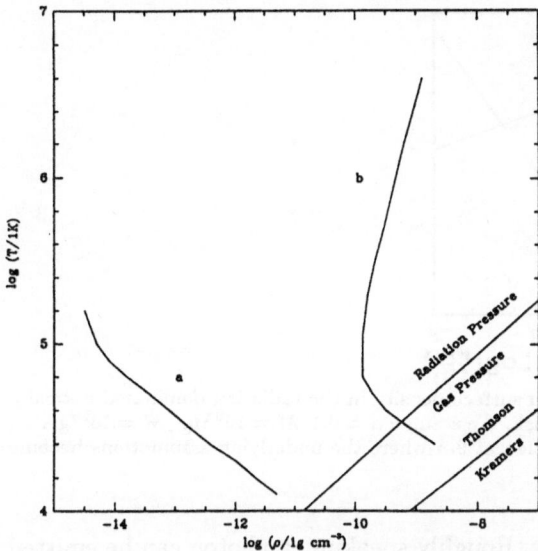

Fig. 4.4. Regions of dominance of gas pressure and radiation pressure and Thomson scattering and Kramers opacity in the density-temperature plane. Two extreme disk models, a)shear stress= p_{rad}, b)shear stress= p_{gas} are also shown [141].

$$\frac{H}{m} = \frac{3\dot{M}\kappa_T \mathcal{J}}{8\pi m} = \frac{3\dot{m}\mathcal{J}}{4} \tag{4.34}$$

where $\dot{m} = \frac{\dot{M}}{M_E}$. As the accretion rate approaches the Eddington rate, the disk thickness becomes comparable with the hole mass and it is no longer self-consistent to assume that the disk is thin in its innermost parts.

We can solve for the flux $F = Q^+/2$, the surface density, the thickness and the temperature if we assume a shear stress of αp.

$$F \simeq 7 \times 10^{18} \dot{m} M_8^{-1} \mathcal{J}(r/m)^{-3} \text{erg cm}^{-2}\text{s}^{-1}$$
$$\Sigma \simeq 2\alpha^{-1}\dot{m}^{-1}\mathcal{J}^{-1}(r/m)^{3/2}\text{g cm}^{-2}$$
$$H \simeq 2 \times 10^{13}\dot{m} M_8 \mathcal{J} \text{cm} \tag{4.35}$$
$$T \simeq 4 \times 10^5 \alpha^{-1/4} M_8^{-1/4}(r/m)^{-3/8}\text{K}$$

The validity of these expressions extends only as long as

$$\frac{r}{m} \lesssim 900\alpha^{2/21} M_8^{2/21} \dot{m}^{16/21} \tag{4.36}$$

which marks the outer boundary of the inner disk (Fig.(4.5).) These Newtonian relations cannot be used close to the central black hole without relativistic corrections. A further complication is that the magnitude of the radial velocity is not small near the inner edge of the accretion disk and, in some models, becomes supersonic.

These formulae implicitly assume that the gas in the accretion disk is able to radiate away all the heat that has been generated. We can give a simple

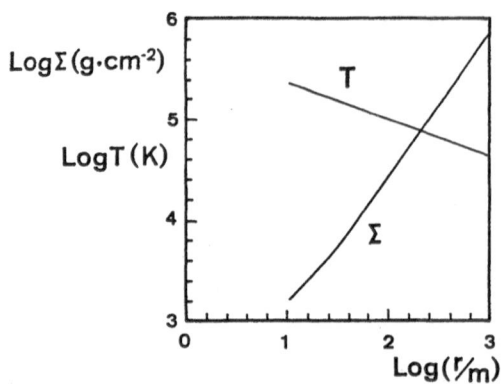

Fig. 4.5. Variation of temperature and surface density in the radiation-dominated optically thick inner region of a thin accretion disk. We assume $\alpha = 0.1$, $M = 10^8 M_\odot$, $\dot{M} = 10^{25}$ g s^{-1}, $r_{min} = 6m$ and terminate the calculation at $9m$ where the underlying assumptions become invalid.

criterion for when this is possible. Roughly speaking, a photon can be emitted if it can be absorbed. Now when Thomson scattering dominates, photons will have to random walk out of the disk and will traverse a mean path length $L \sim \kappa_T \rho H^2$. The photon will be absorbed if $\kappa_R \rho L > 1$. In other words if

$$(\kappa_T \kappa_R)^{1/2} \Sigma > 1 \qquad (4.37)$$

The effective opacity is the geometric mean of the scattering and absorptive opacity[91], [20]. Imposing this condition for our example of an inner accretion disk leads to the the requirement that

$$r/m > 10 \alpha^{1/3} \dot{M}_{25}^{2/3} M_8^{-2/3} \qquad (4.38)$$

Within this radius, we expect that the disk will heat up and become geometrically thick. In fact the disk may get so hot that the electrons become relativistic. Under these conditions it is not even safe to suppose that the ions and electrons have the same temperature [122]. In two temperature accretion disks, the ion temperature exceeds the electron temperature and the rate of cooling is controlled by the rate at which energy is transferred from the ions to the electrons.

4.2.8 Spectrum of emitted radiation. We next consider the atmosphere of the accretion disk. The simplest assumption to make is undoubtedly that each ring of the disk radiates locally like a black body [143]. This implies that, in a steady disk the local flux will scale with radius as $F \propto r^{-3}$, one power of r for the binding energy per unit mass and two powers of r for the emitting area. Using the Stefan-Boltzmann law we deduce that $T \propto r^{-3/4}$. Next integrating over all radii, we form the integrated spectrum

$$S_\nu \propto \int 2\pi r dr F \delta(\nu - KT(r)) \propto \nu^{1/3} \qquad (4.39)$$

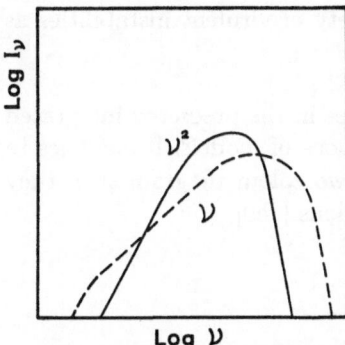

Fig. 4.6. Spectrum emitted from the inner regions of an accretion disk assuming that the local emission is as a black body (bold line) or that scattering suppresses the emission (dashed line).

Unfortunately this does not seem a good match to the observations and there have to be other significant influences on the spectrum.

Further complications that have been included in detailed models are:

(i) *Surface irradiation.* There may be far more radiation incident on the outer parts of a disk than generated locally from the inner parts, especially if the disk flares or is warped[123].

(ii) *Atmospheric effects.* The surface of an accretion disk is no more likely to be a black body than a normal stellar atmosphere. It is possible to treat it as a superposition of early type star atmospheres although the disk surface gravities are typically of lower surface gravity and are more dominated by electron scattering than their stellar counterparts [144]. An electron scattering atmosphere will be partially reflective and will therefore be a poorer absorber than a black body. By Kirchhoff's law it will therefore be a poorer emitter [145]. This causes the spectrum to flatten and extend to a higher frequency (Fig.(4.6)). (A major difficulty is that Ly edge absorption is anticipated though rarely, if ever, seen.) [146]

(iii) *Chromosphere and corona.* When the disk is either irradiated from above, or releases much of its energy in a mechanical form then it will develop an active chromosphere and corona just like the sun [139], [254], [255]. It may be that most of the disk binding energy is released through instabilities in this manner.

(iv) *Polarisation.* If a disk is a flat, scattering surface it ought to produce a measurable linear polarisation [147]. In general the measured polarisations are very small. This may occur because real disks are very thick or their surfaces are highly corrugated. Under extreme conditions, Faraday rotation can depolarise the continuum emission.

(v) *Relativity.* Special relativistic Doppler shifts and general relativistic redshifts can be significant for radiation emitted by the innermost parts of the accretion disk [148]. In addition capture of photon orbits by the central black hole must also be taken into account [149].

(vi) *Instability.* Probably the largest factor, which may invalidate the stationary disk models, is that they are prey to a variety of virulent instabilities as we shall discuss presently.

All of the above items represent major changes in the predicted integrated spectra of accretion disks and have led to a plethora of models. If one tries to represent the predictions of these models on a two colour diagram then they more than encompass the variety in the observations [150].

4.3 Stability of Accretion Disks

4.3.1 Thermal instability. The easiest type of instability to discuss is thermal stability. Consider, for example, a radiation pressure and Thomson scattering dominated inner disk of variable half thickness H. As we are only interested in thermal processes, there will be no time for mass transfer and the surface density Σ can be treated as constant. From equation(4.33)the internal pressure is $p \simeq \Sigma \Omega^2 H \propto H$. Now suppose that the viscous stress is given by αp so that the rate of viscous heating per unit area of disk is given by equation(4.9)as

$$
\begin{aligned}
Q^+ &= \frac{3G\Omega}{4\pi r^2} \\
&= 3\alpha p \Omega H \\
&\propto \alpha H^2
\end{aligned}
\tag{4.40}
$$

However if the disk is optically thick, the rate of radiative cooling will satisfy

$$
Q^- = p\tau_T^{-1} \propto H
\tag{4.41}
$$

Now, consider a disk in thermal equilibrium with $Q^+ = Q^-$ and increase H slightly. If α is constant, the rate of energy production will increase faster than the rate of energy escape and the disk will heat up, increasing its thickness. In other words the disk is thermally unstable. Conversely, if the disk were to shrink, it would continue to cool. More generally, an equilibrium thin disk will be thermally unstable if

$$
\left(\frac{\partial(Q^+ - Q^-)}{\partial H} \right)_\Sigma > 0
\tag{4.42}
$$

[123], [125], [20]. The time scale for thermal instability to develop is the local cooling time

$$
t_{th} \sim \frac{2\pi}{\Omega\alpha}
\tag{4.43}
$$

A thermally unstable disk will develop a corrugated surface, where the amplitude of the corrugations is limited by non-linear effects. Many types of equilibrium disks are thermally unstable. Note that the assumption that the viscous heating is described by the α model with constant α is crucial and different prescriptions can render our particular example stable.

4.3.2 Viscous instability. An alternative class of instabilities can develop more slowly on the timescale associated with mass accretion. Let us suppose that the disk can always adjust to thermal equilibrium so that $Q^+ = Q^-$. Under these circumstances we can evaluate the viscous stress and determine what the mass accretion rate will be for a given surface density $\Sigma(r)$ at radius r using equation(4.22). To simplify a somewhat more general argument [151], let us imagine a disk that accretes mass at a steady rate so that $\dot{M}$ is the same at all radii. Now, make a small, discontinuous increase in the accretion rate at some radius and ask how the disk will respond. The density will also increase and the local mass accretion rate $\propto \nu\Sigma$, (*cf.*equation(4.22)) will adjust. If the accretion rate increases, then the local density can readjust to a new equilibrium; if not, then the disk is subject to viscous instability. The criterion for viscous stability is then given by

$$\frac{d(\nu\Sigma)}{d\Sigma} > 0$$

[125], [20]. When this inequality is reversed, then the disk will depart further from equilibrium and become unstable. If there is viscous instability, then perturbations will grow on the mass accretion timescale

$$t_{vis} \sim \frac{r}{v} \sim t_{th}\left(\frac{r}{H}\right)^2 \tag{4.44}$$

A more complete discussion of time-dependent accretion follows from combining the time-dependent generalisations of the equations of mass and angular momentum conservation (4.7), (4.8). The result is the diffusion equation

$$\frac{\partial\Sigma}{\partial t} = \frac{3}{r}\frac{\partial}{\partial r}\left(r^{1/2}\frac{\partial}{\partial r}(r^{1/2}\nu\Sigma)\right) \tag{4.45}$$

[125], [20]. This equation demonstrates that under normal (stable) conditions, a locally overdense ring will spread as a result of viscous torques, most of the mass migrating inward, but a minority of mass spreading outward so as to conserve angular momentum overall.

The combined thermal and viscous stability of a wide variety of thin accretion disk models has been systematically studied. Rather less attention has been devoted to the difficult question of the non-linear development of these instabilities and to understanding whether or not they are likely to disrupt global mass accretion flow. One exception is a class of disks where the relationship between the mass accretion rate and the surface density has the form shown in Fig.(4.7). If the rate at which mass is supplied to the disk lies within an unstable range, then the disk may alternate between two stable states following a limit cycle through this evolution.

This limit cycle behaviour is held responsible for dwarf nova outbursts in cataclysmic variables. In this case a portion of the disk associated with an ionisation zone is believed to have an unstable $\dot{M}(\Sigma)$ curve. Similar behaviour may occur in AGN accretion disks [152], [153].

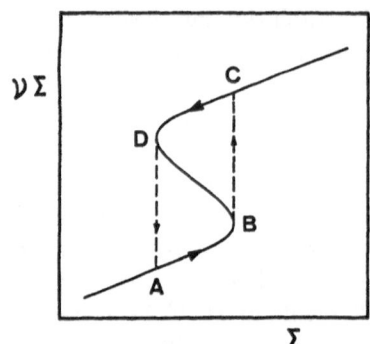

Fig. 4.7. Mass accretion rate at a given radius as a function of the local surface density. When $d(\nu\Sigma)/d\Sigma > 0$, the disk is stable to small perturbations in the surface density. When the inequality is reversed, the disk is unstable. If mass is accreting at an unstable rate the disk may follow the limit cycle variation ABCD instead of accreting steadily.

4.3.3 Dynamical instability. A third type of instabilty, *dynamical instability* is possible in accretion disks. This develops on the orbital period timescale. Thin accretion disks in a Keplerian potential are believed to be dynamically stable. However, when a disk is sufficiently massive that its self-gravitation becomes important, then the disk can become dynamically unstable. We defer discussion to section 6 below. A second class of dynamical instabilities arises in thick accretion disks.

4.4 Thick Accretion Disks

4.4.1 Radiation tori. As we have already mentioned, when the mass accretion rate is close to the Eddington rate, the inner, radiation and electron scattering-dominated disk will thicken. When $\dot{M} \gtrsim \dot{M}_E$, it can no longer be called a disk and is more reasonably described as a *torus*. The equation of hydrostatic equilibrium for the gas contained within this torus can be written, in the Newtonian approximation, as

$$\frac{-\nabla p}{\rho} + g + r_\perp \Omega^2 = 0 \tag{4.46}$$

where g is the gravity and $r_\perp$ is the cylindrical radius [20], [91]. In the simplest model of a torus, it is assumed that the equation of state is *barytropic*, in other words, there is a functional relation $p(\rho)$. It can then be seen immediately from equation (4.46) that the centrifugal force must be the gradient of a potential or that $\Omega = \Omega(r_\perp)$. The angular momentum is constant on cylinders. This is known as *von Zeipel's theorem*. We can then define a family of nested equipotential (gravitational plus centrifugal) surfaces whose shape will depend upon the assumed form of angular velocity variation $\Omega(r_\perp)$. These surfaces will coincide with the isobaric surfaces of constant pressure and density.

The relativistic theory is qualitatively similar [154], [155]. The equation of hydrostatic equilibrium is obtained by setting the divergence of stress energy tensor to zero and can be written

$$\frac{-\nabla p}{w} + \nabla e + \frac{\Omega \nabla \ell}{1 - \Omega \ell} = 0 \qquad (4.47)$$

where w is the enthalpy per unit volume, effectively the rest mass energy density ρ. $e = -u_0$ is the specific energy and $\ell = -u_\phi/u_0$ is the *fluid angular momentum*, which is to be distinguished from the *mechanical angular momentum*, u_ϕ. von Zeipel's theorem is now modified to the statement that a torus described by a barytropic equation of state (*i.e.* one with $p = p(w)$) has its angular momentum ℓ constant on surfaces of constant angular velocity Ω. The angular velocity is defined relativistically by $\Omega = u^\phi/u^0$. Given a specific functional form $\Omega(\ell)$, isobaric surfaces can then be computed. The surface of the torus must then coincide with one of these isobaric surfaces.

One reason for using the relativistic theory [154], [155] is that the non-Newtonian hardening of the potential close to the event horizon causes the equipotential surfaces for a given angular momentum distribution to form a toroidal cusp close to the event horizon. The separatrix passing through this cusp is a limiting surface for matter orbiting in a stationary torus (Fig.(4.8)). If we make the simplest assumption that the angular momentum is constant then the cusp must lie between the marginally bound and marginally stable circular orbits. (A pseudo-relativistic treatment of thick accretion disks is possible [156], [20], and of some heuristic use, but the relativistic theory in a Schwarzschild metric is no more difficult [154].)

If a thick accretion disk develops, it is expected that the torus will expand until its innermost radius approaches the cusp. At this point there are no pressure gradients and the gravitational orbit is unstable. On passing through the cusp, matter will spill through onto the black hole with little further emission. The thickness of a stationary torus increases with the mass accretion rate, and in principle, such structures could extend out to many thousand Schwarzschild radii. However, the specific energy of the surfaces of large tori is necessarily

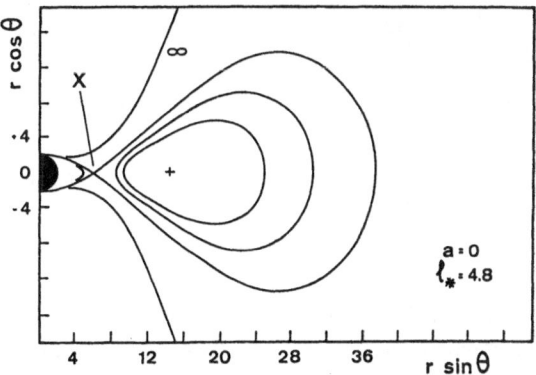

Fig. 4.8. Nested equipotential surfaces for a barytropic radiation torus in orbit about a massive black hole. In this example, it is assumed that the angular momentum is constant. Note the presence of a neutrally stable ring, X. Matter that fills the equipotential surface passing through X, is able to spill through onto the black hole. (Adapted from 155).

small and so the radiative efficiency is correspondingly small making them unacceptable as AGN prime movers.

The pressure in the torus is dominated by radiation pressure and so the radiation flux can be written in the diffusion approximation as

$$F = -\frac{\nabla p}{\kappa_T w} \qquad (4.48)$$

[154]. This allows us to compute the flux emergent from a torus.

Modestly inflated radiation tori are extremely attractive from a phenomenological perspective as they could radiate the intense UV continua characteristic of the brightest quasars. Furthermore, the most intense radiation will be produced in a funnel [157], and will be beamed parallel to the symmetry axis. This can be interpreted is in accord with some observations of Seyfert galaxies [158].

4.4.2 Global instablity of radiation tori. Unfortunately, as Papaloizou and Pringle [159], first realised, radiation tori are prone to global, dynamical instability. This can be described in the Newtonian approximation, with a slender torus [160]. Suppose that there is a non-axisymmetric mode excited in the torus with angular and temporal variation $\propto e^{i(m\phi - \omega t)}$ where ω is the angular frequency of the mode and m is the azimuthal quantum number. We can define a co-rotation radius, r_{cr}, presumed to lie within the torus, where $\Omega(r_{cr}) = \omega/m$. Now the key point is that when $r < r_{cr}$, the mode has negative energy (and angular momentum) density. For $r > r_{cr}$, the energy density is positive. Now suppose that a positive energy wave packet propagates inward to the co-rotation radius. It can be partially transmitted there as a negative energy mode allowing the reflected wave to carry away more energy than the incident wave. The transmitted wave can be reflected from the inner edge of the disk and be re-transmitted as a positive energy mode further increasing the wave amplitude.

More detailed studies suggest that the modes are most damaging when

$$\frac{d\ln\Omega}{d\ln r} < -3^{1/2} \qquad (4.49)$$

As they are dynamical, these modes grow on the time scale of a few orbital periods.

$$t_{dyn} \sim \Omega^{-1} \sim \left(\frac{r}{H}\right)^2 \alpha^{-1} t_{vis} \qquad (4.50)$$

For a thick disk with $\alpha \sim 0.1$, as commonly assumed, they grow somewhat more rapidly than viscous modes.

The consequences of dynamical instability are not well understood. Numerical simulations [161], exhibit the formation of counter-rotating *planets* as the non-linear evolution of this instability. These structures may be highly dissipative and destroy the disk. If this happens, then the concept of a highly optically and geometrically thick torus, may be invalidated. In particular the fluid approximation may no longer be appropriate. Alternatively, if the radial velocity is sufficiently large, particularly in the vicinity of the cusp, the equilibrium flow may convect the waves inwards and inhibits their reflection [162]. The waves might then be maintained at a level of a marginal stability transporting angular momentum just fast enough to limit the wave growth. In this sense, they may act like an effective viscosity and ultimately drive a steady disk flow. More numerical simulations are undoubtedly needed to settle this controversial matter.

There is an alternative type of torus that has been proposed. This is an *ion torus*[122], in which relatively cool electrons are supposed to be supported by the pressure of a hot ion gas with temperature as high as $\sim 100 \mathrm{MeV}$. This coronal gas is quite optically thin. It is also subject to strong radiative losses and is may contain a strong magnetic field. It is not clear whether or not it too is subject to dynamical instability.

5 Jets

5.1 Introduction

5.1.1 Discovery. In 1953, Jennison and Das Gupta [163]discovered that the radio emission from Cygnus A originated from two lobes straddling the associated optical galaxy rather than the galaxy itself. Subsequent observation [164]of other powerful radio sources showed that this was a general phenomenon. Originally it was thought that the powerful radio-emitting lobes had been shot out of the galaxy. This, however, created dynamical problems which prompted Rees [165]to propose that the lobes were instead fuelled continuously along channels emanating from the galactic nucleus. These channels, or *jets* as they came to be known, had already been seen in sources like M87 and 3C273. With the advent of the Cambridge 5km telescope and the VLA, jets were discovered to be associated with large number of double radio sources. The development of VLBI imaging [166]showed that jets were also common among compact radio sources.

5.1.2 Bibliography. Good general reviews of the observations of extended extragalactic radio sources and jets include Bridle and Perley(1984) [93], and the conference proceedings [167], [168], [169], [170], [171]. Compact radio sources are reviewed by Phinney(1987) [172]and the conference proceedings [173], [92]. Theoretical interpretations are reviewed in Begelman, Blandford and Rees (1984) [18], and the above conference proceedings.

5.1.3 Observations of jets. Historically, strong extraglactic radio sources have been divided into two classes, the *extended radio sources* with most of their emission originating from regions more than a kpc from the nucleus of the associated galaxy or quasar, and the *compact radio sources*, for which the opposite is true. The extended sources typically have a steep spectrum ($\alpha \sim 0.5 - 1$), whereas the compact sources have flat spectra ($\alpha \sim 0 - 0.5$). Originally it was thought that these were quite different objects. However, it is now believed that they are essentially two different aspects of the same underlying phenomenon. Part of the evidence for this is that the extended sources invariably have compact central components which facilitate their optical identification, and, conversely, most compact sources exhibit faint halos that could be extended sources seen pole-on.

The extended sources are conveniently divided into two sub-types, the *edge-brightened* and the *edge-darkened* sources[93]. (Sometimes these are known as Fanaroff-Riley [174], class 1 and class 2 respectively.) Members of the former sub-class have their radio emission concentrated in their extremities. They are also luminous ($L_{rad} \gtrsim 10^{42}$erg s^{-1}). They usually have at most one jet with spectral index ~ 0.5. They also exhibit linear polarisation with electric vector perpendicular to the jet. When interpreted on the synchrotron hypothesis, this implies that the embedded magnetic field lies predominately along the jet. (Fig. (5.1).)

The second class, the edge darkened sources, are mostly of lower radio power ($L_{rad} \lesssim 10^{42}$erg s^{-1}). In contrast to the edge-brightened sources, they usually contain two jets emerging along anti-parallel directions from the nucleus. They are also more likely to give parallel polarisation suggesting a field perpendicular to the jet axis. The majority of extended radio sources are associated with elliptical galaxies, though a sizeable minority, selected at a given flux level are identified with quasars.

When we turn to the compact sources, we find a variety of morphological types. Prominent among these are the *core-jet* sources, which comprise a flat spectrum ($\alpha \sim 0$) core and a steeper spectrum, one-sided jet[92]. Features in these jets are often observed to be moving away from the core with apparent *superluminal* speed. Also distinguished are compact double sources where there are two comparable source components and the so called steep spectrum compact radio sources which are like miniature extended sources. Most of these compact radio sources are identified with quasars, though a subset are identified with nearby *BL Lac objects* which are found in elliptical galaxies [177].

BL Lac objects are now considered a part of a large class of objects called collectively *blazars*. These also include the optically violently variable (OVV) quasars, which, like BL Lac objects exhibit rapid variability and high optical linear polarisation. It is becoming increasingly apparent that most compact radio sources are, at some level, blazars [178].

5.1.4 Interpretation. The simple unifying interpretation of many of these observations is that our view of a radio source is very strongly influenced by our orientation with respect to the source axis. Powerful radio sources are supposed to comprise a core, and two jets which feed a pair of radio-emitting lobes. The radio-emitting electrons are supposed to be convected outward along the jet with relativistic speed so that they beam their emission along their directions of motion. The compact radio sources are then mostly intrinsically weaker sources that are beamed in our direction so that their compact jet emission outshines the unbeamed emission from the lobes. The edge- brightened sources

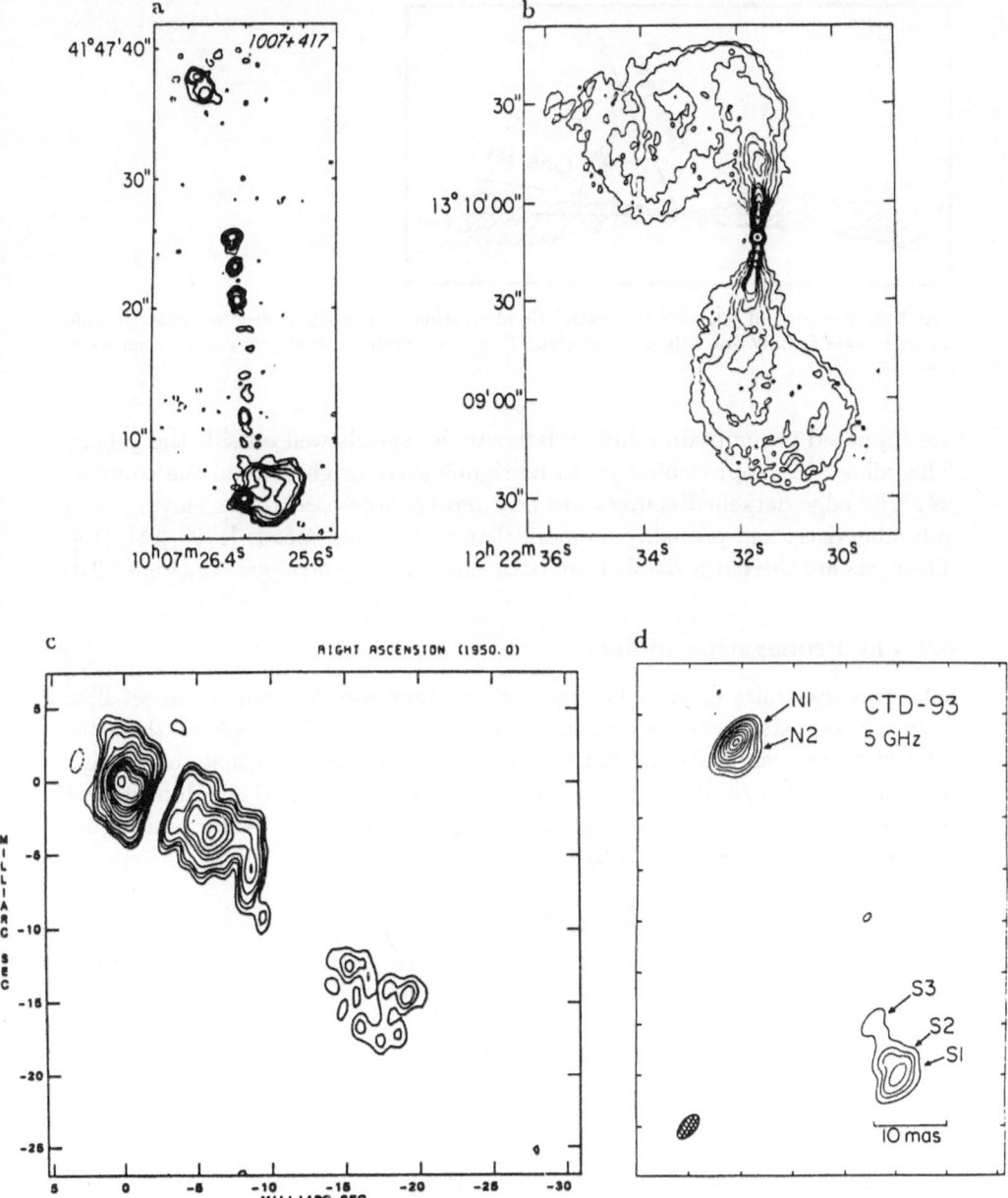

Fig. 5.1. a) Edge brightened extended radio source (Q1007+417[93]). b) Edge-darkened extended radio source (M84[93]). c) Core-jet compact radio source(3C273 [175]). d) Compact double radio source (CTD-93 [176]).

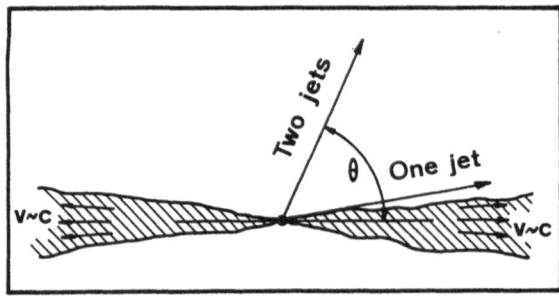

Fig. 5.2. Beaming of jets due to relativistic aberration. When the source is observed from a small angle θ, only one bright jet is seen. Two, comparably bright jets can be seen when $\theta \sim \pi/2$.

are supposed to maintain mildly relativistic jet speeds well outside the galaxy. This allows the approaching jet to be significantly brighter than the counter-jet. The edge-darkened sources are now mostly supposed to be moving with sub-relativistic and probably no more than trans-sonic speeds [179], [93], [18]. Their jets are therefore visible from both sides of the central galaxy (Fig.(5.2)).

5.2 The Propagation of Jets

5.2.1 Gas dynamics of jets. The most elementary way to understand jet flow is to concentrate upon the conservation laws. Let us first look at the non-relativistic case and assume that the gas is ionised plasma so that the specific heat ratio is $\gamma = 5/3$. If we approximate the jet as a collimated one-dimensional flow of variable area A, discharge $\dot{M}$ and speed v then the power and thrust (*i.e.*force) in the jet are given by

$$
\begin{aligned}
L &= \frac{1}{2}\dot{M}v^2 \left(1 + \frac{3}{\mathcal{M}^2}\right) \\
P &= \dot{M}v \left(1 + \frac{3}{5\mathcal{M}^2}\right)
\end{aligned}
\tag{5.1}
$$

where $\mathcal{M} = v/s$ is the Mach number of the flow [180]. The first term in equation(5.1)for the power is the bulk kinetic energy transported by the jet fluid. The second term describes the transport of a combination of the internal energy and the "pV" work, known, together, as the enthalpy. Similarly in the equation for the thrust, the first term is the bulk momentum flux and the second term is contributed by the static pressure. When $\mathcal{M} \gg 1$, the bulk motion dominates. The simplest assumption to make is that a jet is adiabatic and stationary and therefore propagates with constant power and discharge $\dot{M} = \rho A v$ where ρ is the density. We can also suppose that it is in pressure equilibrium with the surrounding gas in which the pressure p presumably diminishes with distance from the nucleus. Now, as the gas is, by assumption, adiabatic, $\rho \propto A^{-1}v^{-1} \propto p^{3/5}$. The Mach number and velocity therefore scale as $\mathcal{M} \propto p^{-4/5}A^{-1}, v \propto p^{-3/5}A^{-1}$. We can evaluate the area as a function of

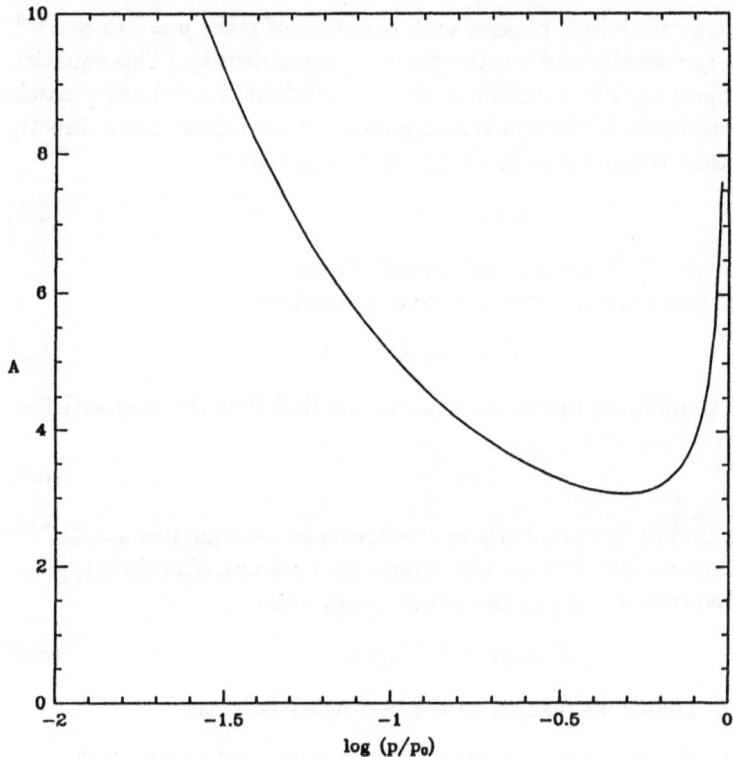

Fig. 5.3. Cross sectional area A as a function of pressure p in units of the stagnation pressure p_0 for a one dimensional, adiabatic jet. Note that, as the pressure diminishes, the area passes through a minimum where the jet speed v is trans-sonic.

pressure for a given luminosity (Fig.(5.3)). The area is seen to pass through a minimum where the jet speed becomes trans-sonic. Therefore if an adiabatic jet accelerates to supersonic speed, it must do so by passing through a *converging-diverging* or *de Laval* nozzle where the pressure is approximately half the stagnation pressure (*i.e.*the pressure where the fluid is at rest)[32]. In the subsonic portion of the flow, the pressure and density are roughly constant and the area diminishes inversely proportional to the velocity. In the supersonic portion, the velocity is mostly constant and the area increases as $A \propto p^{-3/5}$. In a galaxy, the external pressure scales vary roughly with distance as $p \propto r^{-2}$, and so we have approximately that the angle the jet width subtends at the nucleus decreases as $A^{1/2}r^{-1} \propto r^{-2/5}$. Therefore a supersonic jet can be collimated as the pressure diminishes despite expanding in cross sectional area.

In contrast to the power, the *thrust* is not constant in an adiabatic jet. It diminishes during the subsonic portion of the flow and increases in the supersonic regime. This is because the surface of the jet is not exactly parallel to the mean flow velocity and so there is an external pressure force acting parallel to the jet which can change its momentum.

Now let us contrast this non-relativistic theory with a simple relativistic theory[32], [181], [182]. The most elementary assumptions to make are that the jet fluid is an ultra-relativistic plasma with equation of state $p = e/3 \propto n^{4/3}$, where e is the energy density and n is the particle proper density. (This equation of state is also appropriate for a radiation-dominated fluid in which the particle rest masses are negligible.) The relativistic power can be written down directly using the space-time components of the stress-energy tensor

$$L = 4p\gamma^2 vA \qquad (5.2)$$

where $\gamma = (1 - v^2/c^2)^{-1/2}$ is the usual Lorentz factor.

Now particles (e.g.baryons) will also be conserved and so

$$\dot{N} = n\gamma vA \qquad (5.3)$$

is also constant. Combining these two relations we find that the area satisfies

$$A \propto \frac{\gamma^2}{v} \qquad (5.4)$$

A is again minimised at the point where the flow is trans-sonic $v = s = 3^{-1/2}c$ and where the pressure is 4/9 times the stagnation pressure. The thrust, given by the space-space components of the stress energy tensor is

$$P = (4\gamma^2 v^2 + 1)pA \qquad (5.5)$$

and changes in an analogous fashion to the non-relativistic case.

5.2.2 Real jets. Just as was the case with accretion disks, we know both theoretically and observationally that real jets do not satisfy the assumptions we have just made. This is not surprising. There will be significant velocity shear across the jet and probably also turbulence and internal shocks. There are radiative losses and there must be internal dissipation causing relativistic particle acceleration. The mass flux will not even be constant as we expect jets to entrain gas from the surroundings. Entrainment may be particularly important for the low power, edge-darkened sources which may decelerate substantially as they escape the nuclear regions. All of these complicating factors can be exhibited by numerical simulations (Fig.(5.4)).

However, perhaps the most important omission from the preceding discussion is the possibility that the jets are hydromagnetic.

5.2.3 Hydromagnetic confinement of jets. To motivate the proposition that jets are hydromagnetic flows, let us estimate the minimum internal pressure in the jet using the expression given in equation(2.16). Rewriting in terms of the brightness temperature at 5GHz (T_5) and the jet width w, we obtain an expression for the equipartition pressure

$$p \sim 10^{-12} \left(\frac{T_5}{1K}\right)^{4/7} \left(\frac{w}{1kpc}\right)^{-4/7} \text{ dyne cm}^{-2} \qquad (5.6)$$

The first deduction to draw from the observations is that in many well-studied jets, the variation of equipartition pressure along the jet does not conform with

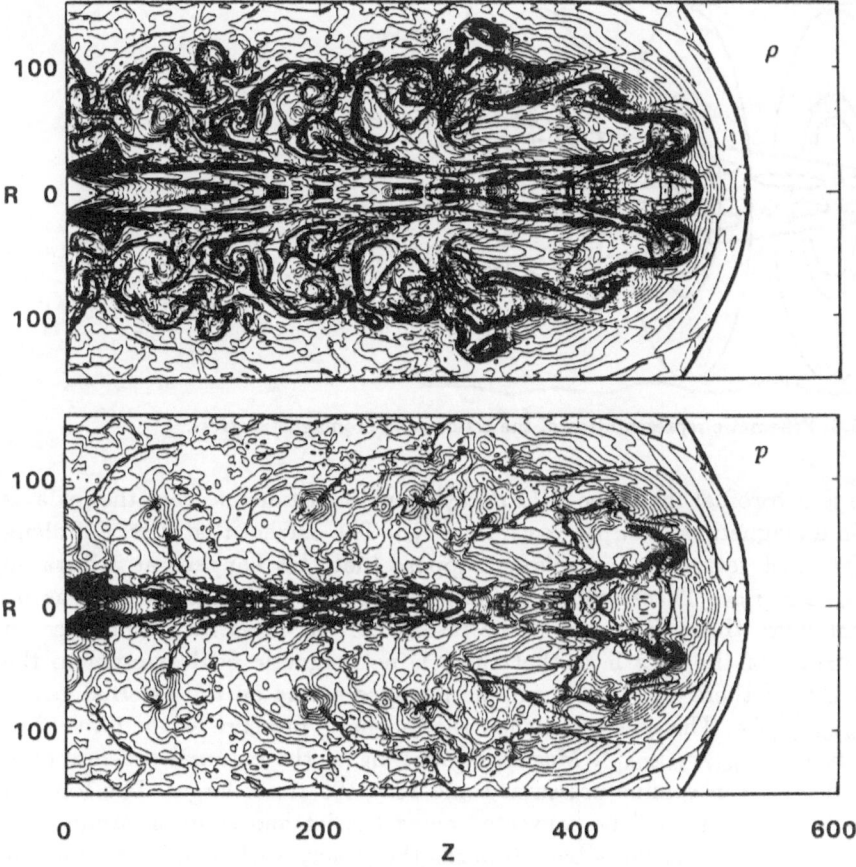

Fig. 5.4. Numerical simulation of a supersonic ($\mathcal{M} = 6$) light jet [183]. Contours of density ρ and pressure p are exhibited.

the law $p \propto w^{-10/3}$ deduced for supersonic, adiabatic, non-relativistic jets with specific heat ratio $\gamma = 5/3$. The pressure falls off much more slowly suggesting that indeed there is internal dissipation and the entropy of the jet fluid increases along the flow[93].

A second inference can be drawn by putting an upper bound on the gas pressure in the circumgalactic gas surrounding the jet from X-ray observations. In several edge-brightened jets this appears to be lower than the lower bound on the internal jet pressure[93]. One way to resolve this discrepancy is to regard the high brightness and, by inference, high pressure regions as transient, perhaps formed by instability or shocks. They would then be freely expanding and cooling. However, some jets show excess pressure over too much of their volume for this to be a viable explanation.

The overpressure in the jets may also be due to the action of a large scale hydromagnetic stress [184], [18]. The manner in which this might happen is outlined in Fig.(5.5). Suppose that the jet carries an electrical current so that

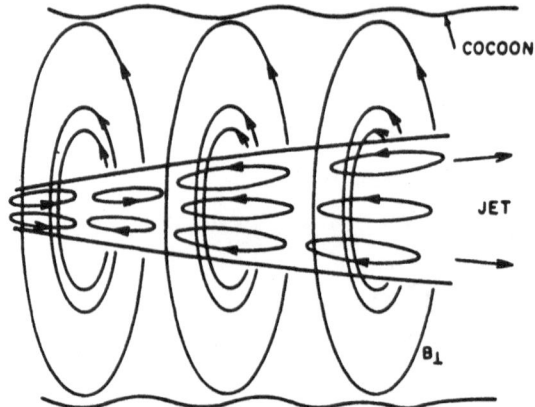

Fig. 5.5. Schematic representation of the action of hydromagnetic stress.

there is a toroidal magnetic field decaying with distance r from the jet axis within a magnetic *sheath*, $B_\phi \propto r^{-1}, r > r_{jet}$. The current that flows out along the jet must return to the galaxy. Suppose that this happens at a distance $r_{sheath} \gg r_{jet}$. There will be no magnetic stress within the sheath as no current flows there. However, if the return current flows in a thin layer on the surface of the sheath so that there is no magnetic field external to the sheath, then there will be a magnetic pressure acting on the external gas of $B(r_{sheath})^2/8\pi$. This is smaller than the magnetic pressure at the jet surface by a factor $(r_{jet}/r_{sheath})^2$. Magnetic focussing of the external pressure onto a slender jet can confine pressures some $10 - 100$ times larger. The toroidal magnetic flux may well be convected along the jet and spun off around the jet as the jet plasma flows back towards the galaxy within the radio lobe at the end of the jet. Numerical simulations of these hydromagnetic flows display rather more complicated behaviour [185], [183]. However, the general principle of magnetic focussing can still be demonstrated.

5.3 Compact Radio Sources

5.3.1 Interpretation of core-jet sources. The earliest models of compact radio sources assumed homogeneous, slowly expanding spheres of magnetised relativistic electrons. [186]. We now know directly from VLBI that sources are not like this and that instead they are inhomogeneous and expand with relativistic speed. (In fact this was partly inferred prior to the development of VLBI.)

Inhomogeneity can have a profound effect on the spectrum. To demonstrate this, let us make a simple model of a non-relativistic, synchrotron-emitting jet. Suppose that the "core" of a core-jet source is the unresolved, optically thick inner portion of the jet. (Fig.(5.6).) Let us suppose that relativistic electrons are accelerated with a power law distribution function so that $N_\gamma = K\gamma^{-2}, \gamma_{min} < \gamma < \gamma_{max}$. The pressure and energy density in the relativistic particles is then contributed roughly equally by high and low energy

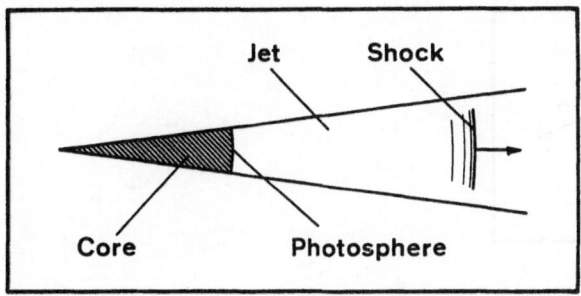

Fig. 5.6. Simple model of a non-relativistic core-jet source. The core is presumed to be formed by the unresolved optically thick inner region of the jet. The resolved jet is presumed to be the the optically thin outer region where relativistic electrons accelerated by shocks and instabilities are convected outward by the flow.

particles. We assume that particle acceleration takes place all the time and that $\gamma_{min}, \gamma_{max}$ remain constant. Let us also, for simplicity, suppose that the jet moves with constant speed within a cone so that its width increases linearly with distance r along the jet. The constant K in the particle distribution function (*cf.*equation(2.12)) will then scale as $K \propto r^{-2}$. The magnetic field will also diminish as the jet expands. Let it do so in such a way that magnetic flux is conserved. The component of magnetic field parallel to the jet will diminish $\propto r^{-2}$. The transverse component will decay more slowly $B \propto r^{-1}$ and this must eventually dominate [18].

Now find how the flux from an octave of radius ($r \rightarrow 2r$) varies with r. Consider first the optically thin flux from a volume $\propto r^3$. In this case, using equation (2.14), we obtain

$$S_\nu(r) \propto KB^{1.5}\nu^{-0.5}r^3 \propto \nu^{-0.5}r^{-0.5} \tag{5.7}$$

The optically thick flux from a jet area $\propto r^2$ is likewise given by

$$S_\nu(r) \propto \nu^{2.5}B^{-0.5}r^2 \propto r^{2.5}\nu^{2.5} \tag{5.8}$$

*cf.*equation (2.19). The true flux will be the minimum of these two estimates. By equating them, we find the maximum flux in the spectrum radiated by an octave of jet is given by

$$S_{\nu max} = \text{const} \tag{5.9}$$

and the associated frequency is

$$\nu_{max} \propto r^{-1} \tag{5.10}$$

Note that the predicted source size scales with the wavelength, just like the resolution. So if the optically thick core is unresolved at one wavelength it will be unresolved at all other wavelengths for which this model is applicable.

If we superpose the spectra formed by superposing several octaves of jet radius (Fig.(5.7)), we find that the integrated spectrum is flat ($\alpha = 0$). This is because the flux at a given frequency is dominated by the contribution from

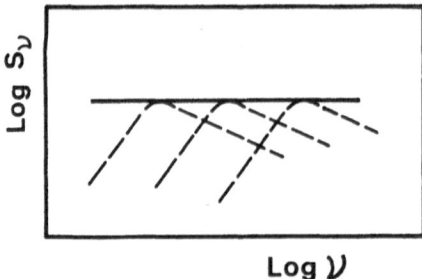

Fig. 5.7. The integrated spectrum (solid line) from an inhomogeneous jet is formed by the superposition of the spectra from individual jet radii (dashed lines).

the radius where this frequency is the local peak frequency. This model is in rough accord with the observations save that it is non-relativistic.

5.3.2 Superluminal motion. One of the first discoveries made about compact radio sources was that they varied surprisingly frequently, on timescales $t_{var} \sim$ several months [187]. This allowed an upper bound of $\sim ct_{var}/D$, (where D is the distance to the source), to be placed on the angular size of the source. This, in turn, gave a lower bound on the brightness temperature of the source which worked out to be in excess of the inverse Compton limit of $\sim 10^{12}$K. Rees [188]realised that this problem could be ameliorated by relativistic expansion and predicted that sources would appear to expand faster than the speed of light. Relativistic speeds are to be expected, if relativistically deep potential wells are present.

The simplest demonstration of superluminal motion is provided by a two source model. One source remains at rest, the other moves toward the observer with speed βc at angle θ to the line of sight (Fig.(5.8)). If we denote time measured in the frame of the stationary source from the time the two sources coincided by t, then the moving component will move a distance $\beta c \sin \theta t$ perpendicular to the line of sight after time t. If the observer receives radiation from the moving source, the time that elapses after it left the stationary source is $t_{obs} = t(1 - \beta \cos \theta)$ making allowance for the motion of the source toward the observer. The measured transverse expansion speed will then be

$$\beta_{obs} = \frac{\beta \sin \theta}{1 - \beta \cos \theta} \qquad (5.11)$$

The apparent expansion speed β_{obs} is maximised when $\theta = \cos^{-1}\beta$ where its value is $\beta_{obs} = \gamma\beta$. This exceeds the speed of light when $\beta > 2^{-1/2}$. Of course there is no material motion faster than the speed of light. Superluminal motion is characteristic of more general flows involving relativistic speed.

Although our model referred to two similar spherical sources, this need not be the case. We can adapt it to modify the core-jet model described above. Let the core comprise a stationary relativistic outflow. Although the fluid is moving, the position of the radio photosphere, where the source becomes optically thick

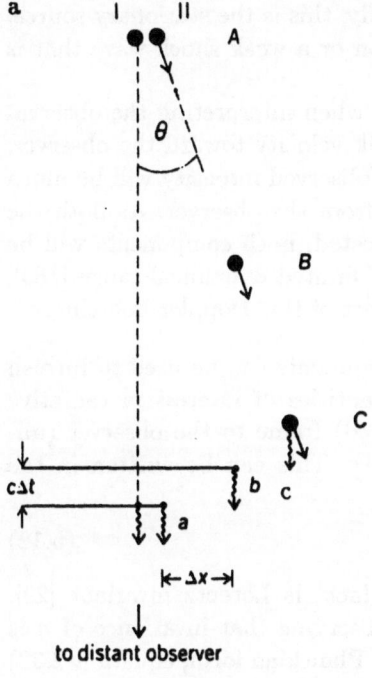

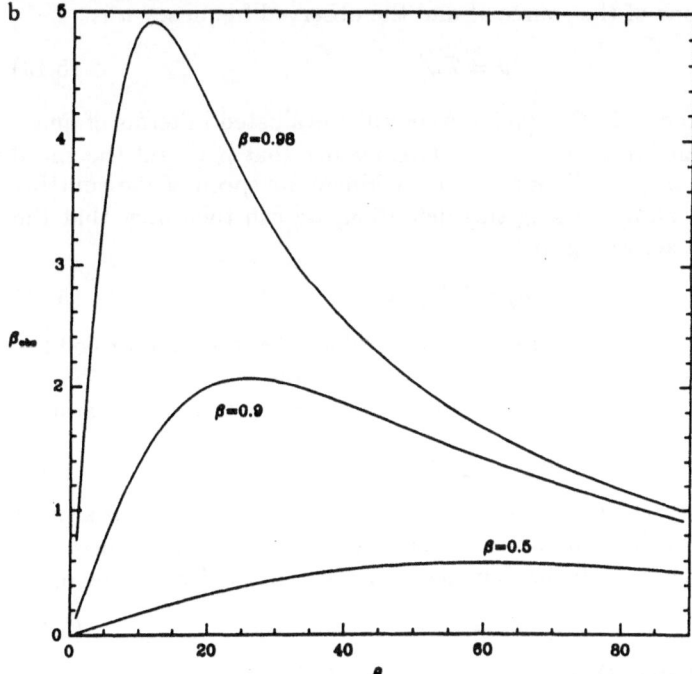

Fig. 5.8. Two source model to demonstrate the kinematics of superluminal motion. a) The moving source approaches the observer with speed βc moving along a direction making an angle θ to the line of sight. b) Apparent transverse expansion speed measured by the observer, β_{obs}.

at a given frequency, will be fixed. Kinematically, this is the stationary source. The moving source can be a density fluctuation or a weak shock wave that is convected outward with the speed of the jet.

This arrangement has a further advantage when interpreting the observations. When the radiating particles have a bulk velocity toward the observer, the radiation will be Doppler-boosted and the observed intensity will be much greater than if the source were moving away from the observer. As both the core and the jet radiation will be Doppler-boosted, both components will be bright and observable in VLBI observations of limited dynamical range [189], [190]. We now explain how to calculate the effect of this Doppler-boosting.

5.3.3 Relativistic radiative transfer. Simple arguments can be used to furnish the necessary Lorentz transformations for quantities of interest in radiative transfer. We transform from the source (primed) frame to the observer (unprimed) frame. Let us begin with the intensity. This can be written in the form

$$I_\nu = \frac{2h\nu^3 n}{c^2} \tag{5.12}$$

where n is the photon occupation number which, is Lorentz invariant [29]. (That this must be the case can be seen by observing that invariance of n is necessary for black body radiation to retain its Planckian form, equation(2.32) under Lorentz transformation.) Now the relationship between the frequency measured in the frame of the source ν' and the observed frequency ν is

$$\nu = \mathcal{D}\nu' \tag{5.13}$$

where the Doppler factor is $\mathcal{D} = [\gamma(1 - \beta \cos\theta)]^{-1}$ evaluated in terms of quantities measured in the observer's frame. (Remember that β is still the speed of the source which may be distinct from the kinematic speed of the emitting region, as discussed above.) Using this definition, we can then infer that the intensity transforms according to

$$I_\nu(\nu) = \mathcal{D}^3 I'_{\nu'}(\nu') \tag{5.14}$$

Note that the left hand side of this equation is evaluated at the transformed frequency which differs from the frequency measured in the source frame. We can also transform the brightness temperature T by observing that the invariance of n requires that

$$T(\nu) = \mathcal{D}T'_{\nu'}(\nu') \tag{5.15}$$

The optical depth, being a probability, is also a scalar invariant. For a small interval of distance in the source frame, this is given by $\tau = \mu' dx'$, where μ is the absorption coefficient. Now dx Lorentz transforms in the familiar manner.

$$dx = \mathcal{D}dx' \tag{5.16}$$

We conclude immediately that

$$\mu(\nu) = \mathcal{D}^{-1}\mu'(\nu') \tag{5.17}$$

Finally, we can transform the emissivity j_ν, (the power radiated per unit frequency per steradian per unit volume) using the equation of radiative transfer. This can be written in the form

$$\frac{dI_\nu}{d\tau} = -I_\nu + \frac{j_\nu}{\mu} \tag{5.18}$$

Lorentz invariance allows us to conclude that the emissivity transforms according to

$$j_\nu(\nu) = \mathcal{D}^2 j'_{\nu'}(\nu') \tag{5.19}$$

When making models of a relativistically expanding source, it is usually most convenient to evaluate the emissivity and the absorption coefficient in the source frame, transform them into the observer frame using equations(5.19), (5.17), and then solve the equation(5.18)of radiative transfer in this frame.

In deriving these formulae, we have ignored the expansion of the universe, a real consideration in their application to AGN. However, the incorporation of cosmological corrections is very simple as analogous transformations to those quoted above [191]can be derived with the substitution

$$\mathcal{D} \to (1+z)^{-1} \tag{5.20}$$

Let us illustrate the use of these formulae by computing the flux from an optically thin source. If the observed intensity is $I_\nu(\nu)$, the observed flux density is obtained by integration over the solid angle subtended by the source.

$$
\begin{aligned}
S_\nu &= \int d\Omega I_\nu \\
&= \int \frac{dA}{D^2} dl j_\nu(\nu) \\
&= \frac{\mathcal{D}^{2+\alpha}}{D^2} \int dV j'_{\nu'}(\nu)
\end{aligned}
\tag{5.21}
$$

where $d\Omega$ is an element of solid angle, dl is an element of length along the line of sight, dA is an element of area normal to the line of sight and $dV = dA dl$ is an element of volume, all quantities being measured in the observer frame. $j'_{\nu'}$ is the conventional emissivity measured in the frame moving with the emitting plasma.

Now suppose that we have two identical, anti-parallel, optically thin radio jets in which the velocity v makes angles, $\theta, \pi - \theta$ to the line of sight. Let the spectral index be $\alpha = 0.5$. The emissivities will be the same in both jets at a given distance from their origins and so the ratio of their fluxes will be given by

$$\frac{S_1}{S_2} = \left(\frac{1 + \beta \cos\theta}{1 - \beta \cos\theta}\right)^{2.5} \tag{5.22}$$

For $\gamma \gtrsim \theta^{-1} \gtrsim 1$, this can approximated as $S_1/S_2 \sim (2/\theta)^5$. This is a very large factor, $\sim 10^3$ for $\theta \sim 30°$. This ratio exceeds the dynamic range of present VLBI observations and routine VLA maps. It should not be surprising that jets are observed to be one-sided. The appearance of real sources, that are partly optically thick, depends sensitively upon aspect and the predicted flux variation can be considerably more or less than that estimated here [192], [193], [172] [194]. The key point is that relativistic beaming is a very powerful amplifier and that intrinsically weak components moving with no more than mildly relativistic velocities in our general direction can outshine the rest of the source.

5.3.4 Distribution of source fluxes. The presence of relativistic beaming has major implications for the distribution of radio source fluxes. Although the details are still a matter of keen debate, we can illustrate its importance with a simple example. Let us suppose that all radio sources comprise identical jets with a fixed outflow speed βc and that they have spectral indices $\alpha = 0.5$ so that their flux densities vary with viewing angle θ according to

$$S_\nu = S_0(1 - \beta \cos\theta)^{-(2+\alpha)}; \quad \theta < \pi/2 \tag{5.23}$$

We ignore the counter jet which will only contribute for $\theta \sim \pi/2$. The angle θ is presumably distributed isotopically so that the fraction of sources observed with viewing angle $< \theta$ is $\propto (1 - \cos\theta)$. Therefore for $\beta \sim 1$, the source count distribution, (the number of sources in a flux-limited survey with flux in excess of S) is expected to satisfy

$$N(> S) \propto S^{-0.4}; \quad S \lesssim \gamma^5 S_0 \tag{5.24}$$

where $\gamma = (1 - \beta^2)^{-1/2}$ [195], [196].

The observed source counts do not look like this [197], [172]. However, we should again not be too surprised. Real jets must be kinematically and dynamically much more complex containing regions moving with different velocities in different directions. Furthermore the morphological differences between the strong and weak extended radio sources lead one to anticipate corresponding differences in the intrinsic source properties and, in any case, the difficulties involved in defining a fair sample are legion[172] [198]. Nevertheless, this example does demonstrate once again the very powerful influence that relativistic beaming can have on the statistical properties of radio sources.

5.3.5 Variability brightness temperatures. We now return to the historical problem of interpreting the rapid variability observed in some compact radio sources. Suppose that an unresolved radio source changes its flux density S in a time $\sim t_{var}$. If its distance is D, (once again ignoring cosmological niceties), then we can assume that its linear size is $\sim ct_{var}$ and define a *variablity brightness temperature*, T_{var} by

$$T_{var} = \frac{SD^2}{2k_B\nu^2 c^2 t_{var}^2} \tag{5.25}$$

In many instances the derived values of T_{var} exceed the inverse Compton (and also the induced Compton) limits.

Now let us suppose that the source comprises a simple radio-emitting cloud moving with speed β so that its Doppler factor is $\mathcal{D}$. The frequency in the source frame is then $\nu' = \mathcal{D}^{-1}\nu$. The brightness temperature will be boosted by a similar factor. If the source has size R, so that its proper variability timescale is $\sim R/c$, then a (model- dependent) estimate of the observed variability time is $t_{var} \sim R/c\mathcal{D}$. Now, the observed flux can be expressed as

$$S = \frac{2k_B\nu^2(\mathcal{D}T')R^2}{\mathcal{D}^2} \tag{5.26}$$

Combining equations(5.15)and (5.26), we obtain

$$T_{var} \sim \mathcal{D}^3 T' \tag{5.27}$$

Doppler factors $\mathcal{D} \sim 10$, consistent with observed superluminal expansion, are adequate to accommodate most compact radio source variability and render it consistent with the inverse Compton limit. However, reported variablity brightness temperatures $T_{var} \sim 10^{18}$K require Doppler factors an order of magnitude larger [46], and may require some alternative explanation.

5.3.6 Relativistic jet models. As we have already remarked, real jets are expected and indeed are observed to be quite complex. To date, there are about 20 known superluminal sources among the brightest known compact radio sources and they exhibit expansion speeds in the range $\sim 2 - 10c$, adopting a Hubble constant of 100km s^{-1}Mpc^{-1} [199]. There is some circumstantial evidence that they are beamed towards us because the jets appear to be quite strongly curved, consistent with strong projection. In addition, those compact jets associated with luminous extended radio lobes, which are presumably less strongly beamed, exhibit lower superluminal expansion speeds.

The fundamental problem in interpreting these observations is that, if we adopt the most naive model introduced above, then, for every source observed to expand superluminally with speed $\beta_{obs}c$, there should be $\sim (\beta_{obs}^2 - 1)/2$, other sources that are less beamed. In addition, their linear sizes should be larger by factors up to $\sim (1+\beta_{obs}^2)/2\beta_{obs}$[196]. Radio astronomers have not been able to locate these unbeamed sources and they have therefore been driven to contemplate more complex models. In particular, these observational difficulties are ameliorated if the source can beam radio emission in a cone with opening angle much larger than $\sim \beta_{obs}^{-1}$. One way to accomplish this and a natural feature of high speed flows is to invoke relativistic shock waves.

5.3.7 Relativistic shock waves. The properties of relativistic shocks are analogous to, and in some respects simpler than those associated with non- relativistic shock waves. Let us consider first an unmagnetised plane shock in the frame in which the shock front is at rest. Let us designate the velocity and pressure ahead of the shock as β_1, p_1, replacing the subscript with 2 for quantities measured behind (downstream from) the shock. (Fig.(5.9).) For simplicity, let us follow our discussion of relativistic jets and adopt an ultrarelativistic equation of state,

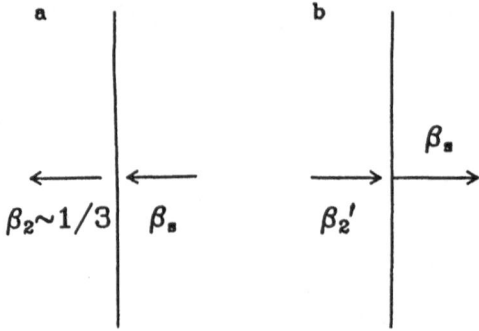

Fig. 5.9. a) Relativistic shock in the frame in which the shock wave is at rest. b) Relativistic shock in the jet frame in which the shock is moving with speed $\beta_s = (1 - \gamma_s^{-2})^{1/2}$.

$$p = \frac{e}{3} \tag{5.28}$$

where e is the internal energy density. The relativistic shock jump (Rankine-Hugoniot) conditions[32], [200], [182] are derived by requiring that the energy flux $4p\beta/(1 - \beta^2)$ and the momentum flux $(3\beta^2 + 1)p/(1 - \beta^2)$ are conserved across the shock. Eliminating the pressure, we deduce that

$$\beta^2 - \frac{4K\beta}{3} + \frac{1}{3} = 0 \tag{5.29}$$

where the constant K is the ratio of the momentum flux to the energy flux. The two speeds, β_1, β_2 are the two roots of this quadratic equation. Therefore, their product is $\beta_1\beta_2 = s^2 = 1/3$, where $s = 3^{-1/2}$ is the speed of sound in an ultra-relativistic plasma. The fluid speed ahead of the shock is supersonic and that behind the shock is necessarily subsonic. In the limit of a strong shock, $\beta_1 = 1, \beta_2 = 1/3$ and $p_2 = 8\gamma_1^2 p_1/3$.

Now let us suppose that there is a strong shock moving with a relativistic speed $\beta_s \sim 1 - 1/2\gamma_s^2$ along a slowly moving jet. We can now use the usual special relativistic velocity addition formula to derive the post shock speed in the jet frame.

$$\begin{aligned} \beta_2' &= \frac{\beta_s - \beta_2}{1 - \beta_2\beta_s} \\ &= 1 - \frac{1}{\gamma_s^2} \end{aligned} \tag{5.30}$$

Hence we find that the Lorentz factor associated with the post shock flow is $\sim 2^{-1/2}$ times that associated with the shock speed. In this example the speed of superluminal motion would be given by the shock speed, β_s, which is distinct from and larger than the speed of emitting fluid, β_2'. The amount of Doppler boosting and the beaming are then smaller than would have been predicted on the basis of the naive model. The changes introduced by distinguishing these two speeds are not great in this simple example. However, it is not difficult to devise models in which they are substantial and which make radically different observational predictions.

240

One way in which this can be achieved is to relax our assumption that the jet fluid is unmagnetised. If the magnetic field ahead of the shock is dynamically significant, then the post-shock speed, β_2 in the frame of the shock can be ultrarelativistic and not much smaller than β_1. In this case, on transforming into the jet frame we find that $\gamma_2' << \gamma_s$. The shock essentially passes by the jet fluid imparting only a small velocity kick but accelerating particles and causing it to radiate, with only weak beaming. The dynamical importance of magnetic field may be indicated by the observation of systematic large swings in linear polarisation in some sources [201], [202], [203].

5.4 Jet Formation

5.4.1 General requirements.
Surely the most tantalising scientific question about jets is how are they made? Many mechanisms have been proposed, all of which present some difficulties[18]. However, there is a wide variety in the observed properties of jets, so there may be a variety of jet collimation mechanisms. For this reason, it is worth considering critically mechanisms which are known to be inappropriate for some jets.

The main constraint on these models is that many jets are well collimated by the time they have propagated to a distance $\lesssim$ 1pc from the nucleus. At this distance, their internal pressures, computed either on the assumption that the radio emission is synchrotron emission and allowing for relativistic beaming or using the power that they are believed to transport, can be quite large, requiring strong forces to shape the flow.

5.4.2 Thermal pressure.
In the *twin-exhaust* model[181], [204], two anti-parallel channels are supposed to propagate adiabatically in anti-parallel directions out of the galactic nucleus forming nozzles where their cross sectional areas are minimised (*cf*.equation(5.4), Fig.(5.10)). In the supersonic part of the flow, the jet will expand transversely as the external pressure decreases but it will also become more collimated. Suppose, with our simple example of an ultrarelativistic equation of state, and an external gas pressure varying with radius as $p \propto r^{-2}$, the jet width will vary as $w \propto \gamma \propto p^{-1/4} \propto r^{1/2}$ and the angle the jet subtends at the nucleus will diminish as $\theta \propto w/r \propto r^{-1/2}$. Even allowing for the reduced collimation caused by internal dissipation, entrainment etc, this mechanism does seem capable of creating well-collimated jets similar to those that are observed.

The reason why it is not believed to be responsible for collimating the powerful radio sources is that VLBI observations tell us that the collimation occurs on scales $\lesssim$ 1pc, where the gas pressure would have to be very large and the density so great that the consequent X-ray emission would easily violate observed upper limits. To be quantitative, suppose that a jet carrying a power of $L = 10^{45} L_{45} \mathrm{erg\ s^{-1}}$ forms a nozzle at a radius $r \sim 10^{18}$cm where its cross-sectional area is $A \sim 10^{34}$cm^2. The pressure at this point would be $p \sim 1$dyne cm^{-2}. If the gas is able to fill the gravitational potential well then its sound speed will be comparable with the central stellar velocity dispersion and its

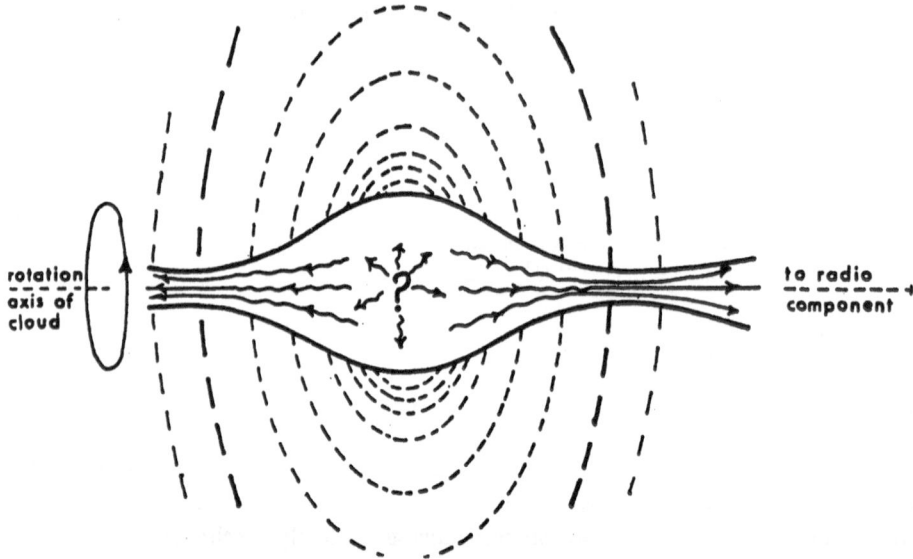

Fig. 5.10. Twin exhaust model for the collimation of jets emerging from an AGN. The hot (probably relativistic), outflowing fluid is in pressure equilibrium with the surrounding plasma which sits in the gravitational potential well created by the galactic nucleus.

temperature will then be $T \sim 3 \times 10^7\mathrm{K}$. The X-ray luminosity of this hot gas is then readily calculated from equation (2.59)

$$L_X \sim 3 \times 10^{48} L_{45}^2 \mathrm{erg\ s^{-1}} \tag{5.31}$$

Only low power radio jets $L \lesssim 10^{43} \mathrm{erg\ s^{-1}}$, could be collimated in this manner without producing excessive X-ray emission.

5.4.3 Radiation pressure-driven jets. Quasars are extremely luminous objects and if they are close to their Eddington limits, then the pressure of radiation acting on electron-proton pairs may be sufficient to overcome gravity along certain directions. This could happen in the funnels formed by a radiation-pressure supported torus orbiting a black hole [205]. (Fig.(5.11).) However, the radiation field in such a funnel is not too anisotropic and a mildly relativistic Lorentz transformation reduces its net momentum density to zero. It is therefore only possible to accelerate the outflowing gas to mildly relativistic speeds, much less than those usually invoked to account for superluminal motion [206]. This explanation is probably also not relevant to the low power, though energetic sources like Cen A which presumably contain giant black holes but appear to be radiating with luminosities far below the Eddington limit. A third, theoretical complication is that these radiation tori are believed to be unstable as discussed in section4.4.2.

In a variant on this mechanism, we suppose that the radiation originates from an accretion disk [207]. Here, somewhat larger outflow speeds are possible although the collimation will be poor. If the outflowing gas arises in an a pair

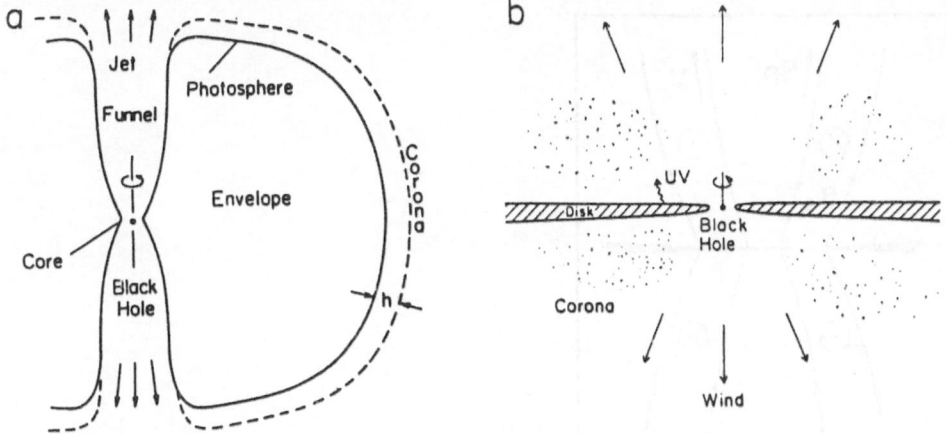

Fig. 5.11. a) Radiation pressure driven outflow collimated by funnels in a radiation-pressure supported torus orbiting a black hole. b) Radiation-pressure driven ouflow from a disk.

plasma then charge neutrality can be maintained with positrons instead of protrons which are 1836 times heavier. The effective Eddington limit for an electron positron plasma can therefore be 1836 times smaller than that for an electron-proton plasma[18], [208]. This allows radiation pressure to drive outflows from thin accretion disks.

5.4.4 Hydromagnetic jets. A third class of jet mechanisms relies upon hydromagnetic stresses exerted by a magnetised accretion disk to fling gas outward centrifugally (*cf.*section 4.2.5) [142], [133], [209], [210]. The gas will be tied to the magnetic field and its inertia will cause the magnetic field lines to be bent backwards creating a toroidal component. Now this toroidal component of magnetic field has an associated "hoop" stress which can act to collimate the poloidal flow of the plasma (see (5.12)). Models can be made of stationary, axisymmetric MHD jets which generalise more familiar models of spherically symmetric stellar winds. As the flows are centrifugally driven and magnetically confined, the gas pressure is not very important except, perhaps, close to the disk.

There are three types of wave mode in MHD, fast waves, intermediate (or Alfvén) waves and slow waves, in contrast to the single sound wave of simple gas dynamics. This implies that three critical points can be found along a flow line where the flow velocity becomes sonic with respect to these three speeds. Real solutions probably have to pass through all three critical points. After passing through the final fast mode point, the magnetic energy and angular momentum fluxes are typically half converted into mechanical energy and angular momentum. The magnetic flux becomes predominately toroidal. This is because the parallel component of magnetic field decays inversely as the square of the jet width, w in a jet with approximately constant outflow speed. However, the toroidal component decays $\propto w^{-1}$ and eventually dominates the parallel component. Under certain conditions, it can be shown that

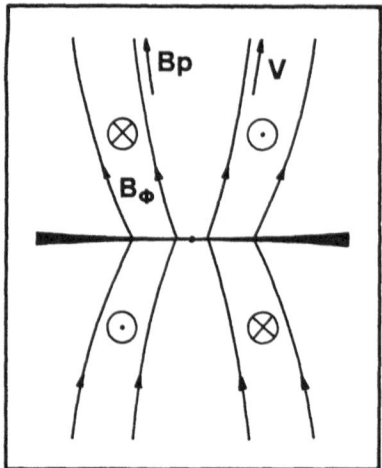

Fig. 5.12. Magnetic collimation of a centrifugal jet from an accretion disk. The inertia of the plasma is responsible for the production of a toroidal magnetic field component which can then collimate the jet.

MHD winds will always be collimated asymptotically, both non-relativistically [211], and relativistically [212].

There are two problems associated with this explanation for AGN jets. The first is that if they are collimated by a toroidal field as suggested here, then we might expect to see this reflected in the polarisation observations and this hasn't happened yet. The second difficulty is theoretical. Magnetically collimated plasma is notoriously unstable , especially to non-axisymmetric instabilities. Until it becomes possible to perform non-axisymmetric hydromagnetic numerical simulations, it will be hard to be sure that hydromagnetic jets are viable.

6 Fuelling of AGN

6.1 Introduction

6.1.1 General considerations. Implicit in the general black hole accretion disk model of AGN is the requirement that gas be supplied to the galactic nucleus at an adequate rate to sustain its bolometric luminosity. There are two problems on which we should focus, what is the origin of the gas and how is it forced into the nucleus? From an observational perspective, this problem is complementary to that of inferring the structure of the flow close to the black hole in the sense that it is possible to resolve much of the activity spatially, though not temporally. From a theoretical standpoint, studying the fuelling of AGN naturally raises the much larger question of the origin and interactions of galaxies.

Some of the principal issues in this subject are dynamical and the best contemporary reference, covering both star clusters and gaseous disks is Binney and Tremaine(1987) [213]. More specific reviews of AGN fuelling include [214], [215], and the conference proceedings [216].

6.1.2 Orders of magnitude. To estimate the required rate of gas supply, let us take a typical quasar with luminosity $L \sim 2 \times 10^{46}$ erg s^{-1}, ($H_0 = 70$km s^{-1}Mpc^{-1}, $\Omega_0 = 1$), at a redshift $z = 2$, when the universe was ~ 2Gyr old. Let us suppose that the quasar radiates at half the Eddington luminosity so that the central black hole mass is $M \sim 4 \times 10^8 M_\odot$. Now if, as suggested by the considerations of section 3, the radiative efficiency is ~ 0.1, then gas must be supplied to the black hole at a rate $\dot{M} \sim 3 M_\odot$yr^{-1}. The black hole mass must then e-fold in $\sim 10^8$yr. There is time for up to 20 e-foldings, long enough for the black hole to have been built up to its inferred mass. Now the actual mass required ($\sim 4 \times 10^8 M_\odot$), is small compared with the mass of gas present in the interstellar medium of our own Galaxy and is presumably even smaller than the mass present in a young galaxy in the process of forming stars. So the problem is not in supplying the fuel but in driving it into the nucleus. Two resolutions of this problem have been proposed; that the mass has been present in the nucleus essentially since the formation of the galaxy and that the gas derives from outside the nucleus and a giant torque is responsible for driving it inward. We consider these in turn.

6.2 Star Clusters

6.2.1 General speculations. Observations of nearby galaxies show that central star clusters with densities in the range $\sim 10^6 - 10^8 M_\odot$pc^{-3} and 1D central velocity dispersions $\sigma \sim 100 - 400$km s^{-1}Mpc^{-1} are typical [217]. If these values are also representative of the central stellar densities in a nascent galaxy, then we can try to determine how the stellar distribution might have evolved. Such a star distribution behaves dynamically like a giant globular cluster and can evolve under the mutual gravitational interactions of the stars to develop a central high density core. The stars in this core can then interact more intimately and collide [218]. This may lead to stellar disruption and the formation of a dense gas cloud. If subsequent star formation is inhibited, then the cloud will collapse and cool to become a radiation-supported supermassive star or "spinar" which itself is subject to dynamical instability. A supermassive black hole is unavoidable as an end result, [89]. If some small fraction of the binding energy released appeared in an electromagnetic channel then we might expect to see a "hypernova". (That none have been reported is an argument against this scenario.)

In an alternative evolution, a modest black hole $\lesssim 10 M_\odot$ is formed initially and gas liberated by stars sinks steadily onto it slowly increasing its mass. Yet a third possiblity is that colliding stars coalesce to build up massive stars which will evolve rapidly, losing most of their mass through winds or supernovae [219].

245

Under these high density conditions, the gas released should be able to cool radiatively before it escapes. Again a massive black hole should form. However this time it might be surrounded by a dense cluster of compact objects, (black holes and neutron stars). Each of these scenarios raises important physical questions.

6.2.2 Stellar cusp. Suppose that a massive black hole of mass M has formed in the centre of a nuclear star cluster. If the star density far away from the black hole is $n_0 = 10^7 n_{07} M_\odot \mathrm{pc}^{-3}$, the stellar mass is $m \sim 1 M_\odot$ and the central velocity dispersion is $\sigma = 300 \sigma_{300} \mathrm{km\ s}^{-1}$, then we can infer the core radius of the cluster r_c, usually defined to be the radius where the surface density falls to half its central value, using the virial theorem and equating the potential energy $\sim 4\pi G n_0^2 m^2 / r_c$ roughly to the kinetic energy $\sim n_0 m \sigma^2 / 2$. A more careful calculation gives

$$r_c = \left(\frac{9\sigma^2}{4\pi G n_0 m} \right)^{1/2}$$

$$\sim \sigma_{300} n_{07}^{-1/2} \mathrm{km\ s}^{-1} \mathrm{pc}$$

(6.1)

[213]. The hole will dominate the mass out to a radius

$$r_h \sim \left(\frac{3M}{4\pi n_0 m} \right)^{1/3}$$

$$\sim M_8^{1/3} n_{07}^{-1/3} \mathrm{pc}$$

(6.2)

which we presume to be smaller than the core radius. (*cf.*Fig.(6.1)). Stars localised within r_h will move in a $1/r$ potential well and may establish a density cusp [220]. (Note though that stars from the outer core moving with speed $\sim \sigma$

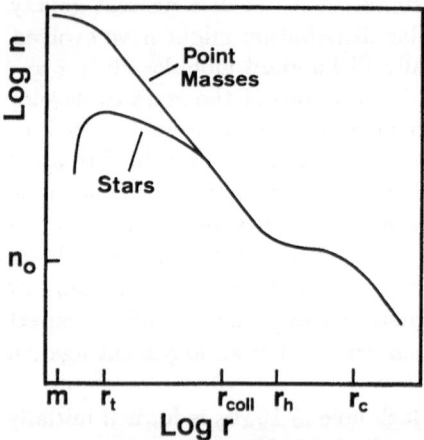

Fig. 6.1. Stellar density distribution within a galactic nucleus. r_c is the core radius of the central star cluster and r_h is the radius within which the mass of the central black hole dominates the gravitational potential. r_{coll}, r_t are the collision and tidal radii.

must pass within a radius $\sim GM/\sigma^2$ for their orbits to be seriously perturbed by the hole.)

We can derive the slope of the density profile in this cusp, by considering the dynamics of stars interacting through their mutual gravitational attraction. Consider two stars approaching each other with relative velocity v. If their impact parameter is b, the impulse will be roughly the product of the force $\sim Gm/b^2$ and the encounter time $\sim b/v$ i.e. $\sim Gm/bv$. They will be deflected through a small angle $\sim Gm/bv^2$ for $b \lesssim b_{\pi/2} \sim Gm/v^2$. (Fig.(6.2).) The mean time for undergoing a $\sim \pi/2$ deflection is $\sim (n\pi b^2 v)^{-1}$ where $n(r)$ is the local stellar density. In fact, smaller angle deflections will occur more frequently and add stochastically so that the timescale for accumulating a total deflection of $\sim \pi/2$ is smaller than the time to achieve this in a single collision by a logarithmic factor. This time scale, which is also roughly the timescale for energy exchange between stars through two body interactions, is know as the *relaxation time*.

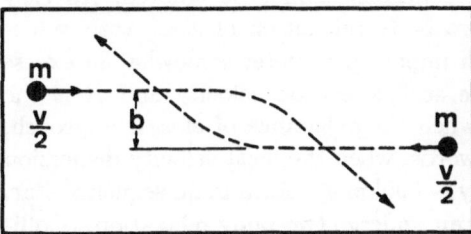

Fig. 6.2. Mutual deflection in centre of momentum frame of two stars with mass m, impact parameter b and relative velocity v.

$$t_r = \left(\frac{2\sqrt{2}\pi n G^2 m^2 \ln \lambda}{3\sigma^3} \right)^{-1} \qquad (6.3)$$

where $\ln \lambda \sim \ln(4\pi n r^3) \sim 10 - 20$ [221]. Numerically

$$t_{r0} \sim 2 \times 10^9 \sigma_{300}^3 n_{07}^{-1} \mathrm{yr} \qquad (6.4)$$

This must be less than the age of the galactic nucleus if a cusp is to form dynamically.

Now suppose that a stationary cusp is established by stellar dynamical processes alone within a central stellar core and that stars slowly accrete onto the central black hole. Unlike in the case of an accretion disk, the binding energy cannot be radiated away, and must be transported outward ultimately to heat the outer parts of the cluster. Now the binding energy residing in the stars found within an octave of radius $r \to 2r$ is $E(r) \sim 4\pi n r^3(-GM/2r)$ and the rate of energy exchange between the stars within this radius range through two body encounters is $\sim E(r)/t_{relax} \propto n^2 r^{7/2}$, where we use the fact that the local velocity dispersion in a $1/r$ potential varies according to $\sigma \propto r^{-1/2}$. As this has to be constant independent of radius, we obtain the density profile

$$n(r) \sim n_0 \left(\frac{r}{r_h}\right)^{-7/4} \qquad (6.5)$$

[221]. Note that it is the flow of binding energy, not stars (which can move separately in and out) that is conserved.

For typical values $M \sim 10^8 M_\odot$, $\sigma \sim 300\text{km s}^{-1}$, $n_0 \sim 10^7 M_\odot \text{pc}^{-3}$, we obtain $r_h \sim r_c \sim 1\text{pc}$ and $t_{0r} \sim 2 \times 10^8\text{yr}$. The stellar density and local velocity dispersion will increase rapidly within this radius. For a cluster of point masses, $(M/m)(r/r_h)^{5/4} \sim 1000$ stars will reside within, say, ~ 10 Schwarzschild radii. However, the local relaxation time at this radius will be $\sim t_{r0}(r_h/r)^{1/4} 2 \times 10^8\text{yr}$. This is not an efficient way to supply mass to the black hole. This is entirely consistent with the majority of galaxies *not* showing nuclear activity. However, there is a more fundamental objection to purely stellar dynamical processes being responsible.

6.2.3 Stellar collisions. The foregoing considerations become irrelevant for main sequence stars because the effects of collisions cannot be ignored. An easy way to see this is to observe that two body relaxation of stars with velocity dispersion σ involve collisions with impact parameter somewhat in excess of $b_{\pi/2} \sim R_*(v_e/\sigma)^2$ where v_e is the surface escape velocity and R is the stellar radius. The stars will collide when their distance of closest approach, $\sim (Rb_{\pi/2})^{1/2} \sim Rv_e/\sigma \lesssim R$. In other words, when the local velocity dispersion exceeds a typical stellar escape velocity $\sim 600\text{km s}^{-1}$ for a main sequence star, then the stars will collide before they can undergo two-body relaxation. (Collisions involving giants will happen even more readily.) Stellar collisions therefore become more important than two body relaxation within the core. It is much harder to predict the density profile of the cusp under these circumstances and the subsequent rate of mass supply to the hole [222].

6.2.4 Tidal disruption by the black hole. Despite these uncertainties, it is likely that stars approach the black hole occasionally. They will then experience strong tidal forces which may be violent enough to disrupt the star. The condition for this to occur is that the tidal gravitational force exerted by the black hole exceed the self-gravitational force of the star, *i.e.*

$$\frac{GMR}{r^3} \gtrsim \frac{Gm}{R^2} \qquad (6.6)$$

This is the usual *Roche limit*; the mean density of the star must exceed the mean density of the interior mass. The associated *tidal radius* is

$$r_t \sim (M/m)^{1/3} R \qquad (6.7)$$

[99], [223], [224].

Tidal disruption of main sequence stars is only important when the tidal radius is larger than the radius of the event horizon. Typically this requires that the hole mass satisfy $M \lesssim 10^8 M_\odot$. Tidal disruption is therefore probably not relevant for the most luminous quasars.

However, it may well be very important for low luminosity AGN with central black holes of mass $M \sim 10^6 M_\odot$, for example our Galactic centre. In this case, the nucleus may have long intervals of quiescence interrupted by occasional outbursts caused by the disruption of individual stars. When this happens, the tidally ejected gas should mostly remain bound to the hole and will eventually circularise to form a torus in orbit about it. It can then accrete from the torus onto the black hole at a rate determined by the effective viscosity. If the accretion rate is large enough, then the power can exceed the Eddington limit and much of the gas may be blown away.

The rate at which stars will be directed within the tidal sphere of influence of the black hole depends a upon dynamical processes occuring within the outer parts of the cluster. Essentially, it is possible to define a *loss cone* at r_h so that any star whose velocity is deflected within this cone will pass within the hole's tidal radius. Now if the stars are simply relaxing dynamically as a result of two body interactions, then it will take a full relaxation time for a stars velocity to random walk into the loss cone. The rate of mass supply will then be somewhat less than the core mass divided by the relaxation time $\sim 0.01 n_{07}^{1/2}$ $M_\odot \text{yr}^{-1}$. However, if some stronger dynamical perturbation is present, for example if a bar forms or a second black hole is captured, then the relaxation time can be significantly shorter and larger rates of gas supply are possible.

6.2.5 Ablation An alternative source of fuel might be *ablation* of stars by the quasar UV or possibly high energy particles [225]. Ablation of the star will occur when it absorbs sufficient power to drive gas away from its surface with the escape velocity. This is more likely to be important for giants which have a large cross section [226]. To get a measure of the likely importance of this process, consider a star of effective temperature $10^3 T_{e3} \text{K}$. Its surface radiative flux is $\sim 10^{-6} T_{e3}^4 \text{erg s}^{-1}$. This can be compared with the radiative flux from the AGN, $\sim 10^{-8} L_{46} R_{pc}^{-2} \text{erg s}^{-1}$. So, presuming the star absorbs the energy, its structure is unlikely to be seriously perturbed outside $\sim 1\text{pc}$ from the central continuum source and is unlikely to contribute significantly to the emission line regions. It is possible that the energy be absorbed within an extensive corona and be carried off as the kinetic energy of the wind. It is then necessary that the gas not cool radiatively in the time it takes to accelerate away from the star. This turns out to be quite restrictive.

6.3 Accretion Disks

6.3.1 Overview. Gas can also be supplied directly from the surrounding galaxy through an extensive accretion disk. This is almost required if the gas originates in the body of the galaxy and it is endowed with angular momentum. Indeed, it may reside in a disk all the way from a few Schwarzschild radii out to several kpc up to ~ 30 octaves in radius. Of course spiral galaxies also have extensive disks which have survived infall for many rotation times. Any explanation of fuelling along these lines must also account for the relative inactivity in most spiral galaxies.

6.3.2 Self-gravitation. One feature of a large accretion disk is that its outer parts are probably self-gravitating [141], [227]. This occurs if gravitational self-attraction across a portion of the disk with size of order its thickness $H \sim s/\Omega$, (with s the sound speed), exceeds the tidal gravitational force exerted by the central mass, *i.e.*

$$G\Sigma \gtrsim \frac{GMH}{r^3} \qquad (6.8)$$

where Σ is the surface density, M is now interpreted as the total interior mass, (not just that of the central black hole if $r \gtrsim r_h$). For a Keplerian disk this can be re-written in the form $(G\rho)^{-1/2} \lesssim \Omega^{-1}$; in words, the gravitational collapse time is shorter than the timescale for density perturbations in the disk to be stretched by shear in the disk's differential rotation. The formal statement of this condition, due to Toomre [228], is

$$Q = \frac{s\kappa}{\pi G\Sigma} < 1 \qquad (6.9)$$

where $\kappa = 2[\Omega d(r^2\Omega)/d(r^2)]^{1/2}$ is the *epicyclic frequency*.

To illustrate the importance of this criterion, let us adopt the α prescription for a Keplerian disk, for which the mass accretion rate is given by $\dot{M} = 3\pi\alpha s H\Sigma$ and $\kappa = \Omega$. In this case, $Q = 3\alpha s^3/G\dot{M}$. Numerically, when

$$s \lesssim 40 \left(\frac{\alpha}{0.1}\right)^{-1/3} \left(\frac{\dot{M}}{3\mathrm{M}_\odot\mathrm{yr}^{-1}}\right)^{1/3} \mathrm{km\ s}^{-1} \qquad (6.10)$$

the disk is likely to fragment and the α prescription is unlikely to be relevant. Disks in quasars cooler than $\sim 10^4$K are therefore unlikely to be describable by the formalism introduced in section 4.

These large scale, cool disks, if they exist, may be more similar to the disks of spiral galaxies than those of X-ray binaries. They may be multi-phase, containing clouds of molecular gas, a hotter intercloud medium and radiatively efficient interfacial regions all coupled together by magnetic field. (Up to $\sim 10^9 \mathrm{M}_\odot$ of molecular gas has been detected in the nuclei of nearby galaxies [229], and our Galactic centre contains a large molecular torus of radius $\sim$ 2pc [230].) Star formation may proceed within the molecular clouds, which may be supported against collapse or even dispersed by the consequent stellar winds, supernovae, planetary nebula formation and novae. Angular momentum may be transported locally through inelastic collisions between these giant "molecules" [231], [232], or globally through non-axisymmetric perturbations in the gravitational potential associated with spiral arms and bars [233]. (A non-axisymmetric perturbation in the gravitational potential, $\delta\phi$, can drive gas inward in $\sim |\phi/\delta\phi|$ orbital periods and so this can be very efficient.) Non-axisymmetric perturbations may either be externally induced or develop internally as in barred spiral galaxies as a consequence of the growth of a dynamical instability. Magnetic fields provide a viable alternative torque just as is the case with protostellar disks.

As well as transporting angular momentum, all of these processes are likely to be highly dissipative. They are therefore likely to heat the disk, not necessarily by simply raising the temperature uniformly, but instead by increasing the velocity dispersion in the molecular clouds. This will increase the thickness of the disk and may be able to keep it at a state of marginal stability with respect to the Toomre criterion.

6.3.3 Observational ramifications. There are some general geometrical features of giant molecular disks that may have important observational consequences. Firstly, as they extend through so many octaves of radius, there is no good reason why they should be coplanar. If gas works its way slowly inwards from the outer parts of the galaxy through the disk, then the plane of the disk at a given distance from the black hole may reflect the history of the gas supply as much as the potential of the galaxy and we already know from the observation of warps in our own and other galaxies that the equipotential surfaces twist. (The characterisation of the gas distribution as a disk is questionable and it may well be more realistic to describe it as flattened distribution of clouds.) This also means that even if a disk is quite thin at a given radius, the whole disk can cover a large fraction of the sky as seen from the black hole. Therefore a large fraction of AGN can be obscured and be only observable through re-emitted or scattered radiation. It is widely suspected that types 1 and 2 Seyfert galaxies are distinguished by viewing angle [231]. In the former case, the broad line and X-ray emission regions are directly visible, whereas in the latter, they are hidden behind a thick disk or torus of molecular gas.

This emission may, however, be observed indirectly through scattering by free electrons located at high latitude (Fig.(6.3)). With the orientation pro-

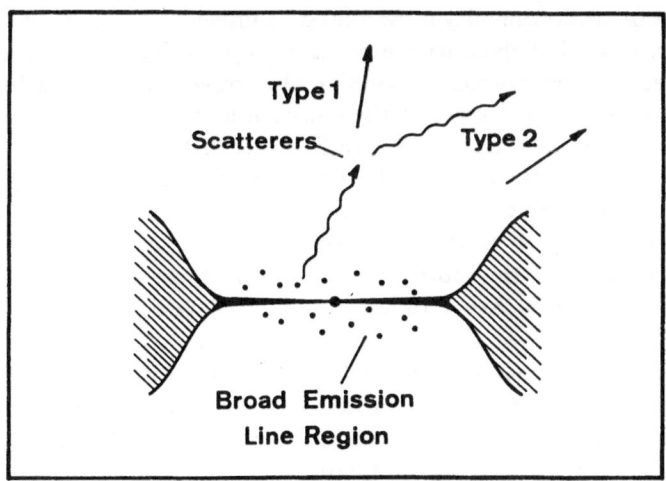

Fig. 6.3. Possible location of Thomson scattering region in a Seyfert galaxy. Low latitude observers classify it as a type 2 Seyfert galaxy and only see the broad emission lines in polarised light; high latitude observers see a type 1 Seyfert.

posed, the scattered radiation will be linearly polarised with electric vector parallel to the major axis of the projected disk.

6.3.4 Infra-red emission. We have argued that cool gas in the outer parts of an accretion disk is unlikely to be confined to a thin equatorial zone with thickness dictated by its temperature. This implies that it is likely to be exposed to the optical, UV and X-ray flux from closer to the central black hole. The cool gas will therefore be heated and should re-radiate the incident power at longer wavelengths [78], [79]. We must determine the conditions under which this re-emission can be efficient enough for the gas to remain cool.

The major coolant for molecular gas under these circumstances is probably dust. If, following the discussion in section 2, we tacitly assume that the dust is not dissimilar from interstellar dust, then we can determine the temperature of a dust grain in radiative equilibrium by balancing the absorbed incident flux F from the quasar with the infra red emission. If we model a dust grain as a sphere of radius a, and for interstellar dust, a lies in the range $\sim 30 - 3000$Å, then

$$\pi a^2 F \sim 4\pi a^2 < Q > \sigma_{SB} T^4 \tag{6.11}$$

where we have assumed that the incident radiation is dominated by UV photons which are absorbed with nearly the geometrical cross section. $< Q > (T)$ is the infra red wavelength-averaged mean absorptive efficiency, which by Kirchhoff's law equals the emissivity relative to a black body at the dust temperature T. Now as we have discussed, $Q(\lambda) \sim 2\pi a/\lambda$ at long wavelengths, and so integrating over the emitted spectrum which peaks at a wavelength $\lambda_{max} \propto T^{-1}$, we have that $< Q > \propto T$ and the dust temperature T satisfying

$$T \sim 1000 L_{46}^{1/5} r_{pc}^{-2/5} \text{K} \tag{6.12}$$

[79] where the constant of proportionality is evaluated by integrating over a distribution of grain sizes. Note that the dust temperature is pretty insensitive to the luminosity. This temperature-to-radius mapping also implies a wavelength to radius mapping in the sense that most of the emission at a given observed wavelength will be associated with a particular radius (Fig.(6.4)).

The integrated infra-red spectrum from the quasar is determined by the covering factor of the dust as a function of radius. The infra-red spectral index is typically observed to be $\alpha \sim 1$ so that the power radiated per octave of frequency is roughly constant. If this power is predominantly re-radiated by dust, then its covering factor per octave of radius would also have to be roughly constant.

It can be see from Fig.(6.4) that dust at radius $r \sim 10$kpc in the outer parts of the surrounding galaxy has a temperature $T \sim 50$K and radiates the $\lambda \sim 100\mu$ emission. Some AGN have been shown to have very steep far infra-red turnovers longward of $\sim 100\mu$ with spectral indices $\alpha < -2$, steeper than a black body [36]. Single temperature dust will radiate a spectrum $S_\nu \propto Q\nu^2 \propto \nu^3$, approximately and so if the distribution of grain temperatures does not

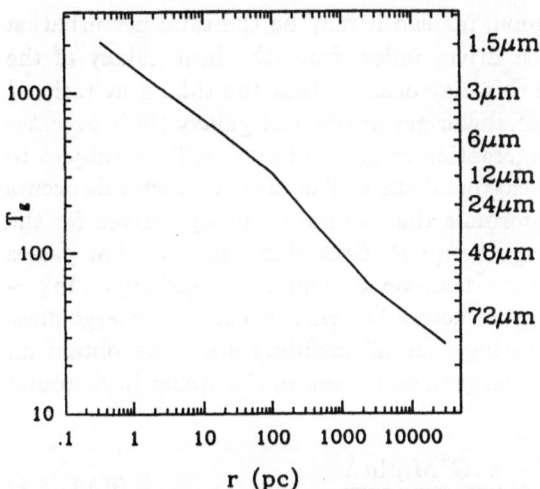

Fig. 6.4. Temperature and wavelength to radius mapping for dust emission from a quasar with luminosity $L \sim 10^{46}$ erg s^{-1}. (Adapted from [234].)

fall below some minimum value, it may be possible to account for steep far infra-red turnovers in dust re-emission model [79].

Of even more interest is the maximum dust temperature. This depends on the composition, but the most refractory graphite grains can apparently resist sublimation for an estimated accretion timescale up to a temperature $T \sim 2000$K. The smallest grains which can be heated impulsively by individual photons may contribute significantly to the emission at shorter wavelengths. The highest temperature dust should be located at a radius $r \sim 0.1$pc (*cf.*equation(6.12)). We might also expect the observed spectra to show some characteristic change at the peak wavelength emitted by dust at this temperature $\sim 1\mu$. When gas is robbed of its dust coolant it will be rapidly heated to a new equilibrium temperature $T \gtrsim 8000$K, where it is cooled mainly by line emission. There will therefore be a relative absence of emission at intermediate temperatures, and a corresponding minimum in quasar spectra at $\lambda \sim 1\mu$, as reported.

An immediate observational implication of this interpretation of quasar infra-red spectra is that they should not be observed to vary at wavelengths longward of 10μ and that their variation at shorter wavelengths should be slow (on timescales $\gtrsim r(\lambda)/c$) and may perhaps track the variation in the UV. Dust emission may also be recognised by the presence of dust-related spectral features, for example a $\sim 3 - 5\mu$ bump and characteristic polarisation.

6.4 Companion Galaxies

6.4.1 Tidally-driven inflow.
Let us now return to the question of the origin of the gaseous fuel. There is some circumstantial evidence that nuclear activity is associated with the presence of a galactic companion. This does not imply that

the gas derives from the companion; instead it may be the tidal perturbation induced by the companion which drives inflow from the host galaxy of the AGN. One mechanism by which this may occur is that the tidal gravitational field may shock and compress interstellar gas in the host galaxy till it becomes self-gravitating [235]. A giant aggregation of gas and stars will be subject to dynamical friction against the background stars. The rate at which this occurs can be estimated modifying the formula that we have already derived for the two body-relaxation timescale, equation(6.3). Each star, mass m, that passes by the large mass, M_t, will induce a transverse momentum exchange $\Delta p_\perp \sim GM_t m/bv$, where b is the impact parameter. The energy lost by the large mass will then be $\sim \Delta p_\perp^2/2m$. Integrating over all colliding stars, we obtain an estimate for the time scale for the large mass to sink in the stellar background under dynamical friction

$$ t_{df}^{-1} \sim \frac{4\pi G^2 M_t \rho \ln \lambda}{v^3} \qquad (6.13)(6.3) $$

[213], where ρ is the density of the background stars and v is the speed of either the stars or the large mass in the galaxy potential well and λ is the Coulomb logarithm (cf.equation(6.3)). Note that the dynamical friction timescale is inversely proportional to the mass. To order of magnitude, this takes $\sim M(r)/2\pi \ln \lambda M_t$ orbital periods, where $M(r)$ is the interior mass in the galaxy. Masses $M \gtrsim 10^{6-7}M_\odot$ can sink into the galactic nucleus over the lifetime of the galaxy. The mass will sink as long as its mean density exceeds that of the background galaxy; thereafter it will be tidally stripped. This amount of mass is sufficient to fuel a major outburst in an active galaxy.

6.4.2 Galaxy mergers. The final possibility that we mention is the most immediate and this is that the fuel derives from a companion galaxy, probably a dwarf, which is actually ingested by the host galaxy [236], [235]. Again it will sink under dynamical friction having its outer parts progressively stripped as it encounters regions of the host galaxy of comparable density. Numerical simulations show that this mechanism is also effective in driving gas from the host galaxy into the nucleus. The association of companion galaxies and morphological distortion with nuclear activity must, in this case, be attributed to galaxy-galaxy-galaxy correlations – where two galaxies are found, it is more likely that there will be a third.

Galaxy mergers provide a possible explanation of double nuclei infra red galaxies, such as that in Arp220 [114]. An even more interesting possibility is that the captured galaxy itself contains a black hole as discussed in section 3.2.6, and that the black hole is actively responsible for driving the fuel inward.

7 Grand Unified Theories

7.1 General Considerations

7.1.1 Precedents and cautions. In the preceding chapters, I have endeavoured to outline some of the physical processes that are believed to be of most relevance to interpreting AGN. I now turn to the problem of assembling these components into a general physical theory of galactic nuclear activity. This turns out to be a rather difficult task. Drs. Netzer and Woltjer, in their lectures, present some alternative solutions and this variety fairly reflects our collective confusion. However, most researchers would agree that this is the long-term goal of AGN research even if some would caution that it is a hopeless task at present. In the context of these lectures, discussion of a particular general interpretation of AGN has the merit of showing *where* the physical principles described above might be applied. In addition, unified models can provide a conceptual framework for describing the observations even when the framework is wrong! The following should be interpreted in this spirit.

The progress in our understanding of AGN, like our understanding of basic physics, has been a story of consolidation and integration. Previously separate phenomena have been shown to be different manifestations of the same underlying physical process, distinguished by luminosity, observer distance, orientation or environment. For example Seyfert galaxies are now believed to be low luminosity radio-quiet quasars, compact radio sources are argued to be mainly modest power extended radio sources seen "pole-on", BL Lac objects are possibly weak emission line AGN in which the Doppler-boosted optical continuum outshines the lines and so on. There is every indication that this progression will continue.

Most prior discussions of a general or unified theory of AGN have taken place and have been assessed in terms of our current understanding of the observations. By contrast, I will attempt a more deductive approach outlining a theory from our fundamental black hole-accretion disk hypothesis. However, before I do this I shall first give an "anatomical" summary of the principal manifestations of nuclear activity working inward from the largest to the smallest scales. I shall then work outward again giving an interpretation of some of the observations, suggesting reasons for the different properties observed in different classes of objects and present a simple classification of AGN. I conclude with some brief speculations about their evolution.

The best existing review of unified models of AGN is due to Lawrence(1987) [237]. Other references specifically devoted to this topic include [238], [195], [239], [240].

7.1.2 "Unification" and "Grand Unification". A distinction should be drawn at this stage. Attempts to unify two classes of active nuclei (most prominently types 1,2 Seyfert galaxies or compact and extended radio sources) are sometimes termed "Unified theories". Attempts to incorporate all nuclear activity within some framework are called "Grand Unified theories". As is the case with fundamental physics, unification is so far more successful than more speculative grand unification.

7.1.3 The primary parameters. The first task in attempting to construct a general theory of AGN is to decide which parameters exert a controlling influence upon their properties. The analogy with stars is helpful. The majority of stars can be organised by their luminosities and effective temperatures into a main sequence, controlled by one primary physical parameter, the star's mass. If we wish to generalise this to embrace pre- and post-main sequence evolution then we must introduce a second parameter, age. There are other parameters, composition, rotation, magnetisation etc. whose effects can be clearly measured. Nevertheless these are only of secondary importance for most phenomena. Still, these parameters are insufficient to describe what is observed. Extrinsic or environmental effects also come into play, notably binarism, reddening, mass accretion, etc. These secondary and extrinsic factors were initially a source of great confusion in taxonomic investigations and remain of dominant interest in contemporary research. Nevertheless, they are secondary.

The position with AGN is less clear. It seems unavoidable that the mass of the central black hole is a primary parameter as it sets the size of the inner accretion disk. However, a black hole cannot radiate at all without accretion and the rate at which gas is supplied to the hole $\dot{M}$, is probably as important as the mass. In addition, we have described several properties of Kerr black holes which distinguish them from non-rotating Schwarzschild black holes and so we might expect that Ω_H would also be a primary parameter. It makes sense to scale the accretion rate and the spin by defining two dimensionless parameters

$$\dot{m} = \dot{M}/\dot{M}_E$$
$$\lambda_H = 2m\Omega_H \tag{7.1}$$

where $0 < \lambda_H < 1$ and $\dot{m} \sim 10$ is necessary to sustain an Eddington-limited luminosity if the radiative efficiency is ~ 0.1. I propose that $M, \dot{m}, \lambda_H$ are the three primary parameters controlling the intrinsic properties of black holes and accretion disks. Note that we are not including age as a primary parameter as the dynamical times around a black hole are short compared with the intervals over which we have been observing them.

Now, as we have already emphasised, AGN exert a powerful influence upon the galactic environment which in turn may affect what we observe, for example, the type of radio source that is formed. A slightly more subtle environmental effect that has been of extreme importance in the historical development of the study of AGN is the luminosity of the surrounding galaxy. Quasi-stellar objects and Seyfert galaxies were distinguished by whether or not the galaxy could easily be seen. This is not of central importance. Location in or out of a rich cluster provides another possible parameter. Also, the recent history of interactions with neighbouring galaxies is undoubtedly important in shaping the observations. However, I shall adopt the convention that all of these effects are extrinsic and secondary and that they have to be removed in much the same way as the effects of interstellar reddening have to be removed from stellar spectra.

7.1.4 The importance of aspect. It will, by now, be clear that the orientation of the observer with respect to the source also has a strong influence on how that source is classified. There are three separate ways in which this can happen. The most dramatic influence is through relativistic Doppler-beaming in a jet. As we have demonstrated in section5.3.3, even mildly relativistic motion can amplify the observed flux by several orders of magnitude and a relatively low power source can appear extremely luminous if we happen to be located within the solid angle illuminated by its twin jets.

The second effect arises from absorption or obscuration, most probably by an equatorial disk or torus. This has been invoked to account for the dissimilarities between type 1 and type 2 Seyfert galaxies[238], [241], [242], those between broad absorption line quasars and normal radio-quiet quasars [243], and, more controversially, for the difference between radio galaxies and radio quasars [244].

The third orientation effect is milder and is associated with a disk geometry. When a disk is viewed at low latitude from close to the equatorial plane, it will appear fainter due to a reduced aspect and limb-darkenening, than when the same disk is viewed pole on. Conversely, if emission lines share this geometry, they will have a broader velocity width at low latitude [245].

Despite their obvious observational importance, I shall also adopt the convention of treating these orientation effects as secondary.

7.2 Anatomy

7.2.1 Radial variation. Nuclear activity is made manifest over a dynamical range of $\sim$ 12 decades of radius. In Fig.(7.1), I show a schematic montage of each decade exhibiting phenomena which are believed to take place on each of these length scales. The phenomena on the large scales are directly observed and we can discern the geometrical arrangement; those on the smaller scales are not and the associated frames represent educated guesses. Let us consider these frames in turn, working from the large scales to the small scales.

7.2.2 1Mpc. Extended radio sources. It is now clear that extended radio sources are jet-fed. It is equally clear that the high power, edge-brightened (type 2) radio sources are qualitatively quite different from the low power edge-darkened (type 1) sources. Physical arguments, buttressed by numerical simulations make it quite likely that the former class contains supersonic, mildly relativistic jets, whereas the latter involves trans- or subsonic flows, whose shapes are sensitive to the circumgalactic environment [93] [18] [179].

7.2.3 100kpc. Radio jets. Radio jets are quite common on this scale. Jets associated with powerful galaxies are relatively fainter than those found in the type 1 sources.

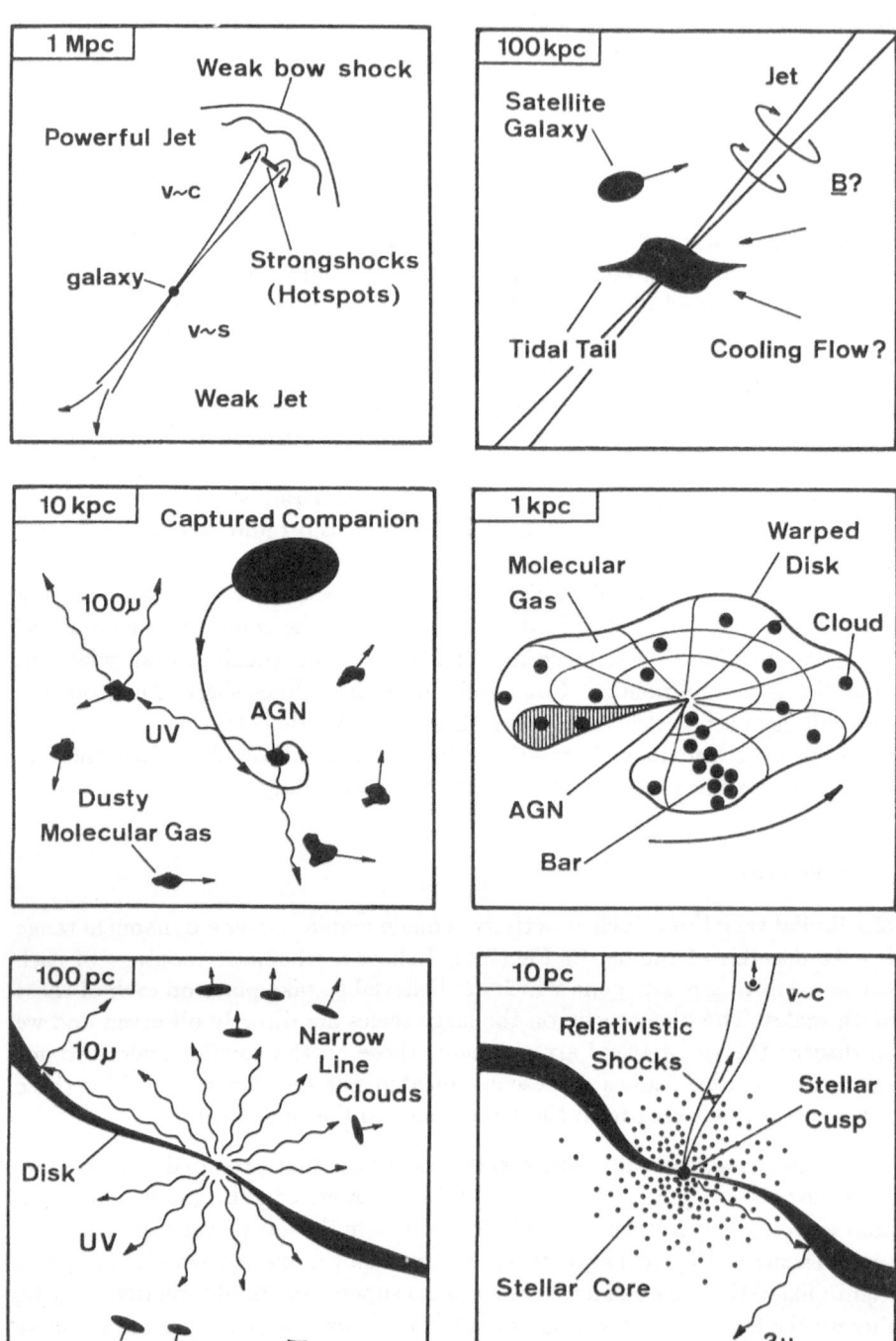

Fig. 7.1. Montage of scales exhibiting some effects of accretion onto a black hole over 12 successive decades of radius ranging from $\sim 1\,\mathrm{Mpc}$ to $\sim 3 \times 10^{13}\,\mathrm{cm}$. This is scaled to a black hole of mass $\sim 10^8\,\mathrm{M}_\odot$. We comment upon the individual frames in the text.

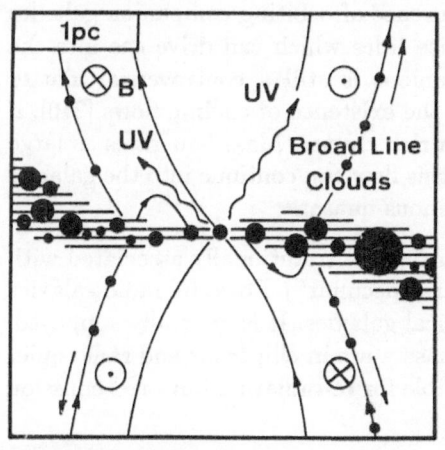

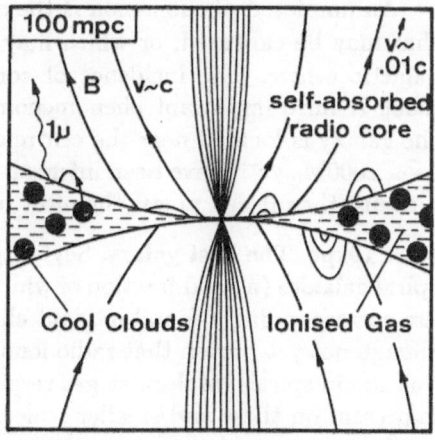

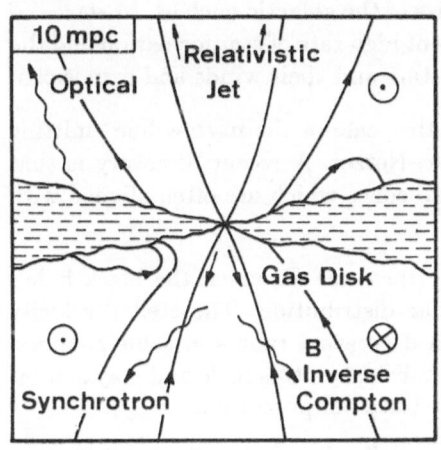

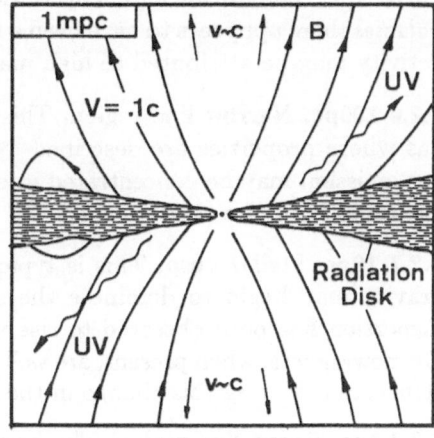

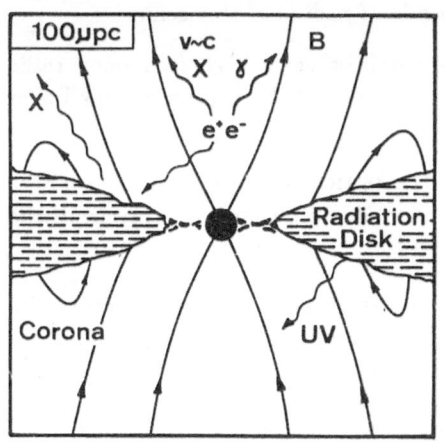

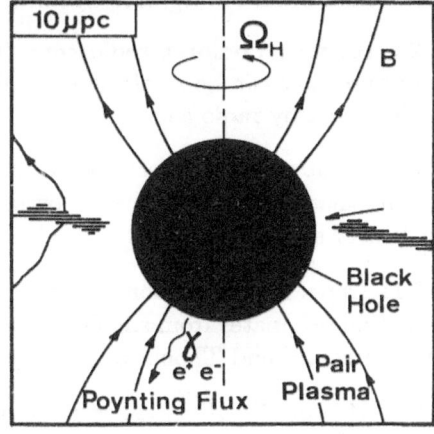

An unrelated phenomenon is the occurence of orbiting companion galaxies that may be captured, or which may raise tides which can drive gas into the galactic centre. The incidence of companions is still a controversial matter [236]. A third important phenomenon is the existence of cooling flows [246], if the galaxy is located near the centre of a rich cluster. Mass flow rates as large as $\sim 1000 M_\odot yr^{-1}$ have been inferred. If this flow can continue into the galactic nucleus, then it can power the most luminous quasars.

7.2.4 10kpc. The host galaxy. Seyfert galaxies are traditionally associated with spiral galaxies (a good fraction of which are "peculiar"). Powerful radio galaxies are correspondingly found in giant elliptical galaxies. It is generally supposed, though not yet proven that radio loud quasars are in ellipticals and radio quiet quasars in spirals. Molecular gas responsible for re-radiating infra red emission is present on these and smaller scales.

7.2.5 1kpc. Galactic nucleus. This is the size of the galactic nucleus. In *starburst galaxies* there apppears to have been a recent high rate of star formation and the activity may be attributed to high mass stars and their winds and explosions.

7.2.6 100pc. Narrow line region. This is the scale of the narrow-line emitting gas whose properties are described by Dr. Netzer. A recent discovery is that the emission may be concentrated into two cones which are often aligned with the radio structure [247].

7.2.7 10pc. Stellar cusp. This is typically the scale at which the black hole's gravity may begin to dominate the stellar distribution. The stellar velocity dispersion has been observed to rise with decreasing radius in some galaxies. Outflowing jets, when present, are well established on this scale and may exhibit structure reflecting this change in the gravitational potential.

7.2.8 1pc. Broad line region. The broad emission lines, also discussed here by Dr. Netzer seem to originate within a decade of radius of this scale.

7.2.9 100mpc. Compact radio core. The smallest scale high frequency radio structure that can be resolved with high frequency VLBI is seen on this linear scale in nearby radio sources.

7.2.10 10mpc. Ionised accretion disk. If the optical continuum in quasars is mainly thermal in origin then it originates on this scale, presumably from the surface of an accretion disk.

7.2.11 1mpc. Radiation-dominated accretion disk. The higher temperature UV radiation originates from smaller scales where an accretion disk would be radiation pressure and Thomson opacity dominated.

7.2.12 100μpc. Relativistic accretion disk. The nature of the gas flow within a decade of radius of the black hole is quite controversial. Observation of highly variable X-ray emission from Seyfert galaxies imply that the X-rays originate from this region.

7.2.13 10μpc. Black hole. Finally, we come to the black hole and perhaps also its magnetosphere.

7.3 Physiology

7.3.1 Black hole. The mass of the hole dictates a characteristic length and luminosity scale for nuclear activity (section 4.1.2), though probably not its qualitative character. The spin parameter, λ_H, may be much more important in this regard. When this is large, spin energy can be extracted electromagnetically in the form of a relativistic hydromagnetic wind (*cf.*section3.3.) The magnetosphere surrounding the black hole is a natural site for a pair plasma, as this is the one place in a galactic nucleus where gas will be efficiently removed! (*cf.*section2.4) Particle acceleration is likely to be very efficient within the magnetosphere and it should be a site of copious γ rays and hard X-rays as well as an outflowing electron-positron, relativisitic electromagnetic wind.

I now make a conjecture. I propose that powerful relativistic jets only form when there is a rapidly spinning black hole. (Slowly spinning holes may not be able to generate sufficient electrical potential difference to make pairs in much the same way as pulsars as supposed to do.) What determines λ_H? If there were just an accretion disk present, then the accreted gas would carry angular momentum as well as mass across the horizon and, in the absence of any decelerating torque, λ_H would eventually increase to nearly its maximum value of unity [124]. However, the strength of the magnetic field responsible for the decelerating torque is more difficult to gauge. It is probably a function of the dimensionless mass accretion rate $\dot{m}$, and the natural prescription to adopt is a generalisation of the α prescription namely that the magnetic pressure is proportional to the pressure, either radiation or gas, in the inner disk. The magnetic energy density near the black hole will be proportional to the shear stress acting in the disk which in turn is proportional to $\dot{M}/mH$, (where H is the height and we have used equation(4.30)), which in turn is proportional to $\dot{M}_E/m^2$. I therefore propose that the electromagnetic power be given by

$$L_{EM} \propto \lambda_H^2 m^2 B^2$$
$$= \alpha_H \lambda_H^2 L_E \qquad (7.2)$$

(*cf.*equation (3.39)). The constant of proportionality α_H is unknown. (See [248], for a related prescription.) The associated torque, is a combination of two terms, that due to the accreted gas, $\sim \dot{M}m$ and the decelerating electromagnetic torque, $\sim \alpha_H \lambda_H m \dot{M}_E$. Balancing these two torques gives an equilibrium angular velocity measured by $\lambda_H \sim \dot{m}/\alpha_H$. Unless $\alpha_H \ll 1$, sub-Eddington mass accretion rates produce slowly spinning holes. The time required to achieve this equilibrium is $\sim t_E/\alpha_H$, which can exceed the timescale on which the accretion rate can change. In this case, the current angular velocity would depend upon the history of the accretion. The main point, though, is that black holes may be spinning rapidly or slowly independent of their current accretion rate and that when they spin rapidly, their electromagnetic powers may greatly exceed the disk bolometric luminosity.

7.3.2 Relativistic jets. The relativistic outflow from the black hole is proposed to be focussed into two jets. As described in section 5.3.3, emission from these jets will be beamed in the direction of motion so that a large flux of polarised synchrotron and inverse Compton radiation will be measured when the observer's direction is roughly parallel to the jet. The jet will be optically thin above infra-red wavelengths but self-absorption will set in at lower frequency and the emission will be seen from a photosphere whose distance from the hole increases with observing wavelength. All of this emission will be spatially unresolved and can be associated with the radio core. Optically thin radio emission from shock waves propagating along the outer jet can however be resolved by VLBI and when a bright jet is observed at a small angle to the line of sight these features move superluminally.

When powerful relativistic jets are formed by the spinning black hole they are supposedly collimated by the surrounding disk outflow. However, if their power dominates that from the disk, and the foregoing considerations imply that this is a strong possibility, then a large scale relativistic jet will be formed. The resulting radio source will be classified as extended, when observed from a general direction, though if entrainment of galactic and circumgalactic gas is minimal and the jet speed remains at least mildly relativistic out to a large distance, only one jet will be seen.

These powerful jets terminate in "hot spots" which are believed to be strong shocks where particle acceleration and magnetic field amplification take place. The shocked jet fluid is supposed to flow back around the unshocked jet to form the lower brightness radio lobes. The whole flow is bounded by a weak bow shock propagating into the circumgalactic medium.

7.3.3 Accretion disk. As described in section 4, an AGN accretion disk is believed to be radiation-dominated in its inner regions, with its thickness proportional to the dimensionless mass accretion rate. Thick accretion disks are subject to dynamical instability, though the ultimate consequences of this remain uncertain. Provided the disk is sufficiently dense to thermalise the radiation (and this requires the Thomson optical depth to be $\tau_T \gtrsim 1000$), most of the radiation will be emitted roughly with peak frequency somewhat in excess of the local effective temperature which lies in the UV. The luminosity will be $L_{UV} \sim 10^{45} \dot{m} M_8 \text{erg s}^{-1}$ and this will dominate the bolometric power when $\dot{m}$ is large.

The unstable accretion disk is likely to be embedded in a very active corona where much of the liberated gravitational binding energy from the disk may be dissipated, probably creating additional UV and X-ray radiation. Roughly half this radiation as well as that from a magnetospheric pair plasma, if present, will illuminate the disk and may be re-radiated as a reflection spectrum, including an iron line. (The iron line cannot be formed too close to the hole, without excessive Doppler broadening.)

Further from the hole, the disk is supported by gas pressure and forms the optical continuum (*cf.*equation(4.32)). The shape of the optical-UV spectrum depends upon the details of the radiative transfer. Observed spectra have spectral indices α greater than the value $-1/3$ predicted for a Keplerian disk radiating black body radiation (equation(4.35)). Radial energy transport either by internal torques or external irradiation and an emissivity that increases with radius as the influence of Thomson scattering diminishes, will both flatten the spectrum.

The disk surface density Σ, depends upon the angular momentum transport which, I argue, is at least partially due to an external hydromagnetic torque. The outflowing hydromagnetic wind collimated by a toroidal magnetic field can collimate the central relativistic jet, when present, even when it transports a significantly smaller power than the relativistic jet.

The outer parts of an accretion disk, beyond $\sim 0.1pc$, may contain dusty molecular gas which can re-radiate incident UV flux from the inner parts of the disk as infra-red radiation (cf.section 6.3.4). This is consistent with the total infra red emission being a significant fraction ~ 10 per cent of the bolometric luminosity, as observed.

7.3.4 Hydromagnetic outflow. Hydromagnetic torques may be very effective in transporting angular momentum from the molecular disk. They can also account for the broad emission lines. As described in detail in Dr. Netzer's contribution, these permitted lines have velocity widths typically ~ 10000km s^{-1} which often have cusps which may be as narrow as ~ 300km s^{-1}, blue-shifted by $\sim 500 - 1000$km s^{-1} with respect to the forbidden lines. Blue asymmetries in the line profile are also common.

Now suppose that the lines come from clouds of neutral gas that are initially in Keplerian orbit in a disk and that they have sufficient ionisation for them to be tied to the magnetic field lines. They will be flung out centrifugally by strong poloidal fields attached to the disk and photoionised by the central UV continuum (Blandford, Emmering and Shlossman, in preparation). A distant observer will only see clouds from a range of radii on one side of the disk. The combination of toroidal and poloidal motion of the clouds will create a line profile that is both blue-shifted in its centre and blue asymmetric in its wings.

However, the observed range of velocity of the line-emitting clouds raises two difficulties. Firstly, if the one and a half decade range in line widths displayed by the lines is really the variation in the virial velocity of the emission line clouds, then these should be located over three decades in radius and exposed to an ionising flux that varies through six decades in intensity. It is then surprising that the line profiles are not more sensitive to the ionisation potential and the critical density for collisional de-excitation. Secondly, if the emitting gas moves with roughly the Keplerian velocity at the distance indicated by simple photoionisation calculations, then the black holes would have masses so large ($M \sim 10^{10}M_\odot$) that they would have been detected in nearby galaxies.

These difficulties can be alleviated if emission lines are broadened by electron scattering. For example, suppose the initial Keplerian velocity of the line-emitting clouds extends from ~ 300km s^{-1} to ~ 3000km s^{-1}. The emission will be dominated by clouds that are moving slightly faster than this. However, roughly half the line photons will illuminate the disk. If this is enclosed in a

$\sim 10^6 \mathrm{K}$ electron-scattering corona, then the back-scattered radiation will be Doppler-shifted by the hot electrons to produce the broad wings. These wings should be linearly polarised and may be more easily seen in polarised light .

One additional advantage of invoking an MHD wind is that it resolves the problem of confinement of the clouds (cf.section2.3.3) as the magnetic pressure will generally be large enough to accomplish this.

The more slowly moving gas clouds originating from the outer disk may also form the narrow emission lines. This is particularly relevant to those Seyfert galaxies in which the spatial distribution of line emission appears biconical[247].

Those AGN that are unable to form relativistic jets by virtue of having slowly spinning black holes, but do form slower hydromagnetic jets, may be responsible for the lower power, edge-darkened (type 1) radio sources. The powers carried by these jets may only be a small fraction of the total bolometric luminosity of the nucleus. These jets may be further slowed by entrainment.

7.3.5 Galactic gas inflow. As reviewed in section 6, gaseous fuel for the AGN must probably be supplied to the nucleus by the host galaxy. Most of the mass should be in molecular form, just as in our Galaxy, and, when it covers a good fraction of the sky as seen by the AGN, will be able to re-radiate the incident UV as long wavelength infra-red emission [79]. Non-axisymmetric gravitational fields, induced externally by satellite galaxies, or internally by bars and warps, should be responsible for driving gas into the central regions. Self-gravitating clumps may sink under dynamical friction [215], [235].

7.4 An AGN Classification.

We can now speculate upon the underlying physical differences between the different types of AGN and attempt to classify them on the basis of their underlying physical properties. I chose the three quantities $M, \dot{m}, \lambda_H$ as primary parameters and I suppose that they vary over the ranges

$$0.01 \lesssim M_8 \lesssim 10$$
$$10^{-4} \lesssim \dot{m} \lesssim 10 \qquad (7.3)$$
$$0 \lesssim \lambda_H \lesssim 1$$

in AGN. Of course, the majority of galaxies are inactive and have parameters lying outside this range. As we also emphasised, secondary parameters, which only characterise our view of the AGN, can have a strong influence upon the observations. The most important is the viewing angle θ between the spin axis of the hole and the observer direction and we shall retain this in our classification scheme. Also important is the covering factor of the AGN by dusty molecular gas; when high a powerful infra red galaxy will be produced. We shall ignore this case.

The resulting classification scheme is summarised in Table(7.1)where we have just characterised the values of the four parameters $M, \dot{m}, \lambda_H, \theta$ by

Table (7.1). AGN Classification Scheme.

AGN	M	$\dot{m}$	λ_H	θ
Extended Radio Quasar	H	H	H	H
Compact Radio Quasar	H	H	H	L
Broad Absorption Line Quasar	H	H	L	H
Radio-Quiet Quasar	H	H	L	L
Narrow Line Quasar	H	L	H	H
High Polarisation Quasar	H	L	H	L
Edge-Darkened Radio Galaxy	H	L	L	H
Core-Halo Radio Galaxy	H	L	L	L
Broad Line Radio Galaxy	L	H	H	H
Compact Radio Quasar	L	H	H	L
Type 2 Seyfert Galaxy	L	H	L	H
Type 1 Seyfert Galaxy	L	H	L	L
Narrow Line Radio Galaxy	L	L	H	H
BL Lac Object	L	L	H	L
LINER	L	L	L	H
LINER	L	L	L	L

whether they are high (H) or low (L). In fact, a continuous variation in these parameters is expected and the bounding values are probably not constant. In general, it is premature to be quantitative.

The radio-quiet quasars and Seyfert 1 galaxies apparently form a single class distinguished by their luminosity which can range from $\sim 10^{42}$erg s^{-1} to $\sim 10^{47}$erg s^{-1}. They probably have mass accretion rates $\dot{m} \sim 1 - 10$ so that their luminosities can approach the Eddington limit. They do not exhibit powerful relativistic jets and supposedly have slowly spinning holes, although radiative drag on the outgoing electrons and positrons can also inhibit the formation of a relativistic jet. The weaker, bipolar radio features that are seen can be associated with MHD outflows from the accretion disks.

The type 2 Seyfert galaxies and perhaps also a class of narrow line quasars may be essentially similar except that the broad line region is obscured by either the more distant, and more slowly moving clouds or an intervening, accretion disk. These objects are presumably viewed close to the equatorial plane. The lower X-ray fluxes from type 2 Seyfert galaxies can have a similar explanation. The linear polarised broad wings reported in Seyfert 2 galaxies may be attributed to electron scattering emission observed from the outer parts of a partially obscured disk, rather than high latitude scattering as described in section6.3.3.

A further class of quasars, the broad absorption line quasars, is most naturally associated under the present unified scheme with large $\dot{m}$ quasars observed roughly equatorially, so that the absorption is caused by the dense, MHD driven wind. As most of these objects are radio quiet, we suppose that their holes spin slowly.

Radio-loud quasars are divided into the steep spectrum, extended sources and the flat spectrum, compact sources. They have broad emission lines and relativistic jets which translates into high accretion rates and spinning holes. Following the considerations of section 5.3 I distinguish these two classes by our orientation, with Doppler boosting raising the fluxes observed in the case of the intrinsically weaker compact sources. In extreme cases where the Doppler boosted optical continuum dominates the emission from the accretion disk, a high power blasar, sometimes called a high polarisation quasar (HPQ) or an optically violently variable quasar (OVV) will be seen.

Radio galaxies have a similar range of radio and emission line properties. The edge-brightened radio galaxies have relativistic jets. The edge-darkened radio galaxies appear to be fuelled by sub-relativistic jets which, on the above hypothesis, derive from the disk rather than the hole. The edge brightened radio galaxies are more likely to have strong emission lines and are further subdivided into broad line radio galaxies and narrow line radio galaxies. [249]. The former appear to be low power, radio-loud quasars; the latter may have less prominent, broad lines because they are accreting less rapidly. (The existence of such correlations demonstrates that the average mass accretion rate is probably stable over long periods of time and is therefore more likely to be controlled by processes occuring outside the galactic nucleus.)

The low power blazars, normal BL Lac objects, are most simply interpreted as narrow line radio galaxies viewed at small angles so that the Doppler boosted continuum from the relativistic jets overwhelms the emission lines. Edge-darkened sources seen at small angles appear as weak core-halo radio galaxies.

This unification of extended and compact quasars [195] has the shortcoming that there appear to be too few extended radio loud quasars to account for the unbeamed compact quasars [172]. However, an alternative suggestion[244], that narrow emission line radio galaxies are all quasars whose broad emission line regions are obscured also has difficulty because their narrow line strengths differ from those of quasars with similar extended radio emission strength [250]. In addition, broad line radio galaxies and quasars are more powerful infra-red sources than narrow line radio galaxies (Browne, private communication). This is inconsistent with [61] but consistent with both the infra-red and the broad line emission being powered by the UV continuum source. The best resolution of these data would seem to be that broad line radio galaxies be considered as low luminosity radio loud quasars, that some (but not most) narrow line radio galaxies be obscured quasars [198], and that the remaining, missing unbeamed compact radio quasars have been classified as infra-red galaxies [251]. Accounting quantitatively for the unbeamed compact radio sources remains a major challenge.

Another intriguing feature of the observations is the association of Seyfert galaxies with spirals and radio galaxies with giant ellipticals. This can be reconciled with the classification scheme if the mass of the black hole is correlated with the mass of the host galaxy. This may be because ellipticals, which are preferentially found in high density galactic environments, form early on and have a large number of dynamical interactions. A low mass black hole in an elliptical galaxy may be heavier than a high mass hole in a spiral galaxy.

The final case to consider is the low accretion rate and low spin objects. These can probably be associated with the *Low Ionisation Nuclear Emission Regions* (LINERS) [252].

7.5 Evolution

7.5.1 Demography of AGN.
A grand unified interpretation of AGN should also address their evolutionary properties. These are summarised by Dr. Woltjer. Although it appears that all nearby quasars are surrounded by low brightness "fuzz", presumably their host galaxies, it is still not certain whether and when these galaxies are elliptical or spiral and what are their luminosities. We can however make a comparison with the space density of bright galaxies, which are mostly spiral. There is a fiducial galaxy luminosity $L^* \sim 10^{10} L_\odot$, comparable with the luminosity of our Galaxy. The number density of galaxies with current luminosity $\gtrsim 0.5 L^*$ is found to be $N_g \sim 10^{-2} h^{-3} \mathrm{Mpc}^{-3}$, where $h = H_0/100 \mathrm{km}$ $\mathrm{s}^{-1} \mathrm{Mpc}^{-3}$. The number density of fainter galaxies increases slowly with luminosity. We don't know directly the fraction of these galaxies which were active in the past. However, we can compute the total energy radiated by AGN per bright galaxy by measuring the local energy density of quasar light by summing all their fluxes and dividing by c. This energy density must be corrected for the redshift of the photons to derive the energy emitted by the quasars. We then divide by the galaxy number density to obtain the energy radiated per bright galaxy [90], [70]. We can follow this simple procedure because the universe is spatially homogeneous and the mean energy density that we measure is typical of the universe as a whole. If we further assume that this radiation was created by black holes with mean blue radiative efficiency, ϵ^B, (*i.e.*the energy of blue light created per unit rest mass energy of accreted gas), then the mean black hole mass required per bright galaxy is given by

$$< M > = \frac{\Sigma F_i^B (1 + z_i)}{\epsilon^B c^3 N_g}$$

$$\sim 3 \times 10^6 \left(\frac{\epsilon^B}{0.03} \right)^{-1} h^{-3} \mathrm{M}_\odot \tag{7.4}$$

where the superscript B refers to the B spectral band. As it is unlikely that the black hole radiative efficiency in the B band greatly exceeds ~ 0.03, there must be, on average, at least $3 \times 10^6 h^{-3} \mathrm{M}_\odot$ of black hole left inside bright galaxies, notably those we observe locally. This is a striking conclusion because we can measure, or place upper bounds upon, the central masses within local galaxies. There is, as yet, no conflict with the observations, but there is not much room for manoeuvre either. In particular, note that the minimal black hole masses increase by a factor 8 if we reduce the Hubble constant to $h = 0.5$. As we remarked in section 3, quasar models with significantly lower radiative efficiency cannot be tolerated.

The number of bright quasars is harder to determine. Let us follow convention and (arbitrarily) define a quasar as having $L_B \gtrsim 10^{44}h^{-2}$ erg s^{-1}. Quasar counts down to $\sim 23^m$ give ~ 100 per square degree or $\sim 4 \times 10^6$ over the sky [21]. Most of these have redshifts $z \lesssim 2.5$ because the quasar luminosity is observed to evolve rapidly back to $z \sim 2$ and appears to be very roughly constant at earlier epochs. (There could be many more faint quasars at larger redshift. If so, this will only strengthen the argument that we are about to make.) Now the accumulated proper volume that we can see back to a redshift $z \sim 2$ in an Einstein-De Sitter cosmology is $\sim 100h^{-3}$Gpc^3, and therefore contains $\sim 10^9$ bright galaxies at the present time. On this basis we can infer that there are roughly 300 bright galaxies per quasar at redshifts $z \sim 2 - 2.5$.

There are two extreme positions to take at this point. The first is that a fraction ~ 0.003 of bright galaxies became quasars and were active for all their lives (up to ~ 1 Gyr at this time). The second is that all bright galaxies were active for roughly 3 million years. Now, in the former case, the minimum black hole mass will be $10^{10}h^{-3}M_\odot$ and the closest dead quasar will be $\sim 30h^{-1}$ Mpc away. Such a galaxy would presumably have an anomalously large central surface brightness and velocity dispersion and we know of no such object. Alternatively, in the latter case, all galaxies would have $M \gtrsim 3 \times 10^6 M_\odot$, which is incompatible with observations of our Galactic centre, unless $h \gtrsim 1$. (In order to comply with the Eddington limit, the most powerful quasars with bolometric luminosities $L \gtrsim 10^{47}h^{-2}$erg s^{-1} must have masses in excess of $10^9 M_\odot$.) Both extreme positions appear to be inconsistent with a simple view of the observations and so we must give a prescription for a distribution of black hole masses among galaxies.

7.5.2 A "Feast or Famine" model.

We can incorporate these evolutionary considerations into the grand unified scheme described above by making one additional hypothesis, namely that when black holes grow, they do so at essentially the Eddington rate [253]. This automatically optimises the radiative ouput per unit mass of black hole. This prescription is appropriate if the mass is supplied to the nucleus dynamically from the host galaxy in large quantities, and radiation pressure prevents the black hole from accepting it faster than the Eddington rate. The hole will then grow in mass with an e-folding time $\sim 0.1t_E \sim 4 \times 10^7$yr, assuming a radiative efficiency ~ 0.1. When the fuel is exhausted or blown away by radiation pressure, the hole will function as a low luminosity AGN or as a radio galaxy if it is still spinning. The brightest quasars will be those galaxies that formed early, allowed big black holes (with $M \lesssim 3 \times 10^9 M_\odot$) to grow at epochs when galaxy interactions were common, and have subsequently become dormant or been transformed into radio galaxies. Lower luminosity Seyfert galaxies are identified with younger, less massive galaxies that have plenty of gas. They should be radiating at roughly the Eddington limit and consequently contain relatively low mass black holes $\sim 10^6 - 10^8 M_\odot$ for $\sim 3 \times 10^{43} - 3 \times 10^{45}$erg s^{-1} luminosity objects. They evolve much less dramatically with cosmic epoch. Low power radio galaxies contain heavier holes but also evolve more slowly than the powerful radio sources. The minimum black hole mass per bright galaxy is distributed roughly logarithmically with most galaxies having $\sim 10^6 - 10^7 M_\odot$ holes, ~ 10 per cent having $10^7 - 10^8 M_\odot$ holes and so on. It is possible to make quantitative models that accommodate our present understanding of evolution and the limits on the masses of relict black holes[253]. However, the results are fairly model-dependent and somewhat sensitive to the assumed Hubble constant.

7.6 Conclusion

The foregoing grand unified theory of AGN is derived primarily from physical considerations and involves many of the processes described in the preceding sections. It combines some generally accepted features with some more speculative and controversial ingredients. When confronted with the observations, it seems to accommodate their gross features but it fails to account for some reported correlations and trends. An exercise such as this is relatively facile and of limited usefulness. What is much more valuable, though considerably harder to achieve, is to refine models like this to the point of making quantitative predictions and to assemble, assess and interpret the observations so as to constrain and refute these theories. If these lectures assist in this endeavour, then they will have fulfilled their purpose.

Acknowledgements

I thank Thierry Courvoisier for his efficient organisation of this school and much helpful guidance on the content of these lectures, Hagai Netzer and Lodewijk Woltjer for their wise and entertaining instruction and several students for their attempts to make the above intelligible. Financial support under NASA grant NAGW1301 is gratefully acknowledged.

References

1. C. M. Gaskell: In *Astrophysical Jets and their Engines* ed. by W. Kundt (Dordrecht: Reidel 1986) p.29
2. S. A. Colgate and A. G. W. Cameron: Nature **200** 870 (1963)
3. G. R. Burbidge and E. M. Burbidge: *Quasi-Stellar Objects* (San Francisco: Freeman 1967)
4. Y. Ne'eman: Astrophys. J. **141** 1303 (1965)
5. F. Hoyle and J. Narlikar: Proc. Roy. Soc. **277** 1 (1964)
6. Ya. B. Zel'dovich and I. D. Novikov: Sov. Phys. Dokl. **158** 811 (1964)
7. E. E. Salpeter: Astrophys. J. **140** 796 (1964)
8. D. Lynden-Bell: Nature **223** 690 (1969)
9. P. Meyer, W. Duschl, J. Frank, and E. Meyer-Hofmeister (ed.): *Theory of Accretion Disks* (Dordrecht: Kluwer 1989)
10. T. M. Heckman: Astron. Astrophys. **87** 152 (1980)
11. D. W. Weedman, F. R. Feldman, V. A. Balzano, L. W. Ramsey, and R. Sramek: Astrophys. J. **248** 105 (1981)
12. H. C. Arp: *Quasars, Redshifts and Controversies* (Berkeley: Interstellar Media 1987)
13. F. Hoyle and W. A. Fowler: Mon. Not. R. astr. Soc. **165** 129 (1963)

269

14. V. Ginzburg and L. M. Ozernoy: Sov. Phys. JETP **20** 489 (1964)
15. J. Arons, R. M. Kulsrud, and J. P. Ostriker: Astrophys. J. **198** 687 (1975)
16. R. Terlevich: In *Evolutionary Phenomena in Galaxies* ed. by J.E.Beckman and B. E. J. Pagel (Cambridge: Cambridge University Press 1989) p.149
17. Ya. B. Zel'dovich and I. D. Novikov: *Relativistic Astrophysics* (Chicago: Chicago University Press 1971)
18. M. C. Begelman, R. D. Blandford, and M. J. Rees: Rev. Mod. Phys. **56** 255 (1984)
19. M. J. Rees: Ann. Rev. Astron. Astrophys. **22** 471 (1984)
20. J. Frank, A. R. King, and D. J. Raine: *Accretion Power in Astrophysics* (Cambridge: Cambridge University Press 1985)
21. D. W. Weedman: *Quasar Astronomy* (Cambridge: Cambridge University Press 1986)
22. D. E. Osterbrock: *Astrophysics of Gaseous Nebulae and Active Galactic Nuclei* (Mill Valley: University Science Books 1989)
23. J. E. Dyson (ed.): *Active Galactic Nuclei* (Manchester: Manchester University Press 1985)
24. J. Miller (ed.): *Astrophysics of Active Galactic Nuclei and Quasi-Stellar Objects* (Mill Valley: University Science Books 1985)
25. G. Guircin, F. Mardirossian, M. Mezzetti, and M. Ramella (ed.): *Structure and Evolution of Active Galactic Nuclei* (Dordrecht: Reidel 1986)
26. G. Swarup and V. K. Kapahi (ed.): *Quasars. Proc. IAU Symposium No. 119* (Dordrecht: Reidel 1986)
27. H. R. Miller and P. J. Wiita (ed.): *Active Galactic Nuclei* (Berlin: Springer-Verlag 1988)
28. D. E. Osterbrock and J. S. Miller (ed.): *Active Galactic Nuclei. Proc. IAU Symposium No. 134* (Dordrecht: Kluwer 1989)
29. G. B. Rybicki and A. P. Lightman: *Radiation Processes in Astrophysics* (New York: Wiley 1979)
30. M. S. Longair: *High Energy Astrophysics* (Cambridge: Cambridge University Press 1981)
31. J. D. Jackson: *Classical Electrodynamics* (New York: Wiley 1975)
32. L. D. Landau and E. M. Lifshitz: *The Classical Theory of Fields* (Oxford: Pergamon 1971)
33. G. R. Burbidge: Astrophys. J. **129** 841 (1958)
34. A. T. Moffet: In *Stars and Stellar Systems. Vol 9* ed. by A. Sandage, M. Sandage, and J. Kristian (Chicago: Chicago University Press 1975) p.211
35. K. I. Kellermann and F. N. Owen: In *Galactic and Extragalactic Radio Astronomy* ed. by G. L. Verschuur and K. I. Kellermann (Berlin: Springer-Verlag 1988) p.563
36. G. Engargiola, D. A. Harper, M. Elvis, and S. P. Wilner: Astrophys. J. Lett. **332** L19 (1988)
37. M. de Kool and M. C. Begelman: Nature **338** 484 (1989)
38. T. J. Turner and K. A. Pounds: Mon. Not. R. astr. Soc. **240** 833 (1989)
39. J. H. Krolik, C. F. McKee, and C. B. Tarter: Astrophys. J. **249** 422 (1981)
40. A. S. Kompaneets: Sov. Phys. JETP **4** 730 (1957)
41. G. R. Blumenthal and R. J. Gould: Rev. Mod. Phys. **42** 237 (1970)
42. S. L. Shapiro, A. P. Lightman, and D. M. Eardley: Astrophys. J. **204** 187 (1976)
43. J. I. Katz: Astrophys. J. **206** 910 (1976)
44. D. L. Band and J. E. Grindlay: Astrophys. J. **298** 128 (1985)
45. A. P. Marscher: Astrophys. J. **264** 296 (1983)
46. A. Quirrenbach, A. Witzel, S. J. Kian, T. Krichbaum, C. A. Hummel, and A. Alberdi: Astron. Astrophys. **226** L1 (1989)
47. D. B. Wilson: Mon. Not. R. astr. Soc. **200** 881 (1982)
48. A. Bazzano *et al.*: In *Proc. Gamma Ray Observatory Science Workshop* ed. by W. N. Johnson (Greenbelt: Goddard Space Flight Center 1989) p.4-22
49. H. Kuneida, T. J. Turner, H. Awaki, K. Koyami, R. Mushotzky, and T. Yoshiyaki: Nature (1990) (in press)
50. G. Ghisellini, C. Done, and A. C. Fabian: Mon. Not. R. astr. Soc. (1990) (in press)
51. V. B. Berestetski, E. M. Lifshitz, and L. P. Pitaevski: *Quantum Electrodynamics* (Oxford: Pergamon 1982)
52. R. D. Blandford and P. Coppi: Mon. Not. R. astr. Soc. (1990) (in press)
53. P. W. Guilbert, A. C. Fabian, and M. J. Rees: Mon. Not. R. astr. Soc. **205** 593 (1983)
54. A. Zdziarski and A. P. Lightman: Astrophys. J. Lett. **294** L79 (1985)

55. A. C. Fabian, R. D. Blandford, P. W. Guilbert, E. S.Phinney, and L. Cuellar: Mon. Not. R. astr. Soc. **221** 931 (1986)

56. R. Svennson: Mon. Not. R. astr. Soc. **227** 403 (1987)

57. S. Bonometto and M. J. Rees: Mon. Not. R. astr. Soc. **152** 21 (1971)

58. C. Done and A. C. Fabian: Mon. Not. R. astr. Soc. **240** 81 (1989)

59. P. Coppi: Mon. Not. R. astr. Soc. (1990) (in press)

60. R. Svennson: Astrophys. J. **258** 321 (1982)

61. M. Baring: Mon. Not. R. astr. Soc. **228** 681 (1987)

62. A. Zdziarski: Astrophys. J. **335** 786 (1988)

63. E. Haug: Astron. Astrophys. **178** 292 (1987)

64. M. Sikora, J. G. Kirk, M. C. Begelman, and P. Schneider: Astrophys. J. Lett. **320** L81 (1987)

65. D. Kazanas and D. C. Ellison: Astrophys. J. **304** 178 (1986)

66. M. Sikora, M. C. Begelman, and B. Rudak: Astrophys. J. Lett. **341** L33 (1989)

67. D. B. Melrose: *Plasma Astrophysics* (New York: Gordon and Breach 1980)

68. R. D. Blandford and D. Eichler: Phys. Rep. **154** 1 (1987)

69. A. F. Heavens and L. O'C. Drury: Mon. Not. R. astr. Soc. **235** 997 (1989)

70. E. S. Phinney: *Unpublished thesis* (Cambridge University 1983)

71. M. L. Burns and R. V. E. Lovelace: Astrophys. J. **262** 87 (1982)

72. K. A. Pounds, K. Nandra, G. C. Stewart, I. M. George, and A. C. Fabian: Nature (1990) (in press)

73. A. P. Lightman and T. R. White: Astrophys. J. **335** 57 (1988)

74. I. M. George, K. Nandra, and A. C. Fabian: Mon. Not. R. astr. Soc. (1990) (in press)

75. P. W. Guilbert and M. J. Rees: Mon. Not. R. astr. Soc. **233** 475 (1988)

76. K. Nandra, K. A. Pounds, G. C. Stewart, A. C. Fabian, and M. J. Rees: Mon. Not. R. astr. Soc. **236** 39p (1989)

77. D. B. Sanders, B. T. Soifer, J. H. Elias, B. F. Madore, K.Matthews, G. Neugebauer, and N. Z. Scoville: Astrophys. J. **325** 74 (1988)

78. M. J. Rees, J. I. Silk, M. W. Werner, and N. C. Wickramasinghe: Nature **223** 788 (1969)

79. D. B. Sanders, E. S. Phinney, G. Neugebauer, B. T. Soifer, and K. Matthews: Astrophys. J. **347** 29 (1989)

80. L. Spitzer: *Physical Processes in the Interstellar Medium* (New York: Wiley 1978)

81. P. G. Martin: *Cosmic Dust* (Oxford: Oxford University Press 1978)

82. B. T. Draine and H. M. Lee: Astrophys. J. **285** 89 (1984)

83. I. Robinson, A. Schild, and E. Schucking (ed.): *Quasi-stellar Sources and Gravitational Collapse* (Chicago: Chicago University Press 1964)

84. I. Iben: In *I. Robinson, A. Schild, and E. Schucking (ed.)* ed. by 64 (Quasi-stellar Sources and Gravitational Collapse 1964) p.I. Robinson, A. Schild, and E. Schucking (ed.)Chicago: Chicago University Press

85. H. J. Smith and D. Hoffleit: Nature **198** 650 (1963)

86. A. C. S. Readhead, M. H. Cohen, and R. D. Blandford: Nature **272** 131 (1978)

87. E. E. Salpeter and R. V. Wagoner: Astrophys. J. **164** 557 (1971)

88. R. H. Sanders: Astrophys. J. **162** 784 (1970)

89. M. C. Begelman and M. J. Rees: Mon. Not. R. astr. Soc. **188** 847 (1978)

90. A. Soltan: Mon. Not. R. astr. Soc. **200** 115 (1982)

91. S. L. Shapiro and S. A. Teukolsky: *Black Holes, White Dwarfs and Neutron Stars* (New York: Wiley 1983)

92. J. A. Zensus and T. J. Pearson: *Superluminal Radio Sources* (Cambridge: Cambridge University Press 1987)

93. A. H. Bridle and R. A. Perley: Ann. Rev. Astron. Astrophys. **22** 319 (1984)

94. A. Dressler: In *Active Galactic Nuclei* ed. by D. E. Osterbrock J. S. Miller (Dordrecht: Kluwer 1989) p.217

95. J. Kormendy: Astrophys. J. **325** 128 (1988)

96. J. L. Tonry: Astrophys. J. **322** 622 (1987)

97. R. Genzel and C. H. Townes: Ann. Rev. Astron. Astrophys. **25** 377 (1987)

98. A. J. Newton and J. Binney: Mon. Not. R. astr. Soc. **210** 711 (1984)

99. M. J. Rees: Science **247** 817 (1990)

100. A. V. Filippenko and W. L. W. Sargent: Astrophys. J. Suppl. **57** 3 (1985)

101. R. D. Blandford: In *300 Years of Gravitation* ed. by S. W. Hawking and W. Israel (Cambridge: Cambridge University Press 1987)

102. Kafatos (ed.): *Supermassive Black Holes* (Cambridge: Cambridge University Press 1988)
103. C. W. Misner, K. S. Thorne, and J. A. Wheeler: *Gravitation* (San Francisco: Freeman 1973)
104. R. M. Wald: *General Relativity* (Chicago: Chicago University Press 1984)
105. K. S. Thorne, R. M. Price, and D. MacDonald: *Black Holes. The Membrane Paradigm* (New Haven: Yale University Press 1986)
106. R. Penrose: Rev, Nuovo. Cimento **1** 252 (1969)
107. S. W. Hawking: Phys. Rev. Lett. **26** 1344 (1971)
108. J. D. Bekenstein: Phys. Rev. D **7** 2333 (1973)
109. S. W. Hawking: Nature **248** 30 (1974)
110. V. B. Braginsky, C. M. Caves, and K.S. Thorne: Phys. Rev. D **15** 2047 (1977)
111. J. M. Bardeen and J. A. Petterson: Astrophys. J. Lett. **195** L65 (1975)
112. S. Kumar: Mon. Not. R. astr. Soc. **233** 33 (1989)
113. M. C. Begelman, R. D. Blandford, and M. J. Rees: Nature **287** 307 (1980)
114. J. R. Graham, D. P. Carico, K. Matthews, G. Neugebauer, B. T. Soifer, and T. D. Wilson: Astrophys. J. Lett. **354** L5 (1990)
115. R. D. Ekers: In *Extragalactic Radio Sources. Proc. IAU Symposium No. 97* ed. by D. S. Heeschen and C. M. Wade (Dordrecht: Reidel 1982) p.
116. B. R. Espey, R. F. Carswell, J. A. Bailey, M. G. Smith, and M. J. Ward: Astrophys. J. **342** 666 (1989)
117. I. Redmount and M. J. Rees: Comm. Astrophys. **14** 165 (1989)
118. R. L. Znajek: Mon. Not. R. astr. Soc. **185** 833 (1978)
119. R. D. Blandford and R. L. Znajek: Mon. Not. R. astr. Soc. **179** 433 (1977)
120. B. Punsley and F. V. Coroniti: Astrophys. J. **350** 518 (1990)
121. E. N. Parker: *Cosmical Magnetic Fields* (Oxford: Clarendon Press 1979)
122. M. J. Rees, M. C. Begelman, R. D. Blandford, and E. S. Phinney: Nature **295** 17 (1982)
123. N. I. Shakura and R. A. Sunyaev: Astron. Astrophys. **254** 22 (1973)
124. I. D. Novikov and K. S. Thorne: In *Black Holes* ed. by C. De Witt and B. De Witt (New York: Gordon and Breach 1973) p.343
125. J. E. Pringle: Ann. Rev. Astron. Astrophys. **19** 137 (1981)
126. A. Treves, L. Maraschi, and M. Abramowicz: Publ. Astr. Soc. Pacific **100** 427 (1988)
127. G. A. Shields: Ann. N. Y. Acad. Sci. **571** 110 (1989)
128. K. O. Mason, M. G. Watson, and N. E. White (ed.): *The Physics of Accretion onto Compact Objects* (Berlin: Springer-Verlag 1986)
129. G. Belvedere (ed.): *Accretion Disks and Magnetic Fields in Astrophysics* (Dordrecht: Kluwer 1989)
130. D. Lynden-Bell and J. E. Pringle: Mon. Not. R. astr. Soc. **168** 603 (1974)
131. J. P. Ostriker: Astrophys. J. **273** 99 (1983)
132. J. E. Gunn: In *Active Galactic Nuclei* ed. by C. Hazard and S. A. Mitton (Cambridge: Cambridge University Press 1978) p.213
133. R. V. E. Lovelace, C. M. Mobarry, and J. Contopoulos: In *Accretion Disks and Magnetic Fields in Astrophysics* ed. by G. Belvedere (Dordrecht: Kluwer 1989) p.71
134. R. D. Blandford: In *Theory of Accretion Disks* ed. by P. Meyer, W. Duschl, J. Frank, and E. Meyer-Hofmeister (Dordrecht: Kluwer 1989) p.35
135. A. R. King: Q. J. R. astr. Soc. **29** 1 (1988)
136. A. Königl: In *Accretion Disks and Magnetic Fields in Astrophysics* ed. by G. Belvedere (Dordrecht: Kluwer 1989) p.165
137. H. C. Spruit: In *Theory of Accretion Disks* ed. by P. Meyer, W. Duschl, J. Frank, and E. Meyer-Hofmeister (Dordrecht: Kluwer 1989) p.325
138. J. C. B. Papaloizou, J. Faulkener, and D. N. C. Lin: Mon. Not. R. astr. Soc. **205** 487 (1983)
139. A. Galeev, R. Rosner, and G. S. Vaiana: Astrophys. J. **229** 318 (1979)
140. D. M. Eardley and A. P. Lightman: Astrophys. J. **200** 187 (1975)
141. P. Sakimoto and F. V. Coroniti: Astrophys. J. **247** 19 (1981)
142. R. D. Blandford D. G. Payne: Mon. Not. R. astr. Soc. **199** 883 (1982)
143. A. Wandel: Astrophys. J. Lett. **316** L55 (1987)
144. K. I. Kolykhalov and R. A. Sunyaev: Adv. Sp. Res. **3** 249 (1984)
145. A. Laor H. Netzer: Mon. Not. R. astr. Soc. **238** 897 (1989)
146. R. R. J. Antonnucci, A. L. Kinney, and H. C. Ford: Astrophys. J. **342** 64 (1989)

147. A. Laor, H. Netzer, and T. Piran: Mon. Not. R. astr. Soc. **242** 560 (1989)
148. C. T. Cunningham: Astrophys. J. **202** 788 (1975)
149. B. Czerny and M. Elvis: Astrophys. J. **312** 325 (1987)
150. W. A. Stein: Astrophys. J. Lett. **351** L29 (1990)
151. T. Piran: Astrophys. J. **221** 652 (1978)
152. D. N. C. Lin and G. A. Shields: Astrophys. J. **305** 28 (1986)
153. C. J. Clarke and G. A. Shields: Astrophys. J. **338** 32 (1989)
154. M. A. Abramowicz, M. Jaroszyński, and M. Sikora: Astron. Astrophys. **63** 221 (1978)
155. M. Kozlowski, M. Jaroszński, and M. A. Abramowitz: Astron. Astrophys. **63** 209 (1978)
156. B. Paczyński and P. J. Wiita: Astron. Astrophys. **88** 23 (1980)
157. P. Madau: Astrophys. J. **327** 116 (1988)
158. C. N. Tadhunter *et al.*: Mon. Not. R. astr. Soc. **235** 405 (1989)
159. J. C. B. Papaloizou and J. E. Pringle: Mon. Not. R. astr. Soc. **208** 721 (1984)
160. R. Narayan and J. Goodman: In *Theory of Accretion Disks* ed. by P. Meyer, W. Duschl, J. Frank, and E. Meyer-Hofmeister (Dordrecht: Kluwer 1989) p.231
161. J. Hawley: In *Theory of Accretion Disks* ed. by P. Meyer, W. Duschl, J. Frank, and E. Meyer-Hofmeister (Dordrecht: Kluwer 1989) p.259
162. O. M. Blaes: Mon. Not. R. astr. Soc. **227** 975 (1987)
163. R. C. Jennison and M. K. Das Gupta: Nature **172** 996 (1953)
164. P. Maltby and A. T. Moffet: Astrophys. J. Suppl. **7** 141 (1963)
165. M. J. Rees: Nature **229** 312 (1971)
166. T. J. Pearson and A. C. S. Readhead: Ann. Rev. Astron. Astrophys. **22** 97 (1984)
167. D. S. Heeschen and C. A. Wade (ed.): *Extraglactic Radio Sources* (Dordrecht: Reidel 1982)
168. A. Ferrari and A. G. Pacholczyk (ed.): *Astrophysical Jets* (Dordrecht: Reidel 1983)
169. A. H. Bridle and J. A. Eilek: *Physics of Energy Transport in Extragalactic Radio Sources* (Green Bank: NRAO 1984)
170. W. Kundt (ed.): *Astrophysical Jets and their Engines* (Dordrecht: Reidel 1986)
171. R. N. Henriksen (ed.): *Jets from Stars and Galaxies* (Toronto:Canadian Journal Physics 1986)
172. E. S. Phinney: In *Astrophysics of Active Galaxies and Quasi-Stellar Objects* ed. by J. Miller (Mill Valley: University Science Books 1985) p.453
173. R. Fanti, K. Kellermann, and G. Setti (ed.): *VLBI and Compact Radio Sources. Proc IAU Symposium No. 110* (Dordrecht: Reidel 1984)
174. B. L. Fanaroff and J. M. Riley: Mon. Not. R. astr. Soc. **167** 31p (1974)
175. J. A. Zensus: In *Superluminal Radio Sources* ed. by J. A. Zensus and T. J. Pearson (Cambridge: Cambridge University Press 1987) p.26
176. M. W. Hodges and R. L. Mutel: In *Superluminal Radio Sources* ed. by J. A. Zensus and T. J. Pearson (Cambridge: Cambridge University Press 1987) p.168
177. L. Maraschi, T. Maccacaro, and M.-H. Ulrich (ed.): *BL Lac Objects* (Berlin: Springer-Verlag 1989)
178. C. Impey: In *Superluminal Radio Sources* ed. by J. A. Zensus and T. J. Pearson (Cambridge: Cambridge University Press 1987) p.233
179. G. V. Bicknell: Astrophys. J. **239** 433 (1984)
180. L. D. Landau and E. M. Lifshitz: *Fluid Mechanics* (Oxford: Pergamon 1959)
181. R. D. Blandford and M. J. Rees: Mon. Not. R. astr. Soc. **169** 395 (1974)
182. A. Königl: Phys. Fluids. **23** 1083 (1980)
183. K. R. Lind, D. G. Payne, D. L. Meier, and R. D. Blandford: Astrophys. J. **344** 89 (1989)
184. K. L. Chan and R. N. Henriksen: Astrophys. J. **241** 534 (1980)
185. D. A. Clarke, M. L. Norman, and J. O. Burns: Astrophys. J. Lett. **311** L63 (1986)
186. K. I. Kellermann and I. I. K Pauliny-Toth: Ann. Rev. Astron. Astrophys. **6** 417 (1968)
187. W. A. Dent: Science **148** 1458 (1965)
188. M. J. Rees: Nature **211** 468 (1966)
189. R. D. Blandford A. Königl: Astrophys. J. **232** 34 (1979)
190. A. P. Marscher and W. K. Gear: Astrophys. J. **298** 114 (1985)
191. S. W. Weinberg: *Gravitation and Cosmology* (191 1972)
192. P. A. G. Scheuer: In *VLBI and Compact Radio Sources. Proc IAU Symposium No. 110* ed. by R. Fanti, K. Kellermann, and G. Setti (Dordrecht: Reidel 1984) p.
193. K. R. Lind and R. D. Blandford: Astrophys. J. **295** 358 (1985)
194. G. Ghisellini and L. Maraschi: Astrophys. J. **340** 181 (1989)

195. M. J. L. Orr and I. W. A. Browne: Mon. Not. R. astr. Soc. **200** 1067 (1982)
196. I. W. A. Browne: In *Superluminal Radio Sources* ed. by J. A. Zensus and T. J. Pearson (Cambridge: Cambridge University Press 1987) p.129
197. P. A. Strittmatter, P. Hill, I. I. K Pauliny-Toth, H. Steppe, and A. Witzel: Astron. Astrophys. **88** L12 (1980)
198. P. A. G. Scheuer: In *Superluminal Radio Sources* ed. by J. A. Zensus and T. J. Pearson (Cambridge: Cambridge University Press 1987) p.104
199. R. W. Porcas: In *Superluminal Radio Sources* ed. by J. A. Zensus and T. J. Pearson (Cambridge: Cambridge University Press 1987) p.12
200. R. D. Blandford and C. F. McKee: Phys. Fluids. **19** 1130 (1976)
201. H. D. Aller, P. A. Hughes, and M. F. Aller: In *Superluminal Radio Sources* ed. by J. A. Zensus and T. J. Pearson (Cambridge: Cambridge University Press 1987) p.273
202. A. Königl and A. R. Choudhuri: Astrophys. J. **289** 188 (1985)
203. T. W. Jones: Astrophys. J. **332** 678 (1988)
204. M. L. Norman, L. L Smarr, J. R. Wilson, and M. D. Smith: Astrophys. J. **247** 52 (1981)
205. M. Jaroszyński, M. A. Abramowcz, and B. Paczyński: Acta Astronomica **30** 1 (1980)
206. M. Sikora and D. B. Wilson: Mon. Not. R. astr. Soc. **197** 529 (1981)
207. G. E. Eggum, F. V. Coroniti, and J. I. Katz: Astrophys. J. **330** 142 (1988)
208. E. S. Phinney: In *Superluminal Radio Sources* ed. by J. A. Zensus and T. J. Pearson (Cambridge: Cambridge University Press 1987) p.301
209. T. J.-L. Courvoisier and M. Camenzind: Astron. Astrophys. **224** 10 (1989)
210. M. Camenzind: Rev. Mod. Astron. (1990) (in press)
211. J. Heyvaerts and C. A. Norman: Astrophys. J. **347** 1055 (1989)
212. T. Chiueh, Z. Li, and M. C. Begelman: Astrophys. J. (1990) (in press)
213. J. Binney and S. Tremaine: *Galactic Dynamics* (Princeton: Princeton University Press 1987)
214. B. Balick and T. M. Heckman: Ann. Rev. Astron. Astrophys. **20** 431 (1982)
215. I. Shlossman, M. C. Begelman, and J. Frank: Nature **345** 679 (1990)
216. A. Toomre and R. Wielen: *Galaxy Dynamics and Interactions* (Dordrecht: Kluwer 1990)
217. T. Lauer: In *Dynamics of Dense Stellar Systems* ed. by D. Merritt (Cambridge: Cambridge University Press 1989) p.3
218. M. J. Duncan and S. L. Shapiro: Astrophys. J. **268** 565 (1983)
219. G. D. Quinlan and S. L. Shapiro: Astrophys. J. (1990) (in press)
220. J. Frank and M. J. Rees: Mon. Not. R. astr. Soc. **176** 633 (1976)
221. A. P. Lightman and S. L. Shapiro: Rev. Mod. Phys. **50** 437 (1978)
222. W. Benz and J. G. Hills: Astrophys. J. **323** 614 (1987)
223. C. R. Evans and C. S. Kochanek: Astrophys. J. Lett. **346** L13 (1989)
224. J. K. Cannizzo, H. M. Lee, and J. Goodman: Astrophys. J. **351** 38 (1990)
225. G. M. Voit and J. M. Shull: Astrophys. J. **331** 197 (1988)
226. C. A. Norman and N. J. Scoville: Astrophys. J. **332** 124 (1988)
227. I. Shlosman and M. C. Begelman: Astrophys. J. **341** 685 (1989)
228. A. Toomre: Astrophys. J. **139** 1217 (1964)
229. N. Z. Scoville, D. B. Sanders, A. I. Sargent, B. T. Soifer, and C. G. Tinney: Astrophys. J. Lett. **345** L25 (1989)
230. R. Genzel: In *The Center of the Galaxy. Proc. IAU Symposium No. 136* ed. by M. Morris (Dordrecht: Kluwer 1989) p.393
231. J. H. Krolik and M. C. Begelman: Astrophys. J. **329** 702 (1988)
232. J. H. Krolik and A. Meiksin: Astrophys. J. Lett. **352** L33 (1990)
233. I. Shlossman, J. H. Frank, and M. C. Begelamn: Astrophys. J. **338** 45 (1989)
234. E. S. Phinney: In *Theory of Accretion Disks* ed. by P. Meyer, W. Duschl, J. Frank, and E. Meyer-Hofmeister (Dordrecht: Kluwer 1989) p.451
235. L. Hernquist: Nature **340** 687 (1989)
236. T. M. Heckman: In *Paired and Interacting Galaxies. Proc. IAU Colloquium 124.* ed. by J. Sulentic (Dordrecht: Kluwer 1990)(in press)
237. A. Lawrence: Publ. Astr. Soc. Pacific **99** 309 (1987)
238. A. Lawrence and M. Elvis: Astrophys. J. **256** 410 (1982)
239. R. D. Blandford: Ann. N. Y. Acad. Sci. **422** 303 (1984)
240. R. D. Blandford: In *Superluminal Radio Sources* ed. by J. A. Zensus T. J. Pearson (Cambridge: Cambridge University Press 1987) p.310

241. R. J. J. Antonnucci J. S. Miller: Astrophys. J. **297** 621 (1985)
242. J. H. Krolik: In *Active Galactic Nuclei* ed. by D. E. Osterbrock J. S. Miller (Dordrecht: Kluwer 1989) p.285
243. D. A. Turnshek: In *QSO Absorption Lines. Probing the Universe* ed. by C. Blades, C. Norman and D. A. Turnshek (Cambridge: Cambridge University Press 1987) p.17
244. P. D. Barthel: Astrophys. J. **336** 606 (1989)
245. G. Miley and J. S. Miller: Astrophys. J. Lett. **228** L55 (1979)
246. C. S. Crawford, A. C. Fabian and R. Johnstone: Mon. Not. R. astr. Soc. **235** 183 (1988)
247. A. S. Wilson and J. A. Baldwin: Astronom. J. **98** 2056 (1989)
248. S. J. Park and E. T. Vishniac: Astrophys. J. **332** 135 (1988)
249. R. G. Hine and M. S. Longair: Mon. Not. R. astr. Soc. **188** 111 (1979)
250. N. Jackson and I. W. A. Browne: Nature **343** 43 (1990)
251. B. T. Soifer, J. R. Houck and G. Neugebauer: Ann. Rev. Astron. Astrophys. **25** 187 (1987)
252. T. M. Heckman: Astron. Astrophys. **87** 152 (1980)
253. R. D. Blandford: In *Quasars. Proc. IAU Symposium No. 119* ed. by G. Swarup and V. K. Kapahi (Dordrecht: Reidel 1986) p.359
254. M. C. Begelman, C. F. McKee, and G. Shields: Astrophys. J. **271** 70 (1983)
255. E. C. Ostriker, C. F. McKee, and R. I. Klein: Astrophys. J. (in press) (1991)

Subject Index

This index complements the table of contents without duplicating it.